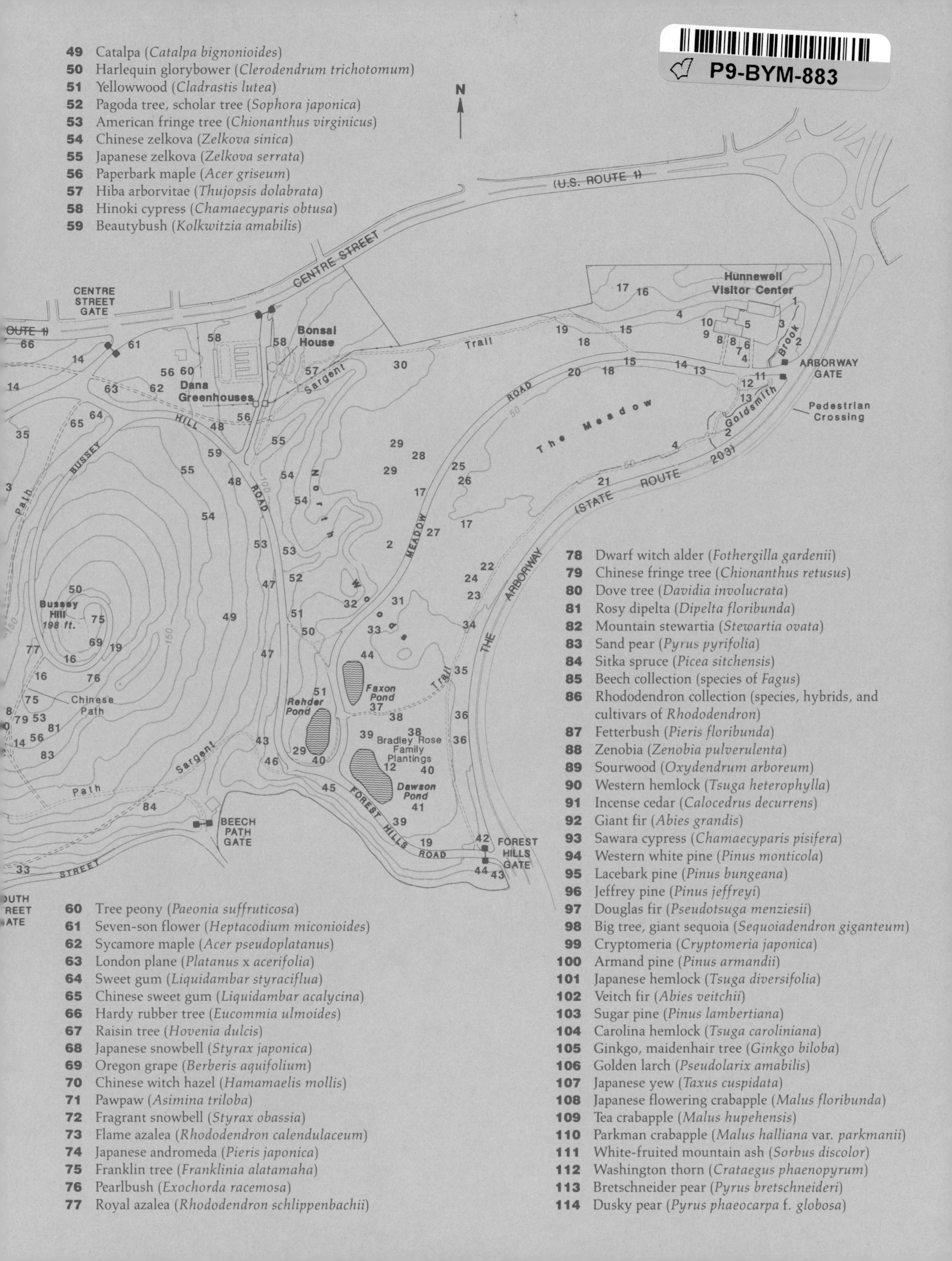

49 Catalpa (*Catalpa bignonioides*)
50 Harlequin glorybower (*Clerodendrum trichotomum*)
51 Yellowwood (*Cladrastis lutea*)
52 Pagoda tree, scholar tree (*Sophora japonica*)
53 American fringe tree (*Chionanthus virginicus*)
54 Chinese zelkova (*Zelkova sinica*)
55 Japanese zelkova (*Zelkova serrata*)
56 Paperbark maple (*Acer griseum*)
57 Hiba arborvitae (*Thujopsis dolabrata*)
58 Hinoki cypress (*Chamaecyparis obtusa*)
59 Beautybush (*Kolkwitzia amabilis*)

60 Tree peony (*Paeonia suffruticosa*)
61 Seven-son flower (*Heptacodium miconioides*)
62 Sycamore maple (*Acer pseudoplatanus*)
63 London plane (*Platanus* x *acerifolia*)
64 Sweet gum (*Liquidambar styraciflua*)
65 Chinese sweet gum (*Liquidambar acalycina*)
66 Hardy rubber tree (*Eucommia ulmoides*)
67 Raisin tree (*Hovenia dulcis*)
68 Japanese snowbell (*Styrax japonica*)
69 Oregon grape (*Berberis aquifolium*)
70 Chinese witch hazel (*Hamamaelis mollis*)
71 Pawpaw (*Asimina triloba*)
72 Fragrant snowbell (*Styrax obassia*)
73 Flame azalea (*Rhododendron calendulaceum*)
74 Japanese andromeda (*Pieris japonica*)
75 Franklin tree (*Franklinia alatamaha*)
76 Pearlbush (*Exochorda racemosa*)
77 Royal azalea (*Rhododendron schlippenbachii*)

78 Dwarf witch alder (*Fothergilla gardenii*)
79 Chinese fringe tree (*Chionanthus retusus*)
80 Dove tree (*Davidia involucrata*)
81 Rosy dipelta (*Dipelta floribunda*)
82 Mountain stewartia (*Stewartia ovata*)
83 Sand pear (*Pyrus pyrifolia*)
84 Sitka spruce (*Picea sitchensis*)
85 Beech collection (species of *Fagus*)
86 Rhododendron collection (species, hybrids, and cultivars of *Rhododendron*)
87 Fetterbush (*Pieris floribunda*)
88 Zenobia (*Zenobia pulverulenta*)
89 Sourwood (*Oxydendrum arboreum*)
90 Western hemlock (*Tsuga heterophylla*)
91 Incense cedar (*Calocedrus decurrens*)
92 Giant fir (*Abies grandis*)
93 Sawara cypress (*Chamaecyparis pisifera*)
94 Western white pine (*Pinus monticola*)
95 Lacebark pine (*Pinus bungeana*)
96 Jeffrey pine (*Pinus jeffreyi*)
97 Douglas fir (*Pseudotsuga menziesii*)
98 Big tree, giant sequoia (*Sequoiadendron giganteum*)
99 Cryptomeria (*Cryptomeria japonica*)
100 Armand pine (*Pinus armandii*)
101 Japanese hemlock (*Tsuga diversifolia*)
102 Veitch fir (*Abies veitchii*)
103 Sugar pine (*Pinus lambertiana*)
104 Carolina hemlock (*Tsuga caroliniana*)
105 Ginkgo, maidenhair tree (*Ginkgo biloba*)
106 Golden larch (*Pseudolarix amabilis*)
107 Japanese yew (*Taxus cuspidata*)
108 Japanese flowering crabapple (*Malus floribunda*)
109 Tea crabapple (*Malus hupehensis*)
110 Parkman crabapple (*Malus halliana* var. *parkmanii*)
111 White-fruited mountain ash (*Sorbus discolor*)
112 Washington thorn (*Crataegus phaenopyrum*)
113 Bretschneider pear (*Pyrus bretschneideri*)
114 Dusky pear (*Pyrus phaeocarpa* f. *globosa*)

A Reunion of Trees

Red
Oak
Chest

A REUNION OF TREES

The Discovery of Exotic Plants and Their Introduction into North American and European Landscapes

Stephen A. Spongberg

Harvard University Press
Cambridge, Massachusetts
London, England · 1990

Printed in the United States of America
10 9 8 7 6 5 4 3 2 1

This book is printed on acid-free paper, and its binding materials
have been chosen for strength and durability.

Library of Congress cataloging information is on the last page.

Frontispiece: An eighteenth-century allegorical engraving representing North America as a warehouse of new plants from which the depleted forests of Europe could be replenished and gardens could be embellished. From F. A. L. von Burgsdorf, *Versuch einer vollstandigen Geschichte vorzuglicher Holzarten* (Berlin, 1783). Courtesy of the Library of the Gray Herbarium, Harvard University.

Contents

Foreword

This book is an invitation to join a series of conversations that have been going on for the past five hundred years. The discussions have not been exclusive but have been carried on among experts and amateurs alike. The participants have been Americans, Englishmen, Germans, Japanese, Chinese, American Indians, explorers, wealthy suburban estate owners, farmers, nurserymen, university scientists, landscape painters, poets, Victorian ladies, families who like to collect things, foresters, plant breeders, fruit fanciers, backyard gardeners, and people who enjoy walking in parks.

Such diverse participants have been drawn toward one another through the twin emotions of curiosity and an appreciation of living things. The particular groups of people that are the subject of this book have focused on the trees and shrubs that so conspicuously compose our human surroundings. Sometimes the exchanges have been slow and distant; other times they have been quick and excited. But whether slow or fast, the skills and resources of any one group have fed the others, so that knowledge and appreciation of trees and shrubs has grown in depth and inclusion.

The purpose of this book is to help all of us, expert and amateur alike, to comprehend our present situation by telling the history of world exploration for woody plants. The history given here is of two sorts: environmental and intellectual. First, the book tells of the transformations of the North American and Eurasian landscapes brought about by the migrations of trees and shrubs through geologic change, exploration, and cultivation. Second, the book tells of some of the shifts in our ideas about woody plants and their meanings for human existence. Neither the explorations nor the biogeographical concepts sprang forth from momentary enthusiasms of bold adventurers or flashes of inspiration of isolated scientists. They were the products of extended and overlapping discussions, correspondence, and publications among generations of amateurs and professionals. This volume should be thought of as part of that ongoing process.

Today we face a crisis of knowledge and understanding. For us to find as a nation, and as citizens of the world, useful ways to comprehend and to safeguard our environments will require the participation of citizens throughout society. Much in these histories demonstrates the efficacy of open exchanges and the utility of the flow of ideas and cooperation among all citizens and across all the nations of the world. Our dependency upon the plants of the earth is no less now than it has ever been, and consequently identifying, classifying, and understanding living things is as urgent today as it has been in the past. Such enterprises also promise to be as rewarding as they have been since the fifteenth century, when the modern botanic conversations began. Of course because our everyday tools are no longer mostly wooden, and because other people supply us with food, drugs, clothing, and fuel, most of us approach the plant world from a distance.

The early explorers stood in quite a different relationship with nature. They experienced the New World as a forest because it was a forest that met them on the Atlantic shores. They hoped that within the forest they might find new plants for food, medicine, and industry. Indeed, the British East India Company made more money importing pepper than any North American trading company made on gold, and surely New World potatoes have nourished more human beings than all the bullion of the Spanish empire.

Yet more was at stake for Europeans than immediate economic profit. The sailing ships enabled Europeans to discover for themselves the rest of the planet, and as they did so the many who looked about experienced the same sort of feelings of wonder that we do today. Contemporary philosophers urged the exploration of the natural world as the study of God's Creation, and in time they elaborated a doctrine that natural science was the handmaiden of good theology. The English poet Alexander Pope (1688–1744) spoke in this voice in his poem *An Essay on Man*. There he wrote of "A mighty maze! but not without a plan . . ."

Fashion also entered the conversations. Princes and wealthy merchants of the fifteenth and sixteenth centuries built ornamental gardens as places of display and entertainment. Such gentlemen, in competition with one another, prized exotic plants to embellish their gardens. Already by 1636 in England one of Charles I's garden consultants had sailed to Virginia to seek fresh plant material for his patron. He returned with the American sycamore, and the happy subsequent outcome was the (*hybrid*) London plane tree that every city dweller today now enjoys.

In these first decades of plant exploration doctors were also important contributors to the intellectual discussions. They established and maintained systematic gardens at university medical schools, gardens like those at Leiden in the Netherlands. Physicians hoped that by the careful study of plants they could find new remedies for human diseases. For example, for a few years after 1577 England experienced a miracle-drug excitement when claims were made that the American sassafras (*Sassafras albidum* (Nuttall) Nees) would cure everything. Alas, the claims proved false, but the search in the plant world for drugs has proved to be an extremely rewarding enterprise.

Openness and universality were the special qualities of these Renaissance botanic conversations, and these twin qualities have proved to be the great Renaissance gift to modern knowledge. The art of printing, a fifteenth-century invention, made it possible for investigators of all kinds to exchange information because for the first time they could look at an illustration and be confident that they were discussing the same plant. Printing made possible, as hand copying had not, an exactly reproducible pictorial statement. Very soon this technical advance came to be joined with the spirit of careful observation.

The author of the first such book, *Gart der Gesundheit*, an herbal printed in Mainz in 1485, wrote with just such a modern point of view: "As man has no greater or noble treasure on this earth than bodily health I came to believe that I could undertake no more honorable or useful or holier work or labor, than to bring together a book in which the virtue and nature of many herbs and other creations of God, with their true colors and form, were made comprehensible for the consolation and use of all the world." In order to provide accurate drawings of more plants than grew at hand in Germany, the author of this herbal traveled with a "painter of understanding and with a subtle and practiced hand" to Italy and Jerusalem so that he "diligently learned the herbs that were there, and had them painted in their true colors and form."

From that time on, the task of exploring the plant kingdom, describing each plant accurately, and classifying and exchanging information has gone forward in ever widening and ever deepening circles. The range of the plant explorers, as is told in this volume, slowly expanded outward to cover both the New World and the Orient. Simultaneously, the number of people who concerned themselves with the study of plants multiplied rapidly. So, for example, a well-known amateur observer like the Reverend Gilbert White (1720–1793) of Selborne, England, corresponded with his colleagues in Italy, and he used the new binomial nomenclature of the Swedish professor Carl Linnaeus (1707–1778). Learned societies for the exchange of scientific information multiplied; the Royal Society (1662) in England was followed by the Horticultural Society of London (1804). Some of the royal gardens in turn—places like Kew Gardens and the Jardin des Plantes—were transformed into places of systematic plant study and the growing of the newly discovered species. The designs of the grounds of palaces and country estates grew ever more ambitious so that an active network grew up among designers of private parks, plant suppliers, and gentlemen gardeners who sought exotic plants and financed world explorations. Much of the history told in the pages that follow is composed of such mixed public and private, professional and amateur, enterprises.

Americans imitated these European practices, but it was not until the middle of the nineteenth century that they could hold up their end of the international conversations. Europeans and Americans had different needs and outlooks. The Europeans saw the United States as a kind of warehouse of new plants. The Americans, because they were a scattered band of settlers establishing new farms and towns, looked to books, correspondence, and exchange as devices for improving their crops and kitchen gardens. Until the middle of the nineteenth century agriculture, for Americans, meant field crops, and horticulture meant fruit trees and vegetables. Thomas Jefferson, an active experimenter and plant correspondent, spoke in the American voice when he wrote on the occasion of receiving a cask of rice seeds from

Africa, "The greatest service which can be rendered any country is to add a useful plant to its culture."

Philadelphians formed the American Philosophical Society in 1743, and a few years later Jared Eliot published the colonists' first agricultural book, *Essays on Field Husbandry* (1743). There followed a flood of American publications, first using the new sexual classification of Linnaeus and then later following the more modern French systems. In Boston the establishment of the American Academy of Arts and Sciences in 1780 was followed by that of the Massachusetts Society for Promoting Agriculture in 1792, then the Massachusetts Horticultural Society in 1829, and the Boston Society of Natural History in 1830. A few decades later the United States had become a full partner in an overlapping series of botanical exchanges that spread outward across the globe, and up and down from botanizing families and backyard gardeners to university scientists.

The Arnold Arboretum, whose critical role in worldwide exploration for plants is described in this volume, was born in that cultural climate. Now as we reassess the Arboretum's tasks in the light of today's environmental concerns it is useful to compare our circumstances to those that existed in the Arboretum's founding years. In the mid-nineteenth century university scientists stood at the apex of the scientific hierarchy, as they do today. They did write popular books and give public lectures, but scholars like Asa Gray (1810–1888) of Harvard University principally occupied themselves with the explorers, with correspondence with other scientists, and with the study and classification of plants that were being sent into botanic gardens and herbaria. Occasionally a stunning hypothesis emerged. As Mr. Spongberg recounts in this volume, Gray's glacial concept gave structure and meaning to the world's natural distribution of plants, and his correspondence with Charles Darwin (1809–1882) proved a significant contribution to Darwin's theory of evolution.

The university scholars connected themselves to the general public indirectly. Their publications and advice informed the standards and goals of the privately financed and government-supported plant explorers and established norms for the new state surveys, works like George B. Emerson's *A Report on the Trees and Shrubs Growing Naturally in the Forests of Massachusetts* (1846). The explorers and surveyors, in turn, relied upon amateur naturalists for local information.

The nineteenth-century amateurs were of many kinds, and their paths overlapped in lectures, learned societies, and exhibitions. Most prominent were the suburban estate owners. Many developed specialties. Horatio Hollis Hunnewell of Wellesley, Massachusetts, collected rhododendrons and conifers; Marshall Wilder of Dorchester displayed camellias; and Judge John Lowell of Roxbury was an orchid grower and pomologist. Charles Sprague Sargent (1841–1927), the first director of the Arnold Arboretum, came from such a family. Its Brookline estate on the edge of Jamaica Pond featured rhododendrons and mountain laurels. These gentlemen were the financial backers of all manner of botanical undertakings, and their gifts subsequently created and maintained the Arnold Arboretum. A number of the suburban estate owners imitated the English style of open "natural" landscape design and helped, thereby, to popularize that style. The large municipal parks of the cities of the United States are essentially public versions of these private preserves. The immensely successful layout and plantings of the Arnold Arboretum are premiere examples of this tradition. They are the product of a collaboration between the park planner Frederick Law Olmsted and the estate owner Charles Sprague Sargent.

The suburban estate owners also patronized the commercial nurseries of their respective cities and thereby encouraged institutions that practiced advanced horticultural technique and taught these techniques to generations of home gardeners. It was the nurserymen of Philadelphia, New York, and Boston who taught horticulture to Americans. They wrote the popular garden manuals, like Bernard M. M'Mahon's *The American Gardener's Calendar* (editions from 1806 to 1857) and Joseph Breck's *The Flower Garden* (1851). The nurserymen's plant lists, and later their illustrated seed catalogues, were both scientifically accurate and written in a style that captured the readers' imaginations.

Other activities were directed toward forestry and conservation of the remaining wilderness. The pioneer in these subjects was the Vermont lawyer and linguist George Perkins Marsh (1801–1882), who published his thoughtful *Man and Nature* in 1864. Marsh called for measures to restore the organic world which had been, he thought, excessively disrupted by farmers and lumbermen. From Marsh, botanic interest branched: one group discussed the management of forest and other natural resources so that they could be sustained with commercial use; the other group sought measures for preservation. John Muir (1838–1914), a friend of Sargent, led this latter group.

Most nineteenth-century authors, however, were family botanizers, people who went out on Sunday to look at, to name, and to collect flowers and plants. Their tastes circled back through the botanic world by means of popular books, prints, and poetry, by voter support for municipal and national parks, and through the seeds and plants they favored. So a popular Victorian shrub like the weigela (*Weigela florida* (Bunge) A. de Candolle) was both the product of exploration in China and an expression of popular acclaim.

By the time Sargent was setting out the trees in the Arnold Arboretum during the late 1880s the institution could take its place amidst these other activities: the explorations, papers, and exchanges among university and botanic garden scientists, the learned societies, the estate owners, the nurserymen, the park superintendents, the landscape architects, the backyard gardeners, and the Sunday botanizers. The topics of concern to these people were similar to those of concern today: cell biology, the processes of evolution, exploration, taxonomy, horticulture, urban, state, and national park management, agricultural technique, and popular appreciation.

Sargent's arboretum served as a public park incorporated within the Boston parks system, as a museum of living trees, and as a base for scientific investigation; these three roles fit easily into nineteenth-century concerns and activities. And owing to the beautiful setting for the trees Sargent and Olmsted designed, the Arnold Arboretum has remained a popular public park ever since. Sargent himself continued the tradition of plant surveys, description, and classification with his *Silva of North America* (1891–1902) and through his sponsorship of successful overseas explorations. The tales of "Chinese" Wilson offered here capture the excitement of that work.

Today, as we drive about the contemporary metropolis with its extensive suburbs, second-growth forests, shopping strips and malls, and office and industrial parks, we no longer imagine the explorations or the plant migrations that have populated these surroundings. Neither in the specialization of our offices and laboratories do we sufficiently appreciate the importance of bridging institutions that both hear the public and speak to it. Yet if science is to continue to contribute to human well-being, it must listen to peoples' experience with the world as well as puzzle out for itself the riddles of nature.

A few arboreta in the United States are currently equipped to participate in all the botanic conversations, from the most difficult genetic analyses to the common complaints of an invasion of aphids or the accurate identification of a garden shrub. The Arnold Arboretum, the Missouri Botanical Garden, and the New York Botanical Garden are such places. Most American arboreta, however, are not so. Most were established in the twentieth century as a means to preserve private estates as public parks.

The founding director of the Arnold Arboretum set out as his goal "to develop an institution which could simultaneously do sophisticated research and express its findings, or the implications of its findings, to a popular audience." Sargent realized many successes, yet thwarting his ambitions to some degree were changes in scientific focus and public habits a generation or two ago, as wealthy suburban estate owners declined in numbers. The American rich moved away from their old English country-house precedents, and the children and grandchildren of the botanizing families turned away from collecting and gardening, choosing to occupy themselves instead with outdoor sports. Wealthy amateur donors were succeeded by government and foundation committees as the source for research funds. Land grant universities like Cornell, Michigan, Illinois, and California assumed central roles in agricultural and forestry research and the dissemination of technique. Landscape architecture and landscape design during the decades from 1930 to 1970 were eclipsed by urban architecture and highway engineering. Home gardening remained popular, but government publications, the Agricultural Extension Service, and the commercial nurseries and seedsmen seemed to meet most gardeners' needs.

Popular concern for the conditions of metropolitan and world environments, however, has brought the botanical conversations to the forefront once more and has magnified the contribution that arboreta can make in our modern society. Scientific investigation of plants, the design of human habitats with appropriate plantings, and the management of human habitats as biological systems have become urgent issues for research, professional technique, public policy, and public consciousness. The old issues of forest and river conservation of Sargent's day are now enfolded in the larger environmental debates. Using plantings and the inherited land form to manage air and water quality seems very much to be a continuation of Olmsted's precedents. Yet, whatever continuities there are, new approaches to science and human settlements now give institutions like the Arnold Arboretum new roles to play.

We have learned that our environmental problems are worldwide, not local, and we have learned that behavior in even one metropolis may affect a whole continent. We have learned that biological diversity may be a central issue in environmental management; and we have learned that different kinds of research yield different questions and different answers. Field research, research on living organisms, and research with laboratory material are distinct undertakings. Finally, we have learned that for modern scientific knowledge to advance human well-being, scientific institutions must not only teach popular audiences but must listen and learn from them.

The Arnold Arboretum is a beautiful place and an institution with a fine tradition and reputation. As such it is a welcome meeting place for people of every level of expertness, from city vacant-lot gardener to genetic biologist and world explorer. It is especially a place where plantsmen, landscape designers, environmental engineers, and private and public practitioners of all kinds can renew their ties to the science that informs their work. By the same token it is a place where scientists might meet and exchange information with those who are working with today's complex metropolitan environment. As in the past five hundred years, the planet earth remains a place that calls for our continued exploration. My hope is that the history presented in *A Reunion of Trees* will stimulate the renewal of conversations that give meaning and focus to these explorations.

Sam Bass Warner, Jr.
Boston, Massachusetts

Preface

Since the very earliest periods of civilization the cultivation of plants has centered primarily on food crops for people and their domesticated animals and on the growth of medicinal herbs. With a few notable exceptions, only in relatively recent periods have plants been grown in gardens and man-created landscapes. And only in the last four hundred years—a period that roughly coincides with the advent of gardening for amenity—have more-or-less continuous, but often incidental, records been kept documenting the human-assisted movement of plants around the globe.

Over the past four hundred years, many adventurers have risked life and limb to procure plants under an astounding variety of circumstances. Taken together, their plant introductions have largely determined the nature of many urban and garden landscapes around the world. Moreover, a large number of the plants that they introduced into Europe and North America from extraterritorial regions now grow in greater profusion in cultivation than in the regions where they occur in nature. And some in North America have become so completely acclimatized that they have invaded forests and woodlands and appear to be native components of the indigenous flora.

A Reunion of Trees has been written with the goal of focusing attention on the wide array of exotic or nonnative trees and shrubs growing in urban and suburban parks and gardens across North America and Europe, and to link those plants with the adventures and vicissitudes of the explorers and travelers who first gathered them and made them known in scientific and horticultural circles. Exploring for plants—ornamental or not—became an interesting and sometimes all-consuming occupation that reached a peak of passion and notoriety during the first years of the present century. Among the first Americans to undertake and champion global plant exploration were the early administrators and plantsmen of the Arnold Arboretum of Harvard University. Their intention—like that of the many naturalists, botanists, and horticulturists who had preceded them—was to increase botanical knowledge and enlarge the selection of plants available to horticulturists and gardeners in temperate regions.

I have used the living collections of the Arnold Arboretum as the backdrop from which examples of introduced species have been drawn, though not all of the exotic species growing in the Arboretum figure as examples in this volume; in fact, the large majority do not. Nor do all of the individuals who have devoted themselves to plant exploration and introduction appear in the pages that follow. I have been forced by limitations of space to be selective, and I have chosen trees and shrubs that appeal to my own aesthetic sense, plants which can also be tied to specific individuals in the sometimes incomplete or sketchy historical record. Suffice it to say that virtually all of the exotic species cultivated in our parks and gardens have direct ties with human endeavor, culture, and history, and the plants and people discussed in these pages represent but a small sampling.

Throughout the text common names of plants have been used, followed by their Latin or botanical names (presented in parentheses) the first time the common name is mentioned. With one exception, the names of the authors of botanical names have been given in full following the Latin binomial. Linnaeus constitutes the sole deviation from this rule, and his name is represented by the accepted abbreviation, "L." The common names I have chosen to use are by and large those which are utilized on the display labels in the Arnold Arboretum; these conform, for the most part, to the most frequently used common names in the horticultural literature and in nursery and trade catalogues. Likewise, Chinese place names have been romanized using the

Wade-Giles system, although the current Pinyin spelling is presented (also in parentheses) the first time the name appears in the text. I have chosen to use Wade-Giles romanizations because this system yields spellings that are either identical to or closer to those used by the majority of plant explorers discussed in this volume.

To aid visitors to the Arnold Arboretum, many of the plants growing in the Arboretum's collections that are featured in this book have been identified by number on the endpaper maps. In addition to the map's numerical key, the boldface numbers in the index can be used to locate particular plants. Other individual specimens not indicated on the map can be found growing in the Arboretum's collections. That these plants are discussed or mentioned in this book is indicated by a small logo that appears on the display label attached to these specimens, alerting visitors to the fact that information concerning these plants can be found in *A Reunion of Trees*.

Today, many of the plants featured in this book are widely grown in North America and Europe, and living examples can be found in many public parks and gardens, arboreta, and botanical gardens. Consequently, this book has been addressed to a wide group of plant enthusiasts, regardless of where they live in relation to the Arnold Arboretum. Moreover, many of the featured plants are available from the nursery industry and help to satisfy the aesthetic sense and requirements of modern-day plantsmen for diverse and interesting ornamental plants that will flourish under a wide assortment of environmental conditions. Other examples remain rare in cultivation, and some have yet to enter the horticultural marketplace. I am hopeful that this volume will encourage more people to cultivate a wider diversity of both native and exotic plants. It is also hoped that this book will enrich the knowledge of people with an interest in their environment and that it will increase their enjoyment of plants.

A Reunion of Trees

CHAPTER ONE

Actively Seeking the Unknown

A Sampling of Exotics

Growing inside the Arborway Gate of the Arnold Arboretum, in the Jamaica Plain section of Boston, Massachusetts, in the low, wet ground at the edge of the meadow, is an unusual moisture-tolerant, cone-producing tree. This tree drops its needle-like leaves in the fall, to expose its spreading branches to winter storms and the cold winter sun. Portions of the branchlets on which the leaves are borne also sever their connections with the tree at the end of the growing season and litter the ground at its base. This annual cycle of needle drop in fall and needle growth in spring is unusual for a conifer, and the common name frequently applied to this species—bald cypress (*Taxodium distichum* (L.) Richard)—refers to its deciduous habit. The bald cypress is not a true cypress (species of the genus *Cupressus* L.), however; true cypresses are evergreen trees.

The bald cypress was one of the first strange and unknown trees encountered by the settlers who attempted to establish the Roanoke Colony. It was noted in written documents by Thomas Hariot on his return from the Roanoke Colony in 1588 and by William Strachey, who visited Jamestown in Virginia in 1610. Native to swamps and bays, where it commonly grows in quiet waters or on low, periodically flooded ground, this tree is characteristic of maritime forests on the Coastal Plain of the southeastern United States. It is not normally encountered in the New England landscape. Its distribution follows the Atlantic coastline from southern New Jersey to Florida and west along the Gulf of Mexico to the Mississippi embayment of Texas. Inland, its range extends northward along the Mississippi River and its tributaries into southern Illinois and Indiana. Throughout much of its native range the bald cypress is festooned with Spanish moss—not really a moss at all but an epiphytic bromeliad—and images of these trees evoke the atmosphere of the antebellum South or of

1. An early illustration of the leaves and cones of the bald cypress that was included in Mark Catesby's *Hortus Britanno-Americanus*, published in London in 1763. Catesby's monogram appears in the lower right-hand corner of the drawing. In the accompanying text it is noted that "the cypress is (except the tulip-tree) the tallest and the largest of all the trees this part of the world produces." Two seeds are associated with each scale, and between eighteen and thirty irregularly shaped seeds are produced in each cone. While cones are usually produced on an annual basis, large numbers are produced only every three to five years.

alligator-infested swamps. While the bald cypress is hardy in southern New England, Spanish moss is not, and the association of the two plants is not encountered north of Virginia. Likewise, in its native habitat the bald cypress often produces curious erect, knoblike, woody growths or knees from its roots. Thought to aid in bringing air to the roots of this frequently aquatic tree, knees are only poorly developed when the species is grown in New England.

A more familiar tree to northeasterners, one which has become naturalized throughout much of New England northward into Nova Scotia, Quebec, and Ontario, is found growing on the slope of the glacial drumlin to the right of Meadow Road as one approaches the three Arboretum ponds. The black locust (*Robinia pseudoacacia* L.) is native to the Appalachian region, from Alabama and Georgia northward into West Virginia, Maryland, and Pennsylvania, and through Kentucky into southern Indiana and Illinois. Its range, however, had been extended eastward into tidewater Virginia by the Indians in pre-Jamestown days because its wood proved to be the finest from which to fashion bows. It is also believed to occur naturally in the Ozark region of Arkansas and eastern Oklahoma; yet the tree has become so completely naturalized in some regions that attempts to determine its original natural range are dependent on historical records. From them we know, for example, that Locust Valley, New York, was named *after* the tree had been introduced and had become commonplace in the woodlands of Long Island.

Its sweetly scented racemes of white, wisteria-like flowers and its picturesque stature have made the black locust a desired ornamental tree. Of equal importance, its durable, hard, and fine-grained lumber has made it one of the most valuable timber trees of the eastern North American forest. During the era of wooden-hulled sailing ships, treenails or trunnels fashioned from black locust wood were in great demand in European shipyards for pegging the planking to the skeletal framework of the hulls; it had been found through trial and error that when the pegs came in contact with water, they swelled and held tighter than iron rivets. Moreover, they did not rust and corrode when exposed to salt water or the rock salt that was packed between the hull and the inner shell as a wood preservative in the process known as "salting down." Between 50,000 and 100,000 black locust treenails were exported annually from Philadelphia alone during the early 1800s.

It was the small, smooth, flat, but many-seeded pods or legumes of the black locust that apparently gave rise to this tree's common name. These seed-filled pods were referred to as "cods" by the English colonists at Jamestown and were confused or fancifully associated with another pod, the carob (the pod of the Mediterranean tree, *Ceratonia siliqua* L.), which sustained St. John the Baptist in the Palestinian wilderness and which today is used to flavor ice cream. In the Old World the carob was called a "locust" (as was the edible insect of that name), and soon the fruits of the North American tree came to be called "locust cods." Despite the inaccuracy of the supposed relationship, the name locust has become permanently affixed to this pod-producing North American tree.

From the slope of the drumlin where the black locust grows with its close relatives—other members of the pea or legume family—visitors to the Arboretum have an unobstructed view across the ponds and the intervening low terrain to the wall along the Arborway that bounds the Arboretum's perimeter. A variety of vines are allowed to climb and scramble over this wall and form a green, living dividing line between the landscape of the Arboretum and the asphalt and concrete of the heavily trafficked roadway. One of these vines is another well-known North American plant—technically, a shrub that climbs by attaching small, aerial rootlets to the object it festoons—the trumpet flower, trumpet creeper, or cow itch (*Campsis radicans* (L.) Seemann). Its bright orange, narrowly funnel-shaped flowers, which easily slip over children's fingers and transform small hands into those of witches or sorcerers or the paws of imaginary beasts, provide a source of sweet nectar for ruby-throated hummingbirds hovering among the flowers. Children, too, are quick to learn of the sticky sweet that can be sucked from the base of the tube. Like the black locust, trumpet creepers (Plate 1) have become naturalized in parts of New England and in other areas of eastern North America beyond the confines of its natural range. Long before the plant was grown in gardens and offered in nursery catalogues, it thrived in hot, humid woodlands, in disturbed areas, and in abandoned fields in the southern United States, in an area that roughly coincides with the natural range of the bald cypress.

In the early spring, weeks before the trumpet flower produces its flamboyant floral display, the small, rounded buds of the cornelian cherry (*Cornus mas* L.) expand in New England, and the plant's small, delicate, pale yellow flowers burst into bloom. One of the first shrubs or small trees to mark the end of winter with its precocious floral display, the cornelian cherry was well known to the ancient Greeks and Romans, who dedicated it to Apollo

2. The black locust (*Robinia pseudoacacia*) is also known as a bee tree, inasmuch as its beautiful, pure white, fragrant flowers are a rich source of nectar. In New England, the short racemes of flowers are produced in May.

3. The delicate charm of the cornelian cherry when it flowers in early spring results from the many clusters of its light yellow flowers that burst from buds along the otherwise naked branchlets. The bright red fruits of the cornelian cherry are sometimes used to make preserves or to flavor sherbets or ices. In hospitals in the Soviet Union the fruits are used as a vitamin C supplement.

and used its extremely hard wood to create lances and javelins. It has been suggested that the Trojan horse was constructed from its timber. The durable wood of the cornelian cherry has been used in the manufacture of wheel spokes and small implements throughout central and southeastern Europe, where the species is native in the region's woodlands. But the common name refers neither to the flowers nor to the properties of its wood; instead, it refers to the beautiful scarlet, cherrylike but ellipsoid fruits that develop from the yellow flowers and mature in late summer. It is disputed whether the name "cornelian" comes from the resemblance of the bright red fruits to the carnelian—a red variety of chalcedony, a semitransparent quartz, which in Europe was often carved into seals for setting in signet rings—or whether the stone was instead named after the fruits of the tree. In any case, these fruits are frequently eaten by birds and, in Europe, were often prepared as a jelly or made into a preserve used in tarts. At the Arnold Arboretum, the cornelian cherry grows at the junction of Meadow and Bussey Hill roads, above the bank of Rehder pond and across from the forsythia collection.

Sometime before the fruits of the cornelian cherry have ripened, the attention of most Arboretum visitors is drawn to a group of far more peculiar shrubs growing along the Meadow Road. Seen from a distance, the rounded contours of these shrubs suggest a cloud of smoke that somehow refuses to be dispersed by summer breezes. Viewed from a lesser distance, their form suggests mounds of pinkish, purplish, or grayish cotton candy or feathery powder puffs. Close inspection will reveal that almost all of the branches of the shrubs terminate in many- and finely-branched inflorescences, and while only a few flowers and fruits are produced, the slender branches of the inflorescences are covered with silky hairs. The abundance of inflorescences coupled with the millions of hairs create the effect of these remarkable shrubs in the landscape. While the hairs are initially pinkish- or purplish-tinged, they soon become smoky gray, and it is not surprising that the shrub is commonly known as smoke tree or smokebush (*Cotinus coggygria* Scopoli); rarely, it is known as Venetian sumac, a name that gives some inkling of the plant's provenance.

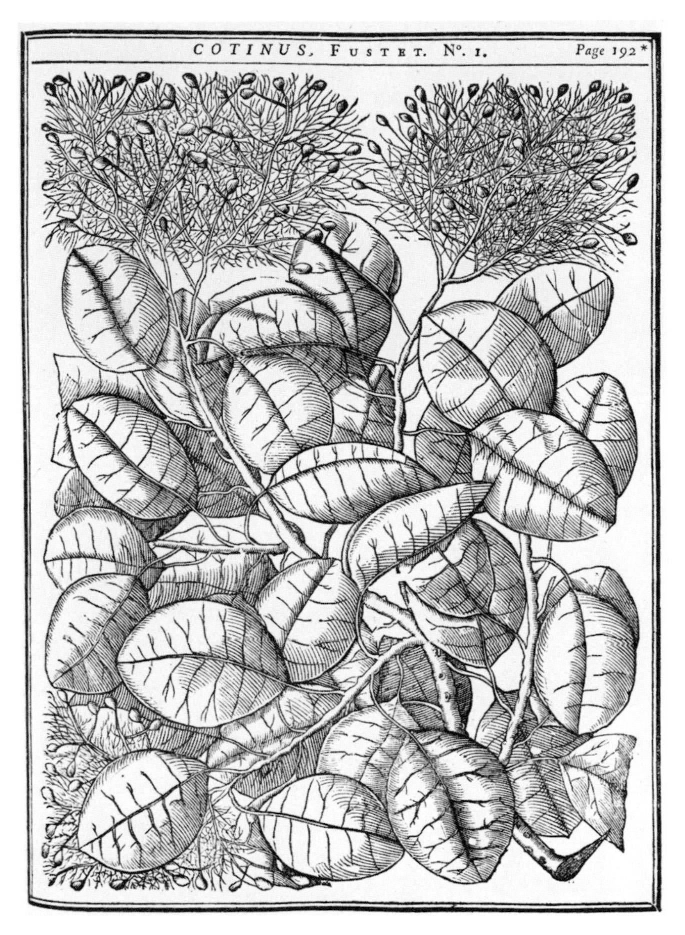

4. "Fustet," the French name of the Venetian sumac, is the title of this woodcut illustration of the plant from Henri Louis Duhamel's *Traite des Arbes et Arbustres*, published in Paris in 1755.

Native to middle and southern Europe from southeastern France to Turkey, the Crimea, and the Caucasus, smokebush also occurs in a broken series of populations that stretch eastward to the Himalayan region and into eastern China. Over this wide range smokebush grows on dry, rocky slopes. Resourceful Greeks and Russians found that it could be used for tanning leather, and its yellowish wood has been the source of a yellow pigment used to dye leather, wool, and even silk.

But what do the bald cypress, black locust, trumpet creeper, cornelian cherry, and smokebush have in common? All five species are able to grow in New England, indistinguishable from native plants when seen in the Olmstedian landscape of the Arnold Arboretum. And yet—as their labels inform—all are aliens in the Boston Basin.

Tradescant, Father and Son

By the year 1640, just twenty years after the Pilgrims had come ashore in New England, all five species described in the opening section of this chapter had been deliberately introduced and were being cultivated as exotics in a London garden in South Lambeth on the right bank of the River Thames. In that garden, these woody aliens attracted interest despite the fact that none had any reputed medicinal value. The unique garden in which they were carefully tended and observed had been established by John Tradescant (d. 1638) and inherited and maintained by John Tradescant the younger (1608–1662). This father-and-son team had eagerly brought together a unique living collection of exotic plants, tapping every possible source and using all the connections available to them to acquire and continuously augment it. Space was allotted not only to the common, well-known herbs, vegetables, and fruit-bearing trees and shrubs required to supply the household medicine chest and stock the kitchen larder. Entirely new plants, some previously never seen in England or cultivated anywhere before, were coaxed and encouraged to establish themselves in English soil. Cornelian cherries and bald cypresses numbered among these plants, as did trumpet creepers, smokebushes, and black locusts.

Gardening in England and in Europe was in the throes of a minor revolution. For centuries the cultivation of plants, other than staple food crops, had revolved around the physic garden, which provided the simples required for the practice of medicine. Largely confined within the walls of monasteries during the Middle Ages, physic gardens slowly became linked with universities and medical instruction. The year 1543 saw the establishment of the first botanical garden in the Tuscan town of Pisa north of Rome. By 1545 similar gardens had been founded in Padua west of Venice and in Florence, also in Tuscany. Other botanical gardens were established in Montpellier, in Paris, and elsewhere, and plants became subjects of experimentation, close observation, and increasingly accurate illustration. One result was that botany became synonymous with the great illustrated herbals, which were compiled from the late fifteenth century well into the late seventeenth century primarily to serve the medical profession as visual aids. But with the advent of gardening on a grand scale, botanical and horticultural knowledge began to divorce itself from medicine and the physic garden. A new body of gardening literature began to emerge; the year 1629 witnessed the publication of John Parkinson's (1567–1650) benchmark book, *Paradisi in Sole Paradisus Terrestris,* that was devoted to garden plants, as opposed to medicinal herbs. The Latin title of the work was a play on the author's name, The Earthly Paradise of Park-in-Sun, and reflected the new perspective in which the garden was viewed. Plants were being placed in the landscape for architectural effects, flowers were being arranged to beautify banquet tables, and gardens were becoming pleasure grounds.

In Elizabethan times Francis Bacon, the noted essayist and sagacious trend-setter, had stimulated the development of gardens by suggesting, in his essay *On Gardens,* that their creation was the highest form of artistic achievement. Men of class, with money and leisure, began to view their estates and country seats in a new way, with an eye toward their beautification, and they loosened their purse strings and directed energies toward creating pleasure grounds as a sign of their social position and their cultured status. The English country house was soon to be surrounded by landscaped parks with vistas framed by avenues of trees and carefully manicured gardens that would provide settings to vie with the architectural genius evident in the houses themselves.

In response to the growing need for knowledgeable plantsmen, James I had issued letters of incorporation to London's Worshipful Company of Gardeners in the late summer of 1605, and gardening in England became recognized as a profession. Statutory years of

5. Title page of John Parkinson's *Paradisi in Sole.* Adam and Eve are shown in the earthly paradise among an assortment of native European and exotic plants, including a vegetable sheep intended to represent the cotton plant.

PARADISI IN SOLE
Paradisus Terrestris.
or
A Garden of all sorts of pleasant flowers which our
English ayre will permitt to be noursed vp:
with
A Kitchen garden of all manner of herbes, rootes, & fruites,
for meate or sause vsed with vs,
and
An Orchard of all sorte of fruitbearing Trees
and shrubbes fit for our Land
together
With the right orderinge planting & preserving
of them and their vses & vertues
Collected by John Parkinson
Apothecary of London
1629
Qui veut parangonner l'artifice a Nature,
Et nos parcs a l'Eden, indiscret il mesure.
Le pas de l'elephant par le pas du ciron,
Et de l'Aigle le vol par cil du mouscheron.
A. Switzer

6. Jean Robin (1550–1629), early member of the network of European plantsmen, was the king's gardener at the Jardin du Roi in Paris from 1590 until his death. Robin and his son Vespasien were the recipients of seeds and plants of many North American species that were sent or brought back to France by travelers in French North America, now mostly Canada.

apprenticeship were established, so that young men eager to enter the occupation would benefit from the experience of old hands at the craft and from the protection of the guild. One of the early members of the company was John Tradescant the younger, whose father, John the elder, had already established his reputation as a gardener. He had laid out the garden at Hatfield House in Hertfordshire—the childhood home of good Queen Bess—for Robert Cecil, first Earl of Salisbury. Other Tradescant clients included Lord Wotton, the first Duke of Buckingham, and, ultimately, Charles I, all of whom required that gardening be done on a grand scale. Not satisfied with the variety of plants that could be obtained in England, the Tradescants seized every opportunity to travel to the continent on behalf of their noble patrons. Only thirty-five different kinds of trees in the native English flora were available for domestication, while on the continent the possibilities far exceeded that number and included plants from the Levant and exotic introductions the French had brought from New France in America.

With opportunities to purchase and collect plants in foreign lands, the Tradescants quickly established a network of continental correspondents with whom unusual plants could be shared and from whom they received shipments of common and not-so-common plants in return. Included in this fraternity were another father-and-son team, Jean and Vespasien Robin, and two brothers, René and Pierre Morin. The Robins were gardeners for a succession of three French kings, and the Paris garden they established would later serve as the nucleus around which the Jardin des Plantes would be developed. The genus *Robinia,* of which the black locust is a member, was named to commemorate their early cultivation of that North American tree in Paris, although it was probably first grown in Europe by the Tradescants. The Morins were well-known Paris nurserymen who grew a wide range of species but specialized in bulbous plants. During the 1630s they were undoubtedly active participants in tulipomania, which captivated growers and speculators in the Netherlands and elsewhere in Europe. In a very real sense, tulipomania epitomized the growing interest in and demand for rare and unusual plants that were moving across seventeenth-century Europe.

The interests of the Tradescants went beyond creating aesthetically pleasing gardens out of trees, shrubs, and herbaceous plants available at home and on the continent. Like an increasing number of their contemporaries, they had become smitten with a consuming interest in novelty plants. They were anxious to obtain seeds, cuttings, and other propagating material of these rarities from any quarter and to experiment with their culture and propagation. To these naturalists and gardeners, Virginia and New England were more impor-

7. Formally established in 1635 by royal edict, the king's garden in Paris, now the Jardin des Plantes, had its origins in Jean Robin's medicinal herb garden, for which he authored a catalogue in 1601. The garden, which now occupies eight hectares (about 20 acres) in the fauxbourg St. Victor, was opened to the public in 1640, at which time 2,300 species were growing in its collections.

8. Charles de l'Escluse (also known as Carolus Clusius, 1526–1609) was one of the first Dutchmen to grow tulips, native to the Near East, in his garden in Holland. This cartoon suggests that the first avaricious plantsmen to be struck by tulipomania were willing to commit a theft in broad daylight to acquire a few of the precious bulbs.

1596. The first aspirant bulbgrower acquires bulbs from Clusius' garden.

9. Thirty-nine years after the scene depicted in the first cartoon, tulip bulbs had become the objects of great speculation, and tulipomania was spreading across Europe. Occurrences such as the one depicted here were not uncommon, and many bulbs bought for speculation were never planted and allowed to flower. They were used only as a medium of exchange.

1635. A house is sold in Hoorn for three tulips.

10. The Tradescant Museum, which was developed as an adjunct to the Tradescant garden collection of plants, likely had an appearance similar to that of the museum established by Ferrante Imperato (1550–1625) in Naples, Italy. In this illustration, from the second edition of Imperato's book, *Dell'historia Naturale* (1672), Ferrante and his son Francesco explain the collection of curiosities to two visitors. Imperato was also responsible for the establishment of the botanical garden in Naples.

tant as potential sources of rarities than as new lands available for colonization. The novelties obtainable there, both plant and animal, were of greater significance than, or at least equal to, the solutions to pressing social and economic problems that the statesmen and politicians for whom they gardened sought in new territories. What was clearly becoming important to a growing number of people was the diversity of nature, the trees, shrubs, birds, reptiles, insects, and other animals that could be studied, named, described, illustrated, classified, and, in the case of plants, used to create and embellish gardens and landscapes. New plants from foreign lands—plants like the black locust and trumpet creeper—as well as plants in the European flora that, like the smokebush, had been overlooked for generations, were tantalizing indicators of additional prizes awaiting discovery.

Not content with overseeing the gardens of his patron, Charles I, the industrious elder Tradescant established his own garden in South Lambeth, where he could grow for himself all of the plants for which he could obtain

11. John Tradescant the younger succeeded his father as gardener to Charles I and continued the horticultural enterprise in the Tradescant garden in South Lambeth. The younger Tradescant also opened his home, a veritable museum of curiosities, to the public.

seeds, slips, or cuttings. His collection, aimed toward diversity, was a precursor of the modern botanical garden, but it lacked the systematic scheme that would characterize the structure of similar comprehensive gardens established in the following centuries.

Like his garden, Tradescant's home was a trove of curiosities from the world over, ranging from birds' nests from China to Edward the Confessor's knit gloves. Appropriately known as "The Ark," the Tradescant collection was the first museum of "rarities" to be opened to a wide, appreciative, and curious public.

John Tradescant the younger was also imbued with a passion for plants. Having worked side by side with his father in the South Lambeth garden, he was fully aware that discoveries of new plants awaited some enterprising plantsman. Not satisfied with unfilled requests for seeds and specimens, and unwilling to depend upon adventurers to foreign lands or continental correspondents for new rarities, the younger Tradescant pushed his enthusiasm to its logical conclusion. He decided to travel to Virginia himself, where he would have personal opportunities to observe and collect all that was of interest. This would be a far better solution than entrusting the mission to another, if an individual could be found, who would likely lack eagerness and a discerning eye for rarities. John Tradescant the younger made three trips to Virginia—in 1637, 1642, and in 1654—and returned to South Lambeth with many Virginian plants, the bald cypress among them.

Numerous other North American and Eurasian trees and shrubs have associations with the Tradescants and their garden of curiosities. The first tulip trees (*Liriodendron tulipfera* L.) and red maples (*Acer rubrum* L.) grown in Europe were brought from Virginia to South Lambeth by the younger Tradescant sometime between 1637 and 1654. Likewise, the Asian persimmon (*Diospyros lotus* L.), the magnificent horse chestnut (*Aesculus hippocastanum* L.), the shrubby bladder senna (*Colutea arborescens* L.; Plate 2), and the mock orange or "syringa" (*Philadelphus coronarius* L.) were southern European or Asian species grown in South Lambeth. The American black walnut (*Juglans nigra* L.) and red mulberry, (*Morus rubra* L.), our native shagbark hickory (*Carya ovata* (Miller) K. Koch), the familiar Virginia creeper (*Parthenocissus quinquefolia* (L.) Planchon), and even poison ivy (*Rhus radicans* L.) were other North American species that flourished as exotics in the Tradescants' garden.

Another well-known tree with Tradescant connections, the London plane (*Platanus* x *acerifolia* (Aiton)

Willdenow; Plate 3), was not brought into cultivation from the wilderness of North America or from a rugged mountain slope in the Levant. To the contrary, this tree, it is believed, first grew in a garden and has become increasingly cosmopolitan ever since. Moreover, its place of origin may have been the Tradescants' South Lambeth garden. When he returned to London from his first trip to Virginia in 1637, the younger Tradescant carried with him either seeds or vegetative propagating material of the button tree or American sycamore (*Platanus occidentalis* L.) and for the first time grew that species as an exotic in Europe. Meanwhile, by 1633 the Tradescants had already been successful in growing the oriental plane (*Platanus orientalis* L.), having obtained seeds or plants of that tree, which is native to southeastern Europe and Asia Minor, from trees already cultivated in England or perhaps from a traveler to the Middle East. With the introduction of the closely related North American species into the confines of the same garden, the prerequisites were satisfied for a chance event that would never have occurred in nature. The two species, previously separated from one another by the expanse of the Atlantic Ocean, were growing together where they might hybridize—the pollen of the oriental species functioning to fertilize the ovules concealed within the flowers of its occidental cousin, or vice versa. The resulting offspring proved unlike any tree then in existence, a tree of remarkably strong and vigorous growth and of easy propagation by cuttings or by layering.

12. German by birth, Jacob Bobart (1599–1680) assembled the first university collection of living plants in the British Isles in the physic garden at Oxford. After his death, Bobart was succeeded as curator of the garden by his son.

Not suspecting that the seeds sown from their plane trees were to give rise to a new type, but being astute plantsmen, the Tradescants would have noted the vigor, strength, and rapid growth of the seedlings. Circumstantial evidence suggests that at least one of the sturdy young saplings was given to Jacob Bobart, one of the members of the expanding circle of English plantsmen. Bobart, who had come from Germany to accept the post of custodian of the Oxford Botanical Garden, planted the sapling where it still grows today, in the garden of Magdalen College. This tree is now of great proportions and is alleged to be the oldest of all the London plane trees.

Thousands of London plane trees have been propagated from the Tradescants' small group of originals. As industrialization altered the environments of cities across Europe and North America during the nineteenth century, the hybrid proved to be exceedingly tolerant of the coal dust, smoke, and compacted soils of cities.

Frequently attaining heights in excess of one hundred feet, this majestic shade tree has added immeasurably to the grandeur of the avenues, boulevards, parks, and squares of almost every European and North American metropolis, and even some urban areas in China, as travelers to the old capital city of Nanking (now Nanjing) will testify. Several of these descendant trees tower above the Arnold Arboretum landscape near the Center Street gate, their trunks remarkable for their girth and the smooth, flaking bark, which is mottled greenish, grayish, and buff. Seldom noticed are the tree's flowers, which appear in spring and are hidden by the expanding leaves, but the annual crop of seed-filled, ball-like fruits that hang from the outspreading branches throughout the winter months never fail to attract attention and help to identify what might qualify as the first hybrid tree of garden origin.

Across the road from the London planes in the triangle of Arboretum land formed by the junction of Bussey Hill and Valley roads is another tree that produces pendulous, spherical fruits, which remain

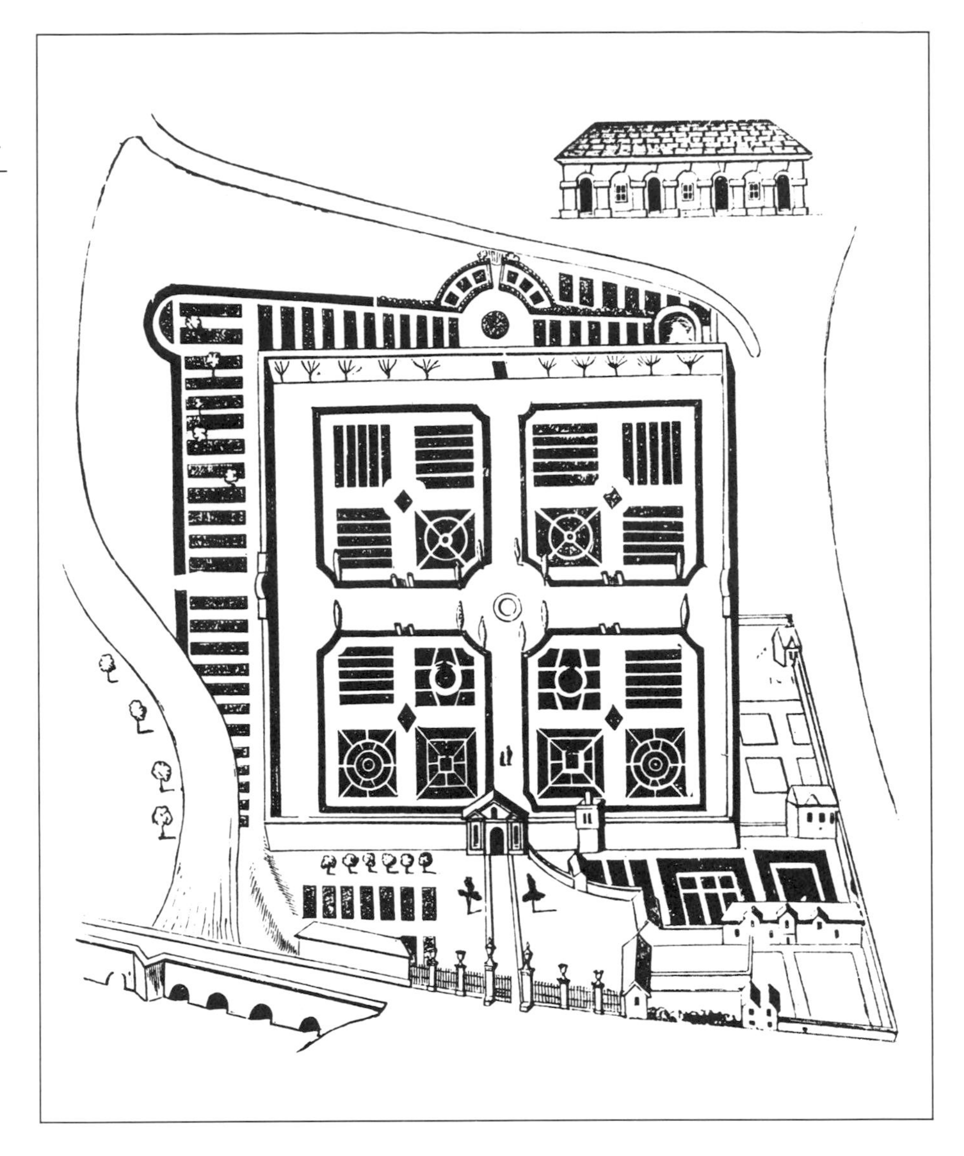

13. Plan of the Oxford Botanical Garden, which was founded in 1621. Now the University Botanic Garden, it still welcomes visitors to its extensive collections.

suspended on the tree's naked branches throughout winter. Commonly known as sweet gum or red gum (*Liquidambar styraciflua* L.), this species produces fruit clusters that become hard and woody, unlike the soft and easily fragmented ones of the London plane. Each fruit cluster of the sweet gum is an aggregate of numerous two-valved capsules, and each capsule valve has a pincerlike beak. It is the numerous beaks that combine to give the clusters their unique appearance, and when the individual capsules have opened in the late fall, the small seeds enclosed within are gradually sifted from the clusters by strong winds over the winter months.

The tree's common name, sweet gum, as well as both its Latin generic name, *Liquidambar,* and its specific name, *styraciflua,* allude to a fragrant, semitransparent, yellowish-brown juice or gum that exudes from fissures in the bark of the tree. When exposed to the air, this resin hardens and takes on an amber color. Despite its sweet balsamlike odor, sweet gum resin has a bitter taste; and yet, ironically, it has been used as a chewing gum to sweeten the breath. The tree is more widely known for its lustrous, green, five-to-seven-lobed, almost star-shaped leaves that turn crimson to purplish in fall, its frequently corky, winged branchlets, and its familiar fruit clusters.

While the sweet gum does not attain the stature of the London plane when cultivated in New England, it does attain heights between eighty and one hundred feet in the woodlands and forests where it occurs naturally. Primarily a tree of the rich soils of moist river bottoms, it

14. The massive trunk of one of the towering London plane trees growing near the Center Street gate in the Arnold Arboretum.

15. A close-up photograph of the mottled bark pattern of a London plane tree, a widely cultivated hybrid with Tradescant connections.

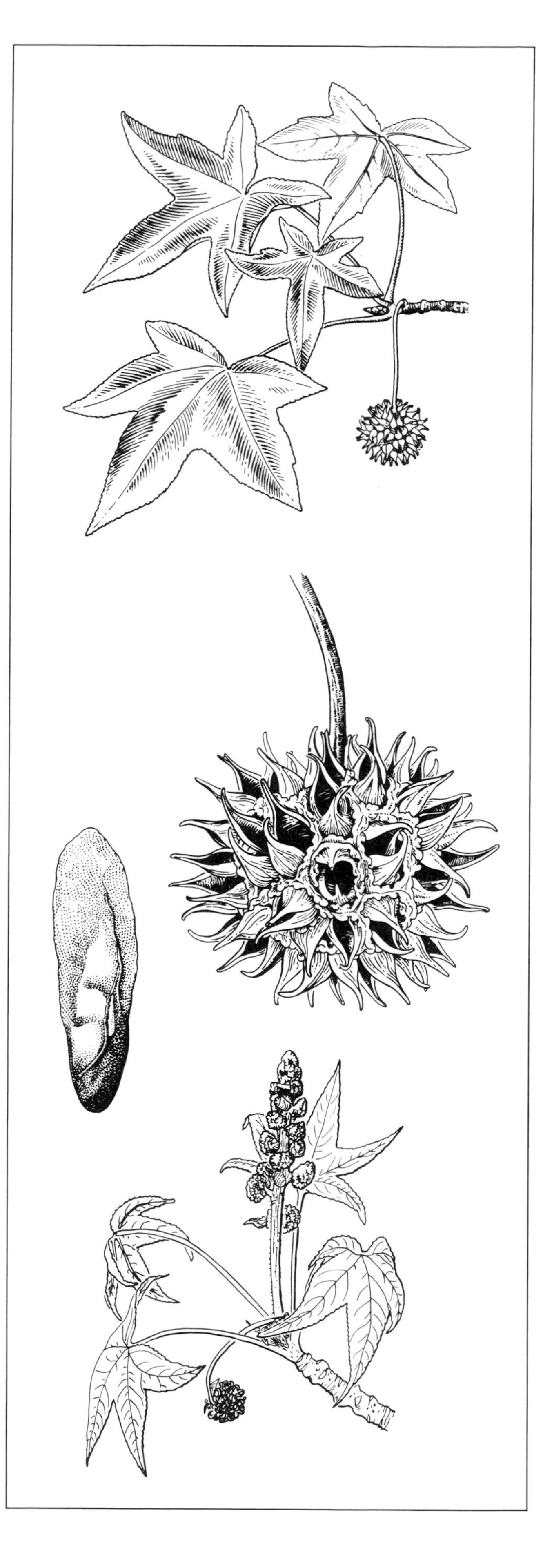

16. As their leaves expand in the spring, sweet gum trees (*Liquidambar styriciflua*) produce staminate and carpellate flowers at the ends of branchlets. The staminate flowers are aggregated into erect inflorescences, while the carpellate flowers are pendulous and develop into the distinctive fruit aggregates. Also illustrated are fruit aggregates of the sweet gum and a fertile seed. Abortive seeds are also frequent.

Plate 1. The bright orange, tubular flowers of the trumpet creeper (*Campsis radicans*) are produced over a long period from early summer until frost. The trumpetlike shape of the corollas gives this climbing shrub its common name.

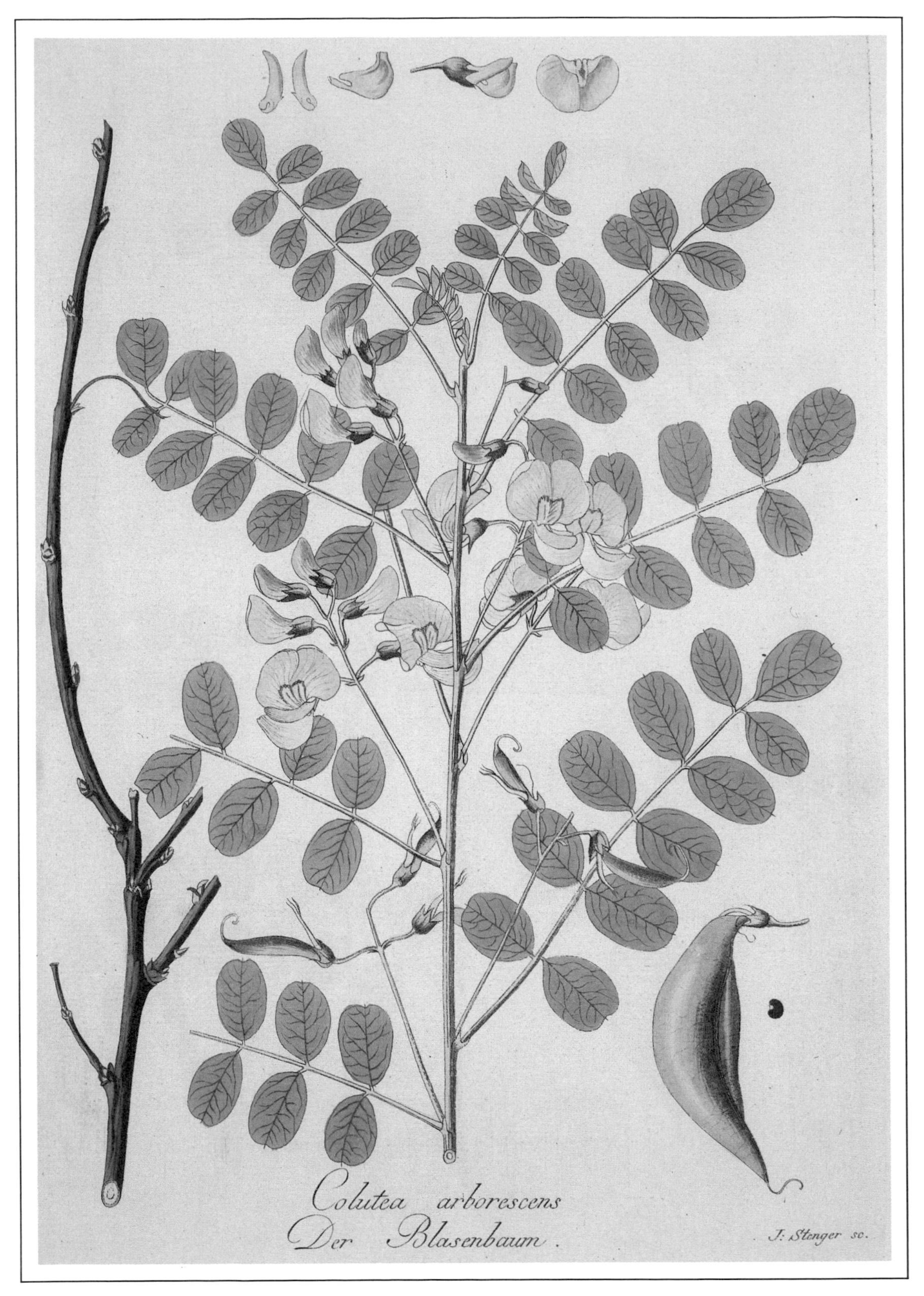

Plate 2. This hand-colored illustration of the bladder senna (*Colutea arborescens*) highlights its bright yellow, pealike flowers. The inflated, bladderlike fruits, which are frequently tinged pink or red, are attractive in their own right and add ornamental interest to these southern European and North African shrubs.

ranges from southwestern Connecticut to Florida and westward into Texas and Missouri, and outlying populations extend its distribution southward into the mountains of eastern Mexico and the highlands of Guatemala. Surprisingly, it was from Mexico—where the gum was used to flavor tobacco—that the first reports of this tree reached Europe. It was described by the Spanish naturalist Francisco Hernandez in 1615. The sweet gum was not grown as a foreign exotic in Europe, however, until around 1688, some twenty-six years after the younger Tradescant had died.

By 1688 the treasures of "The Ark" had been crated by Elias Ashmole, who had come into their possession on the younger John Tradescant's death. Ashmole shipped the rarities to Oxford, where they formed the nucleus of the Ashmolean Museum at the university, a collection that would have been more suitably named for the Tradescants. Artifacts in the collection can still be viewed today, but the South Lambeth garden in which the Tradescants had toiled was already neglected by 1688, and no trace of it now survives. However, in 1688 another garden on the banks of the River Thames was flourishing and had taken the place of the Tradescant garden as an oasis of rare plants from foreign lands. This garden was on the left bank of the river, upstream from Westminster and Lambeth in the then-rural London borough of Fulham.

An Amphiatlantic Diocese

The plants in the garden were carefully cultivated by one George London, who first germinated the small, winged seeds of the sweet gum in European soil and who undoubtedly marveled at the intricate, finely chiseled appearance of the fruit aggregates from which the seeds had fallen. But unlike the Tradescants, London was not tending his personal garden, as the collection of plants for which he was responsible occupied the grounds of Fulham Palace, the residency of the Lord Bishop of London. The Reverend Henry Compton, Oxford-educated and a staunch defender of Protestantism and the Church of England, was the current occupant of the palace and served as Lord Bishop of London from 1675 until his death in 1713. It was this energetic clergyman, simultaneously an enthusiastic plantsman, who had brought together over a thousand species of alien plants to his palace grounds and could claim to be the first to cultivate the sweet gum in Europe.

As Bishop of London, Compton was directly responsible for the Anglican Church in the American colonies. These duties included the appointment of university-trained clergymen who would leave England to establish its Protestant Church in the American wilderness and attend to the spiritual instruction of the colonists. At home, Compton himself was charged with the education of the princesses Mary and Anne. While Compton's career in the Church had been checkered, largely because of the unsuccessful attempt of James II to restore Roman Catholicism in England, Compton had held his post despite reduced church duties. Very likely as a respite from his political activities and ecclesiastical obligations and to fill the time freed during James's brief three-year reign, Compton balanced his life by turning to his fascination with plants and the cultivation and enhancement of the thirty-six-acre palace garden.

One of Compton's appointees to the Virginia Colony was the Reverend Mr. John Bannister (1650–1692), another Oxford-trained cleric who had received a bachelor's degree from Magdalen College in 1671 and who, as a student, may have studied in the cool shade of the young London plane planted in the college garden. Bannister was also at the university when, in 1669, Robert Morrison (1620–1683), who had studied with the Robins in Paris, was named Professor of Botany. This appointment was the first recognition of botany as an academic discipline in England. Bannister, like Morrison and the bishop he served, had also developed an abiding interest in plants and natural history in general, and when he arrived in Virginia in 1678, his intent was to write a natural history of the region. And so, through Bannister, the Lord Bishop of London had a direct source of plants and seeds from Virginia.

From Bannister—who was eagerly engaged in collecting, describing, and attempting to recognize previously discovered plants as well as new curiosities—Compton received the seeds of the sweet gum. Other American plants that came to the Fulham Palace garden by way of Bannister included the sweet bay (*Magnolia virginiana* L.), the first magnolia to be cultivated in Europe, and the swamp honeysuckle or clammy azalea (*Rhododendron viscosum* (L.) Torrey). Both of these "Virginian" plants are wide-ranging in eastern North America, and they grow naturally in swampy woods and along streams in New England. Of the two, the swamp honeysuckle is far more common in the northeast; the sweet bay reaches the northern limits of its natural distribution in one small population a short distance north of Boston, a fact

17. The generic name of the sweet fern, *Comptonia,* was named to commemorate Henry Compton (1632–1713), Lord Bishop of London and early connoisseur of North American plants. This low shrub of sterile soils in woodlands, clearings, pastures, and roadsides is widely distributed in eastern North America. In the Arnold Arboretum it is cultivated along Oak Path near the sweet gums and London plane trees. Among the plants Compton was the first to cultivate was a South African geranium (*Pelargonium inquirans* (L.) Aiton) that served as one of the parents of the common hybrid geranium (*Pelargonium* x *hortorum* L. H. Bailey) so frequently used today in gardens everywhere as a bedding plant or on porches and patios and in window boxes as a pot plant.

reflected in the name of the small coastal community of Magnolia, Massachusetts.

Bannister also communicated with plantsmen and naturalists in England other than Compton. He dispatched seeds, dried plant specimens, and other natural-history objects that proved to be of enormous interest. Seeds were sent to Morrison and Jacob Bobart, the younger, at the Oxford Botanical Garden and to John Ray in Essex. Specimens as well as his carefully delineated drawings of plants were sent to Leonard Plukenet (1642–1706) in London. Of equal significance, Bannister forwarded descriptions of Virginian plants that he had written based on his personal observations. Brought together to form catalogues, some of these descriptions were published by Ray in the second volume of his *Historia Plantarum Generalis,* which appeared in 1687, while Plukenet published many of Bannister's drawings. Bannister's plant descriptions actually consisted of cumbersome Latin phrase names that simultaneously described, classified, and differentiated plants. Thus, the sweet gum was referred to as *Styrax folio Aceris*—literally, the maple-leaved styrax, styrax being the ancient name of a resin-producing tree of the Mediterranean region (now known as *Styrax officinale* L.). Because both trees produced a gum, Bannister thought them related. Appended to the Latin phrase name were observations set down in English. "Here it is called the sweet gumme. It weeps when wounded a white resinous gumme, of a pleasant savour, & some say very balsamick . . . It bears Balls like those of the Plane, but echinated, every prickle being a seed vessel" (quoted in Ewan and Ewan, 1970, p. 250).

John Ray, who had received his education in Trinity College at rival Cambridge University, was also a Protestant cleric of scholarly bent and had achieved an academic position at his alma mater. In 1662, during Charles II's reign, he banished himself from his university post and active participation as a clergyman because he had been unwilling to assent to the Act of Uniformity, Charles's attempt to restore Roman Catholic traditions to church and university. Although of humble means, Ray returned to his Essex birthplace and devoted the remainder of his life to the study of natural history. In conjunction with these studies, he had previously traveled widely over the whole of Britain on natural-history forays and had written the first full-fledged flora of England. By 1682 Ray was engaged in preparing a world flora, his *History of Plants,* and was delighted to receive information concerning American plants. His goal was to account for all plants previously described and to include as many new plants as possible. At the same time, Ray was deeply concerned with the artificiality of the plant classification systems of the day and realized that any improved system would be dependent upon the use of characteristics from all aspects of a plant's structure—not just a few selected ones. His emphasis was on the overall similarities of one plant to another, a concept that in practice he believed would lead to a natural classification, not an artificial one.

Bannister's intentions had undoubtedly been inspired in part by Ray's example. But his Virginia natural history was not to be, nor was he to continue to provide plant materials and first-hand information from Virginia. In May of 1692, while exploring for plants along the Roanoke River in Henrico County, Virginia, Bannister met an untimely death when, at the age of forty-two, he was accidentally shot and killed. The Lord Bishop of London mourned deeply when he received word of Bannister's demise and quickly sought as a replacement another cleric with a bent toward natural history. Ray and Plukenet were equally grieved at the grim news. To Compton, Bannister had been a faithful servant of the Church as well as a source of new plants for the bishop's horticultural endeavors; to Ray and Plukenet, Bannister had been a faithful scientific colleague who had provided new and otherwise unobtainable botanical information. During his lifetime Bannister had seen botany take its place in the academic curriculum, marking its initial distinction from horticulture. Yet he had served both horticulture and botany diligently, as would others who

18. John Ray (1627–1705), Anglican clergyman and member of the Royal Society, has been called the "father of English natural history."

followed him to Virginia and North America shortly after the turn of the new century.

Of the plants described by Bannister, one—listed in his catalogue under the awkward Latin phrase name *Laurus tinus floribus albidis eleganter bullatis*—was not included in Lord Compton's famous garden collection of American plants. However, a misleading and inaccurate illustration of the plant, perhaps based on a drawing by Bannister, was published by Plukenet. As one of the most beautiful of all the small trees or shrubs of the eastern North American forest, it seems surprising that the fringe tree (*Chionanthus virginicus* L.), sometimes known as old man's beard, was not grown in England by the turn of the eighteenth century. Two factors, either alone or in combination, may have conspired to frustrate attempts to obtain and cultivate this charming plant.

While the fringe tree (Plate 4) is widely distributed along the eastern seaboard from New Jersey and Pennsylvania southward into Florida and through the Gulf states into Arkansas and Texas, it is nowhere common, and the flowers of an individual plant are functionally either male or female. As a consequence, a crop of fruits is not always produced unless carpellate plants are growing in the same locale as staminate plants. If this condition is satisfied, pollen from the staminate flowers must be successfully transferred, by pollinating insects, to the carpellate flowers to ensure fertilization and the production of seeds. Moreover, if fruits and seeds have developed by late summer, the collector who wishes to attempt their germination and cultivation must gather them before they have been devoured by flocks of migrating birds.

The germination of these seeds is not a straightforward matter, either, as they must be exposed to alternating periods of warmth, cold, and warmth in order to break their inherent dormancy. Under natural conditions, these shifting temperature requirements may not be met until two years after the fruits have been dispersed from the parent plant. In the garden, unless the plantsman who has sown the seeds is familiar with these prerequisites—or perhaps forgetful or cautiously patient—the experiment could easily be given up as a failure after one growing season. Such were the unknown and unexpected biological idiosyncrasies faced by John London and other gardeners and nurserymen who were experimenting with seeds that had never before been coaxed to germinate in cultivation.

The late spring flowering of the fringe tree creates a display that even an uninterested, unobservant urbanite would ogle. In the Arnold Arboretum these spectacular plants can be seen growing on either side of Bussey Hill Road at the upper end of the lilac collection. Their delicate, four-petaled white flowers literally cover the plants in late May and early June and appear in sharp contrast with the intensely green leaves that emerge and expand as the flowers appear. In nature, where the plants frequently grow on bluffs and rock outcrops above forested stream banks, a flowering specimen looks like a shrub or small tree festooned with snowy tinsel. And if fruits are produced, the plant has a brief, second period of ornamental beauty. These fruits, technically called drupes, assume a dark blue or purplish color and an olivelike shape at maturity. The solitary pit or seed embedded in the fruit's fleshy walls is another telltale indication that *Chionanthus* is classified as a member of the Oleaceae or olive family.

The Colonial Audubon

Mark Catesby (1683–1749), an English naturalist who visited his sister and brother-in-law in Williamsburg, Virginia, between 1712 and 1719, was sure to have noticed the fringe tree. In 1722, when he returned to Charles Town (now Charleston), South Carolina, for a briefer but more productive sojourn, he again encountered this species. On his second trip, Catesby journeyed to North America as a fully equipped naturalist, complete with collecting paraphenalia and the apparatus of an artist. His kit included diggers and quires of brown paper in which plant specimens were to be pressed, quantities of bottles and jars of various sizes with corks and sealing wax to make them tight, and plenty of wine, rum, or brandy in which insects, reptiles, and small

19. As the desire for curiosities and natural-history artifacts increased, publications and broadsides like this one issued by James Petiver served to alert sea captains, ships' surgeons, and intelligent travelers in foreign lands that they could perform a valuable service by collecting materials and samples and sending or bringing them back home. Petiver (1663/4–1718) was an apothecary who became a demonstrator at the Chelsea Physic Garden and a Fellow of the Royal Society. An indefatigable collector, he established a personal cabinet of natural history, which was purchased by Sir Hans Sloane in 1718. Consequently, many of Petiver's curiosities are today preserved in the collections of the British Museum.

Brief Directions *for the Easie Making, and Preserving* Collections *of all* NATURAL Curiosities.

For IAMES PETIVER *Fellow of the Royall Society* LONDON.

All *small* Animals, *as* Beaſts, Birds, Fiſhes, Serpents, Lizards, *and other* Fleſhy Bodies *capable of* Corruption, *are certainly preſerved in* Rack, Rum, Brandy, *or any other Spirits; but where these are not eaſily to be had, a ſtrong* Pickle, *or* Brine *of Sea Water may ſerve; to every* Gallon *of which, put 3 or 4 Handfulls of* Common *or* Bay Salt, *with a Spoonful or two of* Allom *powder'd, if you have any, and ſo ſend them in any* Pot, Bottle, Jarr, *&c. cloſe ſtopt, Cork'd and Roſin'd. N.B. You may often find in the* Stomachs *of* Sharks, *and other great Fiſh, which you catch at Sea, divers ſtrange* Animals *not eaſily to be met with elſewhere; which pray look for, and preſerve as above.*

As to Fowls, *those that are large, if we cannot have their* Caſes *whole, their* Head, Leggs, *or* Wings *will be acceptible, but ſmaller* Birds *are eaſly sent entire, by putting them in Spirts, as above, or if you bring them dry, you must take out their* Entrals; *which is beſt done by cutting them unde^r^ their Wing, and then ſtuff them with* Ockam *or* Tow, *mixt with* Pitch *or* Tar; *and being thorouh= ly dried in the Sun, wrap them up cloſe, to keep them from Moiſture, but in long Voyages, you must Bake them gently, once in a Month or two, to kill the* Vermin *which often breed in them.*

All large pulpy moist Fruit, *that are apt to decay or rot, as* Apples, Cherries, Cowcumbers, Oranges, *and ſuch like, must be sent in* Spirits *or* Pickle, *as* Mangoes, *&c. and to each Fruit, its desired you will pin or tye a ſprig of its* Leaves, *and* Flowers.

All Seed *and dry Fruit, as* Nutts, Pods, Heads, Huſks, *&c. these need no other Care, but to be ſent whole, and if you add a* Leaf *or two with its* Flower, *it will be the more instructive, as alſo a piece of the* Wood, Bark, Root, *or* Gum *of any Tree or Herb that is remarkble for its* Beauty, Smell, Uſe, *or* Vertue.

In Collecting PLANTS, *Pray observe to get that part of either* Tree, *or* Herb, *as hath its* Flower, Seed, *or* Fruit *on it; but if neither, then gather it as it is, and if the* Leaves *which grow near the* Root *of any Herb, differ from those above, be pleaſed to get both to Compleat the Specimen; these must be put into a* Book, *or* Quire *of* Brown Paper *ſtitch'd (which you must take with you) as ſoon as gathered; You must now and then ſhift these into freſh Books, to prevent either rotting themselves or Paper. N.B. All* Gulph-Weeds, Sea-Moſſes, Coralls, Corallines, Sea Feathers, Spunges, *&c. may be put altogether into any old* Box, *or* Barrel, *with the* Shrimps, Prawns, Crabs, Crawfiſh, *&c. which you will often find amongst the* Seaweeds, *or on the Shoar with the* Shells, *which you may place in layers; as we do a* Barrel *of* Colcheſter *Oysters.* *All* SHELLS *may be thus ſent as you find them, with or without their* Snails *in them, and wherever you meet with different sizes of the ſame sort, pray gather the fairest of all Magnitudes; the* Sea ſhells *will be very acceptible, yet the* Land, *and* Freſh-water *ones, are the most rare and deſirable. In Relation to* INSECTS, *as* Beetles, Spiders, Graſshopper, Bees, Waſps, Flies, *&c. these may be Drowned altogether, as ſoon as Caught in a little wide Mouth'd Glaſs, or Vial, half full of* Spirits, *which you may carry in your Pocket. But all* Butterflies *and* Moths, *as have mealy Wings, whose Colours may be rub'd off, with the Fingers, these must be put into any ſmall Printed Book, as ſoon as caught, after the same manner you do y^e^* Plants.

All Metals, Minerals, Ores, Chryſtals, Spars, *Coloured* Earths, Clays, *&c. to be taken as you find them, as also ſuch formed* Stones, *as have any resemblance to* Shells, Corals, Bones, *or other parts of* Animals, *these must be got as intire as you can, the like to be Observ'd in* Marbeld Flints, Slates, *or other* Stones, *that have the Impression of* Plants, Fiſhes, Inſects, *or other Bodies on them: These are to be Found in* Quarries, Mines, Stone *or* Gravel *Pitts,* Caves, Cliffs, *and* Rocks, *on the Sea ſhoar, or wher= ever the* Earth *is laid open.* NOTE *If to any* ANIMAL, PLANT MINERAL *&c. you can learn its* Name, Nature, Vertue *or* Use, *it will be still the more Acceptible.*

N.B. As amongst Forreign Plants, *the most common* Graſs, Ruſh, Moſs, Fern, Thiſtle, Thorn, *or vileſt* Weed *you can find, will meet with Acceptance, as well as a ſcarcer Plant; So in all other things, gather whatever you meet with, but if very common or well known, the fewer of that* Sort, *will be accep= tible to*

Y^r^ most Humble Servant

Alders gate ſtreet
LONDON.

IAMES PETIVER.

mammals could be preserved. Nests of pill boxes and several sizes of so-called deal boxes constructed of pine or fir would be necessary for transporting seeds, fruits, and samples of minerals and ores; and, naturally, plenty of drawing paper was included on which sketches of birds, fish, reptiles, mammals, and plants could be made. Like Bannister, Catesby came to the untamed North American wilderness enthusiastically prepared to obtain the materials necessary to compile a natural history.

Catesby was fortunate that when he arrived in South Carolina—still a pristine and virgin land compared with long-settled tidewater Virginia—he already had an established friendship with the colony's new governor, Francis Nicholson. The governor, as Catesby's colonial patron, offered him logistical assistance, letters of introduction, and an annual stipend to boot. This welcome support augmented the financial backing of a circle of eleven patrons in England and the unofficial blessings and good will of the Royal Society of London. Chief among his patrons at home was the influential Sir Hans Sloane (1660–1753), then President of the Royal Society, who was in the midst of preparing for publication the second volume of his *Natural History of Jamaica.* The first volume of that work had appeared in 1707 and stood as the prototype against which Catesby could measure the value of his own collections and observations. Also included in the list of patrons was William Sherard (1658?-1728), Oxonian botanist and fellow of the Royal Society, who became England's leading botanical academician after the death of John Ray in 1705.

For four years Catesby traveled in the Carolinas and Georgia, moving his equipment from one place to another as he pushed through the wilderness. In addition to recording his observations, he collected and prepared botanical and zoological specimens for shipment to his patrons. These shipments included fruits and seeds and, when it could be managed, living plants in tubs of soil. Friendly Indians—then at peace with the English—helped with his cumbersome load and on several excursions guided the naturalist up the dark waters of the Savannah River to the outpost of Fort Moore near the present site of Augusta, Georgia. These journeys allowed for collections and observations in the rougher terrain along the fall line—the region where the Coastal Plain abuts the Piedmont, the broad belt of foothills east of the Appalachian escarpment. This inland region was virtually unexplored and unpopulated by the English colonists, who preferred the tidewater regions of the Coastal Plain north and south of Charles Town for settlement. Travel across the flat but frequently swampy

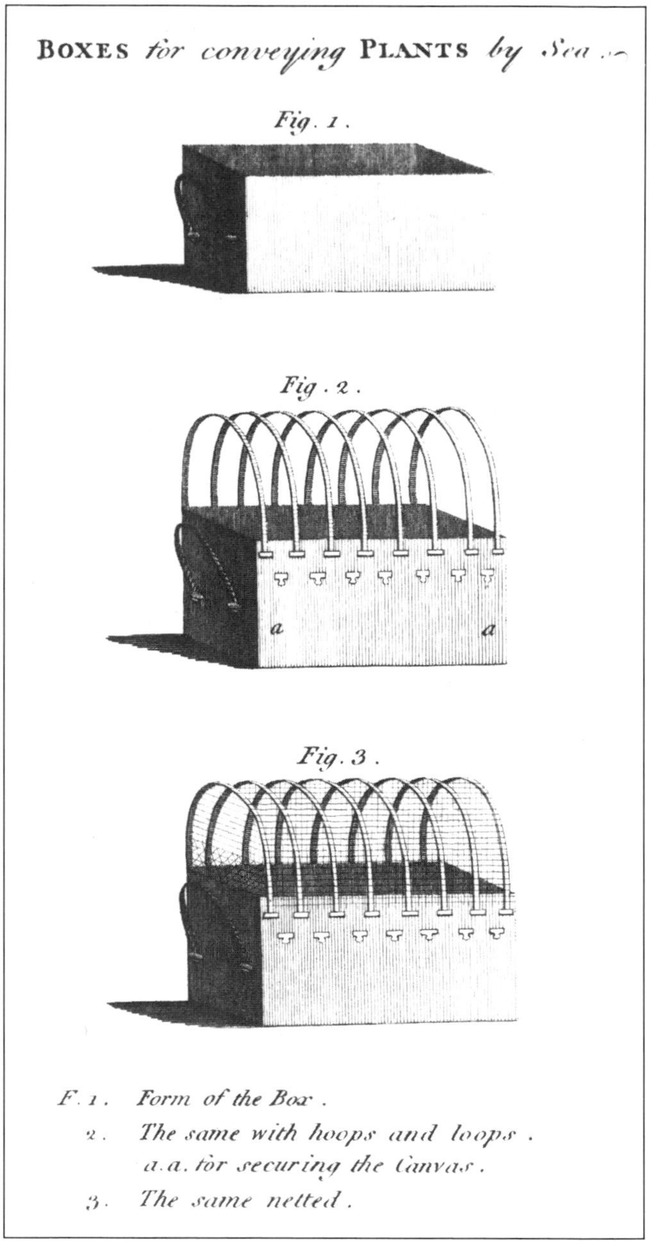

BOXES *for conveying* PLANTS *by Sea.*

Fig. 1.

Fig. 2.

Fig. 3.

F. 1. Form of the Box.
2. The same with hoops and loops.
a. a. for securing the Canvas.
3. The same netted.

20. Boxes of the sort used to transport living plants on transatlantic crossings during Catesby's era. Fastened to the deck of a ship, the plants growing in these containers frequently suffered from salt spray and a lack of fresh water.

land adjacent to the coast was difficult at best, but widely scattered settlements did afford shelter and hospitality, particularly with the letters of introduction written by the governor. Further inland, travel was more troublesome, and shelters of Indian fashion had to be constructed to ensure that collections and equipment were kept dry.

Fortunately, Catesby brought enthusiasm to his task, as his was a prolonged and arduous adventure that was not without danger, disappointment, and frustration. For one three-month period he was unable to make any collections at all, as he was bedridden with a swollen infection of his face. While on a trek in the vicinity of Fort Moore, his party was confronted by a band of sixty Indians who, thankfully, proved friendly. In another incident, a rattlesnake slithered from beneath the counterpane one morning as Catesby's bed was being made. Apparently the reptile had spent the night under the covers, warmed by the recumbent body of the fatigued naturalist. Despite the fact that Catesby was thousands of miles from London, letters from his patrons containing demands for additional materials and complaints about quantities and quality were anything but encouraging. Little did his patrons understand the difficulties the naturalist faced in poorly explored country inhabited by venomous serpents and potentially hostile Indians. But most discouraging of all, Catesby learned of the possible wholesale destruction of one of his shipments of specimens by plundering pirates. This disheartening news came when the ship on which the collections had been loaded for transport to England was forced to return to the Charles Town harbor to repair damage sustained in the attack.

En route back to England, Catesby spent additional time in 1725 and 1726 collecting in the Bahamas, where the beautiful fish of the warm subtropical waters were of particular interest. Once re-established in England, the naturalist set to work to organize, name, describe, and illustrate for publication the results of his four-year enterprise, a monumental undertaking in itself that was to require twenty years to reach completion. Catesby also turned to gardening, probably in association with Christopher Gray, a nurseryman in Fulham who was growing many of Catesby's plant introductions. As a consequence, Catesby began requesting additional seeds and plants from friends he had made in Carolina and from his sister, who still resided in Williamsburg.

Two large folio volumes comprising 220 personally engraved and individually hand-colored plates, coupled with descriptions in Latin and observations in English and in French, eventually resulted from Catesby's labors. The plates vividly displayed the botanical and zoological richness of the lands, waters, and skies Catesby had explored. Plate 86 in his *Natural History of Carolina, Florida, and the Bahama Islands* included an illustration of the fringe tree, the first botanically accurate drawing of the plant. And in plate 46 Catesby illustrated the Carolina spicebush or Carolina allspice (*Calycanthus floridus* L.), seeds of which he had brought with him from Carolina on his return to England in 1726.

Plants of Carolina allspice are included in the collections of the Arnold Arboretum, where they grow beneath the towering tulip trees near the junction of the bridle path and Meadow Road; other specimens grow across the road in the azalea border at the edge of the meadow. This low, rounded, dense shrub, which rarely attains six to eight feet in height, does not compare with the fringe tree as an ornamental. Yet it is a charming plant in its own right, noted for its dark green, lustrous leaves, which are borne in pairs along the branches and twigs. The branches themselves are paired at the nodes. And its flowers—produced freely in late spring and into the summer months—emit a pineapplelike fragrance and consist of numerous petal-like structures termed tepals, which vary in color from a deep brownish maroon to a rich coppery red. Occasional plants even produce greenish-white flowers, but these were a later discovery and unknown to Catesby. In commenting on this plant, Catesby wrote, "The bark is very aromatic, and as odoriferous as cinnamon. These Trees grow in the remote and hilly parts of *Carolina*, but no where amongst the Inhabitants" (Catesby, 1731–1743, 1:46). In a later period, the bark was used as a substitute for cinnamon, and the camphorlike odor of the leaves, wood, and bark were responsible for its common name.

In referring to these shrubs as trees, Catesby's memory had failed him, but in recording their habitat, he was correct in stating that it did not occur in the vicinity of Charleston (Charles Town in Catesby's day). Carolina allspice is primarily a shrub of rich woods on the Piedmont and in the mountains from Georgia northward into Virginia and West Virginia. It seems probable that Catesby encountered the plant on one of his treks to Fort Moore, when he had opportunities to sample the Piedmont flora.

Catesby indicated that another plant, the Indian bean or common catalpa (*Catalpa bignonioides* Walter), was also not known to the colonists of coastal Carolina "till I brought the seeds from the remoter parts of the country." He continued, "And tho' the inhabitants are little

21. This charming drawing of Carolina allspice (*Calycanthus floridus*) appears as plate 46 in Mark Catesby's *Natural History*. The bird, called the chatterer by Catesby, is undoubtedly a cedar waxwing.

curious in gardening, the uncommon beauty of this Tree has induc'd them to propagate it; and 'tis become ornament to many of their gardens" (Catesby, 1731–1743, 1:49). Catesby reputedly first found the tree growing in the fields of Cherokee Indians, and it was their name for the plant, catalpa, that he adopted. He carried seeds of this species, or perhaps the slender, elongated pods or "beans" in which they are produced, back to England in 1726, and the tree has been an object of cultivation in both Europe and America from the time of Catesby's second sojourn in the Southeast.

Specimens of these broadly rounded trees, which attain heights of thirty to forty feet, also grow in the Arnold Arboretum on the slope of Bussey Hill above the lilac collection, a site far removed from their native habitats along rivers and streams in Georgia, Florida, and westward into Louisiana. But the large, erect clusters of yellow and lavender spotted, tubular white flowers that cover these trees in June suggest a tropical lushness and connote their southern origin.

In London, Catesby was impatient to converse with his patrons—the long delay in the transatlantic delivery of correspondence behind him—and to meet and to express his gratitude to Peter Collinson, a Quaker merchant and keenly eager plantsman who had raised additional funds to support the naturalist's activities in America. In these conversations, news of the failure of Thomas More (ca. 1670-ca. 1720), who had become known as the Pilgrim botanist, reached Catesby's ears. More had been sent to New England shortly after Catesby left for Carolina, to undertake a mission similar to Catesby's. Basing himself in Boston, More became alarmed by the general disregard of the New England colonists for His Majesty's forests and undertook unsuccessful political maneuvers to be appointed Royal Forester in New England. He returned to England without satisfying his patrons' longing for specimens and seeds, and the only plant he is credited with introducing into English gardens is the white ash (*Fraxinus americana* L.).

Catesby was also eager to visit the Physic Garden in the London borough of Chelsea. While the garden had been established by the Worshipful Company of Apothecaries in 1673, it was not until 1722, the year Catesby left England for the wilderness of America, that Sir Hans Sloane placed the collection on a firm financial basis and thereby ensured its future. In return for his support, Sloane's only proviso was that the garden staff prepare dried specimens of fifty plants growing in the garden each year for transmittal to the Royal Society.

22. Peter Collinson (1694–1768) gardened at Peckham until 1749, when he established a botanical garden at Mill Hill, where he specialized in cultivating North American plants. Collinson organized a seed trade in North America, in which John Bartram played the essential role of collector. Once shipments from Bartram arrived in England, Collinson would divide and distribute the collections to the various subscribers. In 1766 Collinson wrote that no more would the seed distribution be complete than he would receive requests from the subscribers begging, "Pray Sir how and in what manner must I sow them,—pray be so good, Sir, as to give mee some directions, for my Gardener is a very Ignorant Fellow."

23. In this view of the Chelsea Physic Garden from the banks of the River Thames, two large specimens of the cedar of Lebanon dominate the scene. Cedars of Lebanon (*Cedrus libani* A. Richard) are also hardy in the Boston region, and large specimens grow on Bussey Hill in the Arnold Arboretum.

Moreover, on Sloane's recommendation, a new gardener and administrator, a Scots horticulturist by the name of Philip Miller (1691–1771), arrived in 1722 to steer the course of the garden until 1770.

The changes that Miller had begun at the Physic Garden and Miller's first book, *The Gardeners and Florists Dictionary, or A Complete System of Horticulture*, which was published in 1724, greatly impressed the naturalist who had recently returned from Carolina. And Collinson's remarkable garden at Peckham convinced Catesby that his efforts in plant introduction were well worth the inconveniences and difficulties he had experienced in the wilds of Carolina. The topic of conversation on his visits to these gardens focused on Catesby's experiences and intimate knowledge of American plants. He was frequently called upon to identify glaring gaps—plants missing from the collections of "Americans"—and to suggest correspondents in the colonies who might be persuaded to gather propagating material for dispatch to England.

Peter Collinson continued to serve as Catesby's patron during the years of labor required to publish the *Natural History of Carolina*, and it was Collinson who was finally successful in growing the fringe tree in England. Collinson was also to succeed in cultivating one of the most curious plants to be received from the English colonies in America, one that had figured in Catesby's *Natural History.* A large shrub or small tree with ten-to-twelve-inch obovate leaves, which have an alternate arrangement along the branchlets and emit a disagreeable odor when bruised, the pawpaw (*Asimina triloba* (L.) Dunal) is most noted for its unusual flowers and equally unusual but sweet—some say insipid—edible fruits. The flowers, which appear as the leaves of the plant are expanding in spring, occur individually and are borne on stout pedicels or stalks that arise from buds on two-year-old twigs. Each consists of three greenish sepals, which are densely hairy on the outer surface, and six, thick, heavily and deeply veined petals, which are initially green but, like the tepals of the Carolina allspice flowers, gradually turn purplish or maroon. Enclosed within the central portion of the flower is a ball-like aggregate of stamens that surround the erect protruding carpels. The fruits that develop and ripen in the fall emit a cloying aroma that suggests a cheap perfume. Suspended individually or paired (or rarely in groups of three or four together) from the branches of the tree, they somewhat resemble small Idaho potatoes or strange, fat, underlength sausages. Initially green, the smooth skin of these fruits turns yellow-brown. When fully ripe in September or October, the yellowish pulp, which surrounds the numerous flat, brown, and rather large watermelonlike seeds, is watery and can be sucked from the leathery walls of the fruit.

The pawpaw is most common in the forests of the Mississippi River valley, but it occupies a broad range from eastern Pennsylvania west to Michigan and Illinois and eastern Kansas and southward into Florida and Texas. In this region the plant is an understory tree or shrub that sometimes forms dense thickets in the rich, moist soil of river bottoms and along stream banks. In the Arnold Arboretum also, specimens of the pawpaw grow as understory trees in the shade of larger trees adjacent to the Center Street beds behind the hickory plantation.

A far more handsome small tree, which was also illustrated by Catesby, was first successfully grown by Philip Miller in the Chelsea Physic Garden. Because of its attractive inflorescences of numerous white flowers and the color its leaves assume in fall, plants of the sorrel

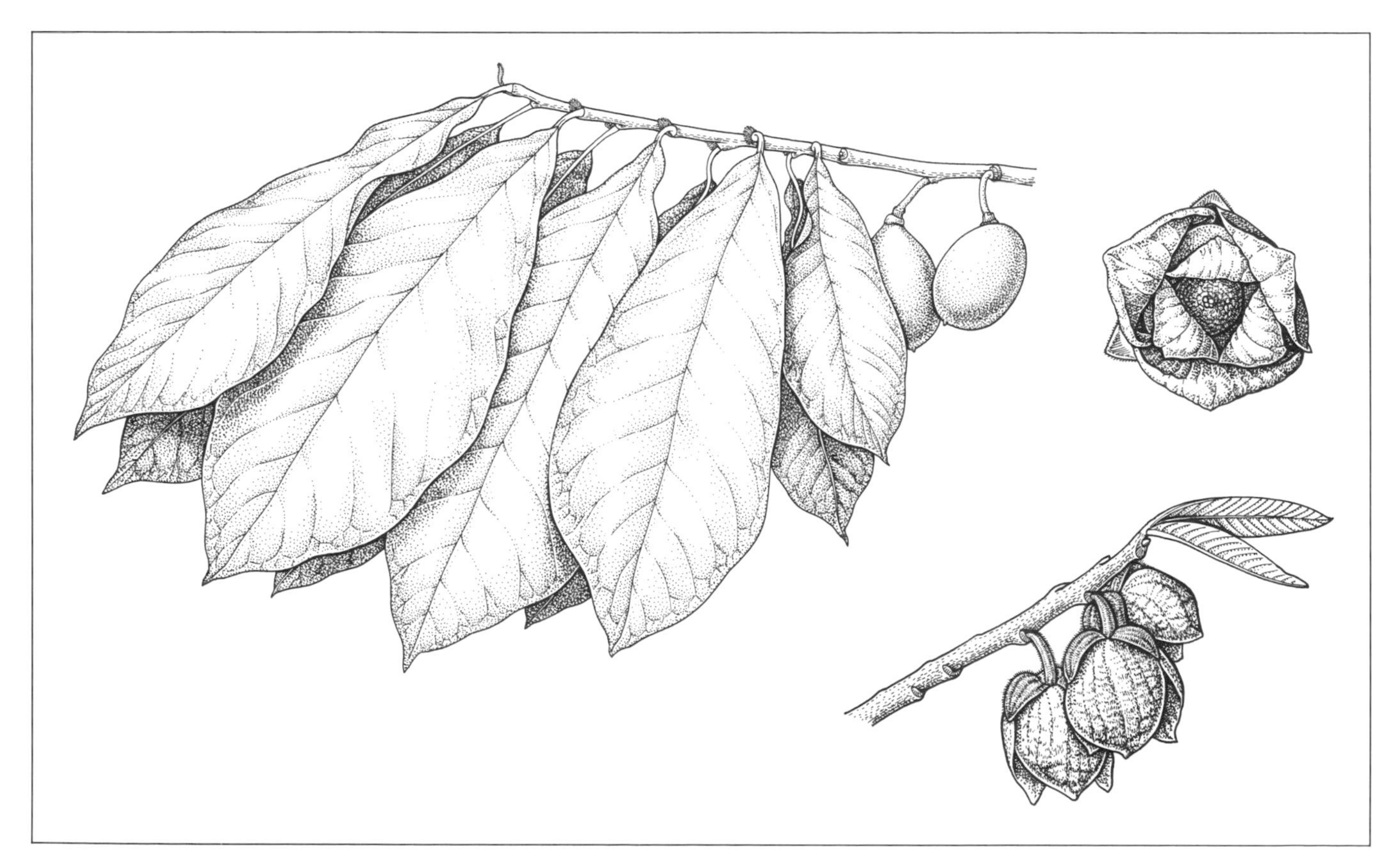

24. Peter Collinson cultivated the pawpaw (*Asimina triloba*) in his garden at Mill Hill. These drawings illustrate the floral morphology as well as a branch with mature fruits. The large leaves lend an almost tropical appearance to the trees.

tree or sourwood (*Oxydendrum arboreum* (L.) De Candolle) have been included in several locations in the Arnold Arboretum. The species is best seen along Valley Road against the backdrop of the hemlocks at the base of Hemlock Hill; from that vantage point, an isolated sourwood can be seen on the open slope above Bussey Brook at the edge of the conifer plantation.

Common in woodlands from Pennsylvania southward through the mountains into western Florida and west into Louisiana, Tennessee, and Indiana, sourwood usually grows in well-drained, gravelly soils on slopes and ridges. The tree is easily distinguished during the summer months, when hundreds of small white flowers are produced in wandlike racemes at the ends of the branches. The individual flowers are smaller but resemble those of the lily-of-the-valley and hang downward, each attached to the inflorescence axis by a small stalk or pedicel. The flowers are extremely attractive to bees, and sourwood honey is a specialty in the southern Appalachians and other regions where the tree is common. After the corollas have fallen from the flowers, the developing capsules retain a whitish color, and the inflorescences continue to provide ornament late into the summer and particularly in the fall, when their color is in sharp contrast with the crimson to burgundy color of the foliage. Close inspection of the fruits will show that the capsules are held erect along the branches of the infructescence, the pedicels having curved upward during their development.

Seeds of the pawpaw had been received by Collinson in 1736; in 1738 Philip Miller was growing an American hawthorn (*Crataegus phaenopyrum* (L. f.) Medicus), now known as the Washington thorn (Plate 5); and by 1752 Miller was also cultivating the sourwood in Chelsea from propagating material brought back from the American colonies. The Washington thorn, while it had been known in Europe before Miller was successful in growing the plant, was little known in gardens. This hawthorn was to obtain wide popularity because of the elegant appearance of its habit, the beautiful coloration of its leaves in fall, its brilliant display of fruit, and its late spring and early summer flowering period. A small tree, rarely attaining thirty feet in height, the Washington thorn has a slender trunk but a wide, symmetrically rounded crown densely clothed with glossy green, three-lobed leaves, which resemble those of some maples. Its extremely numerous clusters of white flowers are produced when the tree is in full leaf in June, after all other species of hawthorn have flowered. In the fall, the tree holds its rich orangish-red foliage until late in the season, after other hawthorns have shed their leaves. Against the bright and cheerful autumn coloration of the leaves are abundant bunches of bright red fruits near the ends of the branches. These clusters persist throughout much of the winter, when the intricate branching pattern of the leafless crown is evident.

Of local occurrence from Pennsylvania south to Florida and west into Arkansas and Missouri, Washington thorn is particularly abundant in the Washington region and in the Potomac River valley. Its occurrence in this area has given rise to its common name, but the species epithet, *phaenopyrum,* to shine, was coined in reference to the bright red, shining fruits.

Quaker Farmer, Royal Botanist

It is likely that Collinson and Miller obtained the seeds of these three American plants from the same source. John Bartram, a Quaker farmer born near Darby, Pennsylvania, in 1699, had been recommended to Collinson by John Logan, William Penn's secretary, as a person who might be depended upon to provide annual shipments of seeds, plants, and other collections. After four or five years of schooling at Darby Meeting School, Bartram had worked on his father's farm but then left his family home in 1728 to purchase his own land along the Schuylkill River at Kingsessing near Philadelphia. Bartram, who had become interested in plants as a youth, established his own garden adjacent to his new home. The correspondence between Collinson and Bartram began in 1733 and continued unabated for the next thirty-five years. During this period Bartram supplied Collinson with quantities of seeds, bulbs, root pieces, cuttings, and living plants from the forests, marshes, and meadows of eastern North America. Collinson, in turn, provided Bartram with a share of his livelihood. Bartram's income was further bolstered through subscriptions from an increasingly large number of Collinson's gardening friends, Miller included. Shipments were received in London by Collinson, a mercer by trade, whose profitable company had an active business with the American colonies and whose offices were located close to the busy London docks and custom houses. Collinson then served as agent and middleman, distributing the chests and crates of seeds and plants sent by Bartram to subscribers, taking requests for future orders, and corresponding frequently with the native-born American naturalist.

As an active member of the Royal Society, Collinson

25. A drawing of John Bartram (1699–1777) by Howard Pyle. Although a portrait has been said to represent Bartram, William Bartram is recorded as having stated that no likeness of his father was painted or drawn during John Bartram's lifetime. Pyle used this drawing to illustrate his article entitled "Bartram and His Garden," which appeared in *Harper's New Monthly Magazine* in February 1880.

did not limit his correspondence to Bartram but also actively exchanged seeds and plants, garden information, and news of scientific advancements with a wide circle of American residents. Included were John Custis of Williamsburg, as well as Bartram's close friend and mentor, Benjamin Franklin, to whom Collinson reported the initial discovery of electricity. Collinson also served (free of remuneration) as the London agent for the Library Company of Philadelphia, an organization that had been established as the brainchild of the "Junto," a discussion group of intellectuals and civic-minded Philadelphia citizens brought together by Franklin. In this capacity Collinson purchased and shipped to Philadelphia the most current and highly respected publications of London and continental book dealers. Every edition of Philip Miller's *Gardeners Dictionary* was received by the Library Company, and copies were also sent as gifts to esteemed colonial correspondents with botanical and horticultural interests. Bartram received a personal copy from Collinson, and by 1755, when the *Dictionary* was in its sixth edition, Miller and Bartram were in direct contact with one another.

Miller, while enthusiastically tending his responsibilities at the Chelsea Physic Garden, stole every opportunity to personally observe new exotics that had been successfully introduced into cultivation, in order to update his *Dictionary.* He must have been delighted to correspond directly with Bartram, the fountainhead of North American novelties. In 1752, along with sourwood seeds, Miller also received seeds of the umbrella tree (*Magnolia tripetala* L.), undoubtedly from Bartram, and by 1755 it, too, was growing in Chelsea. Miller must have closely followed the annual growth increments of his living specimens of this fascinating plant.

A small, often several-stemmed, sometimes shrublike tree with ashy-gray, almost pewter-hued bark and stout branches with large leaf scars, the umbrella tree (Plate 7) is most remarkable for the size and arrangement of its leaves. These are broadly lance-shaped—pointed at both ends but widened near the middle—and can attain lengths of up to two feet. Groups of leaves are closely clustered on the branchlets, and the thick petioles and leaf midribs radiate from the branches like the spokes of a wheel. The leaves are held more or less outright, drooping downward toward their apices, and when viewed from beneath (or above), the crown of the tree appears to consist of hundreds of small, open, green umbrellas!

The flowers of the umbrella tree—produced at the ends of the branches—are scaled in proportion to the

A Draught of John Bartram's House and Garden as it appears from the River 1758
Sent to P Collinson
1. my Studey
2. Common Flower Garden
3 upper Kitchen Garden
4 the Lower Kitchen Garden
5.6. Walks 150 Yards long of a moderate decent
A new flower Garden 25 yards long & 10 Broad
A Pond or Spring head
convaid under ground to the Spring or milk House
The Course of this Fence is Northwest & Southeast
Schuilkiln River 400 Yards wide

size of the whorls of leaves that occur beneath them. A fully opened flower, consisting of three greenish-white outer tepals and six or nine creamy white inner tepals, can measure up to ten inches in diameter. Despite their large size and waterlily-like appearance, the flowers are malodorous, and pollination is mainly affected by small, inconspicuous beetles, hardly the agents that might be expected to play this essential role in such an extravagant floral display.

Native to rich eastern North American woodlands from southern Pennsylvania and West Virginia south into Florida and westward through Ohio and Kentucky into Arkansas and Missouri, the umbrella tree must have been a common sight to John Bartram. The Quaker farmer turned naturalist and botanist, who had hired the local schoolmaster to tutor him in Latin, the botanical language, introduced the tree, along with pawpaws and sourwoods, into his own garden on the west bank of the Schuylkill, upstream from its confluence with the Delaware. This practice became customary with the majority of plants he encountered on his collecting trips, and his American garden began to rival those of his patrons in England. Moreover, the traffic in plants was not a one-way arrangement, as Collinson and other English subscribers sent plants and seeds to Bartram to be grown in his garden and to be shared with other interested colonists (Plate 6).

Books and other botanical works published in England and Europe, including Catesby's *Natural History* and the eighth edition of Miller's *Gardeners Dictionary,* which came off the presses in 1768, continued to be addressed to the Library Company and to Bartram. The eighth edition of the *Dictionary* must have been received by Bartram with great anticipation. For in that edition—the last one personally revised and edited by the Chelsea gardener—Miller adopted the binomial system of nomenclature that had been devised by the renowned Swedish naturalist, Carl Linnaeus. In his two-volume work entitled *Species Plantarum,* which was published in 1753, Linnaeus (1707–1778) laid down the rules of botanical nomenclature that are still followed today. Gone forever were the cumbersome and frequently confusing polynomials or phrase names that historically had been used to refer to plants. In their place stood a much simplified and streamlined system for naming plants, whereby each species was referred to by a generic name coupled with a specific name or epithet, one that usually connoted some identifying attribute or the geographic origin of the plant in question. Thus, *Laurus tinus floribus albidus eleganter bullatus*—John Bannister's polynomial for the fringe tree—was supplanted by *Chionanthus virginicus,* literally, the snow flower tree of Virginia. The new system was equivalent to saying "Bartram, John," although the names or epithets were in Latin form.

While Miller continued to adhere to an alphabetical arrangement for the entries in his *Dictionary,* Linnaeus had revolutionized plant classification as well. His "sexual system" was admittedly artificial, but it nonetheless allowed for the placement of any plant in his classification by simply counting the number of stamens and carpels or styles present in the flower of the plant in question. Utilizing this easy means of botanical arithmetic, any new or unknown plant could be distinguished and classified or grouped with its fellows, and any observant investigator with an ability to count could consider (or deceive) himself a naturalist.

By 1760 Bartram had scoured much of eastern North America on horseback, collecting as far north as Lake Ontario in New York, as far west as Pittsburgh and the Ohio River, and as far south in the Carolinas as Charles Town, Catesby's former base of operations. His contributions of plants and specimens to his European colleagues had gained him a reputation as the foremost American botanical authority, and his circle of correspondents had been enlarged to include Linnaeus himself. In remarking on Bartram's abilities, Linnaeus pronounced him "the greatest natural botanist of his time" (quoted by H. Pyle, 1880, p. 321). Another accolade was conferred in 1765 when, at Collinson's urging, King George III appointed Bartram Royal Botanist in the American colonies. Soon after, Bartram—at age sixty-six—set out on yet another collecting expedition, one on

26. *A Draught of John Bartram's House and Garden as it appears from the River, 1758,* drawn by Bartram himself and sent to Peter Collinson, his friend and correspondent in London. The garden and adjoining property passed out of the hands of the Bartram family in 1850, when the estate was purchased by Andrew M. Eastwick. Eastwick died in 1879, and in 1891—at the urging of Thomas Meehan, a prominent Philadelphia nurseryman and politician who had served as Eastwick's gardener early in his career—Bartram's garden was purchased by the City of Philadelphia and preserved for posterity. A master plan for the restoration of the garden was approved in 1984; the pen and ink drawing shown here provided great aid in ensuring the historical accuracy of the restoration.

THE

GARDENERS DICTIONARY:

CONTAINING

The BEST and NEWEST METHODS

OF

CULTIVATING and IMPROVING

THE

Kitchen, Fruit, Flower Garden, and Nurſery;

As alſo for Performing the

Practical Parts of AGRICULTURE:

INCLUDING

The MANAGEMENT of VINEYARDS,

WITH THE

Methods of MAKING and PRESERVING WINE,

According to the preſent Practice of

The moſt ſkilful Vignerons in the ſeveral Wine Countries in *Europe.*

TOGETHER WITH

DIRECTIONS for PROPAGATING and IMPROVING,

From REAL PRACTICE and EXPERIENCE,

ALL SORTS OF TIMBER TREES.

THE EIGHTH EDITION,

Reviſed and Altered according to the lateſt SYSTEM of BOTANY; and Embelliſhed with ſeveral COPPER-PLATES, which were not in ſome former Editions.

BY PHILIP MILLER, F. R. S.

Gardener to the Worſhipful Company of APOTHECARIES, at their Botanic Garden in *Chelſea*, and Member of the Botanic Academy at *Florence.*

.... *Digna manet divini gloria ruris.* VIRG. Georg.

LONDON,

Printed for the AUTHOR;

And Sold by JOHN and FRANCIS RIVINGTON, at No. 62, *St. Paul's Church-yard*; A MILLAR, J. WHISTON, W. STRAHAN, J. HINTON, R. BALDWIN, B. WHITE, L. HAWES and W. CLARKE and R. COLLINS, W. JOHNSTON, T. CASLON, S. CROWDER, T. LONGMAN, B. LAW, C. RIVINGTON, J. DODSLEY, W. GRIFFIN, T. CADELL, T. LOWNDES, S. BLADON, G. ROBINSON and J. ROBERTS, and T. PAYNE.

M. DCC. LXVIII.

27. Frontispiece and title page of the eighth edition of Philip Miller's *Gardeners Dictionary*, published in London in 1768. The binomial nomenclature for species proposed by Linnaeus was adopted for the first time in this edition of Miller's highly popular and widely used work.

which he had the companionship of his talented son, William (1739–1823), who had also become smitten with exploration and botanizing. In large measure, this enterprise, which commenced with an expensive passage by ship down the coast to Charles Town, was made possible by the annual stipend of fifty pounds sterling that was stipulated in Bartram's royal appointment. It was also undertaken at Collinson's suggestion that Bartram should provide materials for his new patron, the King. After all, horticultural novelties from the North American colonies would be most welcome for inclusion in the Royal Gardens at Kew. These extensive gardens had been under development for almost three decades by the King's mother, the Princess Augusta, Dowager Princess of Wales. The Dowager Princess had been aided in these horticultural improvements of the royal estates in the vicinity of Richmond, west of London, by John Stuart (1713–1792), the third Earl of Bute, and William Aiton

28. Only twenty-nine copies of the original edition of Linnaeus's *Systema Naturae* are known to exist today. However, sixteen reprint or new editions followed the original publication of 1735, and of these twelve were prepared under Linnaeus's personal supervision. It was in the pages of the *Systema* that Linnaeus outlined his classification of the three kingdoms of nature and proposed his sexual classification system for plants.

(1731–1793), one of Philip Miller's former pupils at the Chelsea Physic Garden. And Augusta's son, frequently referred to by his subjects as "farmer George," took a personal interest in their development.

Regardless of royal patronage, John Bartram needed little encouragement to start on a new adventure. He was intent upon exploring new territory in East Florida, an area that had been ceded to England by Spain in 1763 in exchange for Cuba. In addition, once in Charles Town, Bartram could renew his acquaintance with Alexander Garden. A Scots physician, Garden (1730–1791) had settled in Charles Town in 1752 and had immediately begun to investigate the natural history of the region. He quickly achieved a place and was welcomed in the network of enlightened colonial correspondents who exchanged scientific views with naturalists and physicians in England and elsewhere in Europe and who, like Bartram, was responsible for dispatching seeds and spec-

29. A view of the lake and island at Kew, seen from the lawn, with the bridge, the temples of Arethusa and Victory, and the Great Pagoda. By comparison with the king's garden in Paris, the gardens at Kew occupied extensive acreage and could be adapted to the style of the English landscape garden. As knowledge of China became more commonplace in Europe, the adoption of Chinese art forms became very fashionable, and existing styles were often modified to add an oriental motif. The pagoda at Kew was not a solitary expression of interest in things Chinese.

imens of the rich and diverse southeastern flora to correspondents abroad.

Linnaeus recognized Alexander Garden's contributions to botany when he named the genus *Gardenia* in his honor. Ironically, this group of highly ornamental evergreen shrubs, well known for their sweetly fragrant white flowers, are of Old World origin and their discovery had nothing to do with the Charles Town physician. The specific name of another shrub, however, the witch alder—first discovered in the Carolinas by Garden and introduced into European gardens when John Bartram sent seeds to Peter Collinson—links Garden with an American plant and commemorates the role he played in the botanical exploration of the American colonies. Strangely enough, when he sent witch alder seeds to Collinson, Bartram suggested that this plant, should it prove to be the basis of a new genus, be named *Gardenia*. But the generic name of this plant, *Fothergilla*, honors John Fothergill, who, like Peter Collinson, stimulated horticultural and botanical exploration of eastern North America and was one of the supporters of John Bartram's activities in the colonies.

Two species of witch alders are now known, and both are elegant ornamental shrubs worthy of wide cultivation. The dwarf witch alder (*Fothergilla gardenii* Murray), the species named in recognition of Alexander Garden's contributions to science, is a low-growing shrub that rarely exceeds three feet in height but often attains a greater spread. In nature, this species occurs at the edges of swamps on the Coastal Plain from North Carolina into eastern Alabama, while the second species (*Fothergilla major* Loddiges) occurs naturally on wooded

30. The flowers of the mountain witch alder (*Fothergilla major*) are freely produced and are showy not because of perianth parts but because of enlarged, creamy white stamen filaments.

slopes along the Blue Ridge and on the adjacent Piedmont from the Carolinas into central Alabama. On Bussey Hill in the Arnold Arboretum, both witch alders grow together and can easily be distinguished from each other by their size alone, for plants of the mountain witch alder frequently attain heights of eight to ten feet and form large clumps because of the proliferation of underground stems. While there are several technical characters that separate the two species, dwarf witch alder appears to be a miniature version of its montane counterpart.

In spring, toward the end of April and into May, the buds on the branches of the witch alders swell and expand, and the young, embryonic leaves begin to emerge and enlarge. Simultaneously, dormant flower buds break into growth to cover the shrubs with white or creamy white bottle-brush-like inflorescences. Close examination of these curious clusters of flowers will reveal that their beauty and appearance are due to the large number of stamens, the filaments or stalks of which are club-shaped and surmounted by small, yellow, pollen-bearing anther sacs. The flowers completely lack petals, and their unique display is totally due to the numerous stamens.

By the time the flowers have withered and the stamens have fallen, the leaves of the shrubs have achieved their full size. The common name occasionally applied to these plants, witch alder, resulted from the similarity of the somewhat leathery leaves to those of species of alder (of the genus *Alnus* L.). And "witch" was used for the same reason that it was applied to a native New England shrub, the familiar witch hazel (*Hamamelis virginiana*

L.). The witch alders and the witch hazels all belong to the same family of plants, the Hamamelidaceae, and they share an identical mechanism of seed dispersal. When the capsules in which the seeds have developed are mature, the capsule walls open, forcing the expulsion of the seeds held within. The seeds literally fly through the air, and their propulsion suggested witchcraft to those who first observed the phenomenon.

The fall coloration of witch alder leaves is reason enough for the shrubs to attract attention at that time of year. The dark green of the leaves in summer gradually gives way to brilliant yellows, oranges, and reds ranging from scarlet to subtle burgundy. Several hues can combine on a single leaf, and individual shrubs frequently progress in stages from yellow through orangish hues to scarlet as the fall season advances.

In Charles Town, Bartram and Garden spent long hours discussing recent botanical advances that had been communicated from England, and excursions were taken to view and study the plants cultivated in gardens in the neighborhood of that growing southern metropolis. Forced to remain in the city because of his active medical practice—and perhaps because of his aversion to the intense summer heat and high humidity—Garden did not accompany Bartram on several collecting forays on the Coastal Plain. When William Bartram arrived in Charles Town in late summer to join his father, Garden again remained in the city but sent the two naturalists off on their southern expedition with a sheaf of letters of introduction to his numerous acquaintances in scattered settlements further south.

From Charles Town the Bartrams went on horseback to Savannah and proceeded southward, reaching the Altamaha River, which then constituted the southern frontier of Georgia, on the first of October. Losing their bearings, they came upon the banks of the river about four miles downstream from Fort Barrington, their intended destination, where the river could be easily crossed. Once on the south bank of the river, they planned to continue into and explore the Indian territory farther south.

After going astray, the Bartrams spent the night of October 1, 1765, somewhere along the Altamaha River. The next day they eventually found their way again, arriving upstream at Fort Barrington. In his journal for October 1, 1765, John Bartram was to note in passing, "This day we found severall very curious shrubs one bearing beautiful good fruite" (J. Bartram, 1942, p. 31). William Bartram, recounting the incident twenty years later, wrote that "at this place [Fort Barrington] there are two or three acres of ground where it grows plentifully" (quoted in F. Harper, 1958, p. 296). One shrub in particular, alluded to by John Bartram and referred to by William, proved to be among the most beautiful small trees or shrubs of the North American flora. It was destined to be named in honor of John Bartram's close associate, Benjamin Franklin, and it has continued to the present day to whet the appetites of botanists and naturalists eager for exploration in the vicinity of Fort Barrington.

Specimens, fruits, and seeds of the Franklin tree (*Franklinia alatamaha* Marshall) were apparently not gathered by the Bartrams in the fall of 1765. It remained for William, on a second journey to the area in 1773, to relocate the small population of trees and to obtain propagating material of the plant for shipment to his patron in England. On this extended, four-year sojourn William was partially supported by Dr. John Fothergill, the successful and wealthy London physician, a fellow member of the Quaker fraternity, an old and close friend of Peter Collinson, and a long-time correspondent of William's father. Fothergill must have been satisfied with his new, second-generation horticultural and botanical agent and correspondent in the colonies, for in 1774 he was able to present seedlings of the Franklin tree to the Royal Gardens at Kew.

Striking out on his own under Fothergill's auspices, William was to travel over a thousand miles in Indian country—the territories of the Cherokee, Creek, Chickasaw, and Seminole nations—recording his observations in his journal and on his sketch pad and documenting his botanical discoveries with dried specimens. Periodically descending from the mountains and Piedmont to Coastal Plain settlements—Charles Town and Savannah on the Atlantic, and Mobile on the Gulf—he dispatched his collections of specimens, seeds, plants, and drawings to Fothergill in London. When he finally headed back to Pennsylvania in January of 1777, once again traveling via Fort Barrington on the Altamaha, he carried with him additional propagating material of the Franklin tree in order to establish the plant in his father's flourishing botanical garden. When he reached Philadelphia he discovered—to his dismay—British troops occupying the city. The battles of Lexington, Concord, and Bunker Hill had been fought, the Declaration of Independence had been signed in Philadelphia the previous summer, and the American Revolution was well under way.

On his return to the farm and botanic garden on the banks of the Schuylkill, William had the pleasure of giving his father a fully detailed description of the beautiful

31. Green and white jasper portrait medallion of Dr. John Fothergill (1712–1780) modelled in about 1785 by John Flaxman for the famous pottery firm of Wedgwood. Fothergill, close friend and associate of Peter Collinson, helped to finance William Bartram's collecting trips through the southeastern United States and was the recipient of Bartram's collections of seeds, specimens, and many of his botanical and zoological drawings. Bartram's plants were cultivated by Fothergill in his extensive garden at Upton in Surrey.

flowers of the Franklin tree they had discovered years previously. John Bartram, then an old man, relived through his son's recountings the joys and thrills of exploration and botanizing. John Bartram was to spend one last growing season in his garden, marveling at the plants in his collection, which by that time included rhubarb, seeds of which Benjamin Franklin had sent from France. In September of 1777, after a brief illness, John Bartram died, and with his demise—hastened perhaps by his concern and anxiety for his garden at the hands of British troops—the brief era of a royally appointed botanist in His Majesty's North American colonies ended abruptly.

A descendant of the Franklin tree that grew from the seeds that William Bartram carried in his saddle bags from the banks of the Altamaha River in Georgia to the garden on the banks of the Schuylkill River in Pennsylvania still flourishes in its transplanted habitat in the industrialized, urban city of brotherly love, Benjamin Franklin's adopted home. Other specimens of the Franklin tree, direct descendants of the Bartram garden plant, also grow in urban Boston, the city of Franklin's birth, on the southeast-facing slope of Bussey Hill in the Arnold Arboretum, a site that provides a beautiful vista of the Blue Hills to the south. Despite repeated attempts to relocate the plant along the banks of the Altamaha River in the vicinity of Fort Barrington, particularly an intensive exploration stimulated by Charles Sprague Sargent at the close of the nineteenth century, the conclusion has been reached that the Franklin tree is extinct in its native habitat. Botanically inclined travelers continue, however, to seek the plant in the sandy soils along the river where the Bartrams discovered the tree in 1765.

The plant that has inspired these repeated searches is well represented by the venerable old specimens growing in the Arnold Arboretum, where the plants have developed multiple stems from the base and grow as large, upright shrubs. The limbs are covered with a smooth, dark, slate-gray bark with longitudinal fissures of a silverish hue, a combination that makes the naked branches distinctive during the winter months. The bright green leaves, arranged alternately along the branches, are generally oblong in outline, tapering gradually to the petiole at the base, and the margins are sharply serrate. The large, attractive flowers are camellia-like in appearance and are clustered, each on a short, stout pedicel, in groups of four or five in the axils of leaves toward the ends of the branches. The rounded flower buds open to disclose five white, scallop-shaped petals that surround a large boss of numerous yellow

32. William Bartram's drawing of the Franklin tree (*Franklinia alatamaha*), which has been annotated, "A beautiful flowering tree discovered growing near the banks of the R. Alatamaha in Georgia. William Bartram Delin. 1783."

stamens. These occur in five groups and in turn surround and conceal the conspicuously ridged and hairy ovary, but the solitary, slender style protrudes from among them. The fruits that develop after pollination require a full year to mature, and their structure is unique within the Theaceae or Tea Family—the family that includes both the Franklin tree and species of the well-known genus *Camellia*. Each rounded, woody capsule splits longitudinally from the base to near the middle as well as from the apex to near the middle when it is fully ripened, allowing for the dispersal of the numerous seeds that have developed within. The walls of a fully dehisced or opened capsule present a curious zigzag appearance not seen in the fruit of any other native North American species.

The flowers of the Franklin tree do not appear in spring but are delayed in their development until late August and early September. As a consequence, the shrub is one of the last ornamentals to provide a floral display toward the end of the growing season. Once flowering has commenced, flowers are continually produced until a killing frost brings the growing season to an end. The first flowers are produced while the foliage retains its summer green; later flowers appear progressively as the leaves assume autumn tints of reddish-orange and burgundy. When the large white flowers are viewed against the background of the plant's fall foliage, the Franklin tree provides a sight unlike any other shrub or tree cultivated in New England.

CHAPTER TWO

Establishing Traditions in the Young Republic

Revolutionary Aftermath

On the wooded, west-facing slope of Peters Hill in the Arnold Arboretum, under the dappled shade cast by towering oaks and tulip trees, unmarked graves share the hillside with a few ancient headstones and provide mute testimony to lives lost in the American War of Independence. The small church that once stood nearby has vanished, but a commemorative plaque informs those who visit the area of the existence of the burial ground and serves to recall to memory the conflicts and battles fought almost one hundred years before the surrounding land became the urban forest of the Arnold Arboretum. During the Revolution, colonial lives and livelihoods were disrupted, commerce was brought to a virtual standstill, hard battles were fought against troops from the mother country, and the colonists, regardless of their origins, became Americans. Under some circumstances friends and family members became foes, and allegiances changed; yet in the end a new Republic was born as the thirteen British Colonies became independent and forged a federation.

By the time the fighting ended in 1781, the network of colonial correspondents and their English patrons had fallen apart. John Bartram was dead, Philip Miller had died a decade earlier, and Peter Collinson and John Fothergill had been recently mourned. In 1778 the great Linnaeus died, and five years later his primary colonial contact, Alexander Garden, who had survived the British occupation of Charleston, felt unable to support the new federation. Returning to London in 1783, he left his adopted city of thirty years, his medical practice, his home and garden, and the surrounding countryside, which he had come to know so well.

William Bartram, with little prospect for further patronage from men of science in England, had joined his brother, John, Jr., in seeking a living as nurserymen, centering their business amid the rich and diverse collections in the botanical garden established by their father.

33. Engraving of William Bartram after a portrait by Charles Willson Peale. The sprig of flowers in Bartram's buttonhole appears to be one of *Porterantherus trifoliatus*, bowman's root, a widespread species in the mountains and Upper Piedmont of the Appalachian mountain system, the region William Bartram explored and about which he wrote in his famous *Travels*.

Shipments of American plants and seeds to England and elsewhere in Europe were expected to recommence as soon as the peace was secured and commerce resumed. Given the successful cultivation of numerous American species in Europe during the colonial era and the increasing interest of an expanding number of Europeans in the art of landscaping, the Bartrams expected the demand for plants to be great. Europeans had come to perceive America as a richly stocked warehouse and inexhaustible source of new plants, and even in the young Republic a growing number of gentlemen, notably in the Philadelphia region, were developing an interest in rural improvement and a taste for gardening.

During the years of revolution, William Bartram—a pacifist, in Quaker tradition—spent long hours tending the plants in the botanical garden on the banks of the Schuylkill. He also found time to prepare a written account of his travels and adventures in the North American wilderness, and by 1781 his manuscript was essentially complete. Missing, however, was the list of names for the new species of plants that he had discovered and collected, names which he was waiting for botanical authorities in England to supply, based on the dried specimens he had so carefully prepared, packed, and shipped to Fothergill in London. However, on Fothergill's death in 1780, his estate—library and herbarium included—had been put up for sale at auction. Through his successful bidding, Sir Joseph Banks, doyen of English scientists and recently elected president of the Royal Society, had acquired Fothergill's herbarium, along with William Bartram's North American plants and drawings.

A zealous and influential collector and promoter of natural-history investigations and the development of all branches of science for public benefit, Banks was dynamically positioned at the center of practically all scientific undertakings in England. He had spent time exploring in Newfoundland and Labrador in 1766 and 1767, charting the coastal waters and personally making extensive collections. The jack pine (*Pinus banksiana* Lambert), a widespread species that grows on the barren, sandy soils of the boreal regions in North America that he visited, commemorates his name. And when he returned to England, he brought with him seeds of the rhodora (*Rhododendron canadense* (L.) Torrey; Plate 8) and the bog laurel (*Kalmia polifolia* Wangenheim), two other northern species, for trial in the Royal Botanic Gardens at Kew.

Banks's interest in the royal gardens increased during the next decade, for in 1772, when Princess Augusta died, the estate passed to her son, George III. "Farmer George," recognizing Sir Joseph's organizational abilities, appointed the scientific entrepreneur royal adviser to direct the botanical development of the gardens. The king's choice could not have been more fortunate. In his pivotal position as president of the Royal Society, a post he assumed in 1778 and held for forty-two years, Banks saw to it that large numbers of plants were sent to England, destined for Kew, as His Majesty's navy actively expanded the British empire and English influence around the globe.

After Banks acquired Bartram's specimens, Daniel Solander (1733–1782), librarian and botanical curator of Sir Joseph's extensive and wide-ranging collections,

34. After his father's death, William Bartram continued to live in the Bartram homestead, although it was inherited by his brother, John, Jr.

35. Engraving of Sir Joseph Banks (1743–1820), staunch promoter of scientific investigation and active participant in shaping the science policy of eighteenth- and nineteenth-century England.

agreed to supply William Bartram with the long-awaited names and accurate descriptions of his new species. But Solander, Swedish born and the favorite pupil of Linnaeus, became sidetracked working on other, apparently more interesting collections from the South Seas. For Solander himself had served as principal naturalist on the scientific expedition that had gathered these materials. In 1768, the British Admiralty, in conjunction with the Royal Society, had determined to undertake an exploratory voyage to Tahiti in the South Pacific, the first expedition of its kind to set sail from a European port. Unlike the solitary naturalist of earlier days who set out on his own with support from a small circle of patrons, Solander—together with a retinue of naturalists that included Banks—boarded a ship specially equipped for the voyage, the *Endeavor*, under the command of Captain James Cook. The undertaking was ostensibly intended to observe the transit of the planet Venus across the face of the sun, an astronomical phenomenon calculated to occur only twice in 110 years. Scientific data collected would allow for great advances in celestial navigation. But Banks seized the opportunity to expand the scope of the expedition and to ensure that natural-history interests were well represented on board.

The expedition was successful in many respects, and future scientific exploration undertaken by European naval powers and eventually the United States would be modeled after Cook's epic voyage of discovery. When the expedition returned to London, the botanical specimens gathered at every port of call, particularly in Australia and New Zealand—plants of unexpected form and variety never before seen by European naturalists—absorbed Solander's attentions until his death in 1782. Bartram's dried plants lay tied in bundles in Banks's herbarium, where they remained unstudied until years later, after their transferral to the British Museum following Sir Joseph's death.

By contrast, on the opposite side of the Atlantic the Bartram collection of living plants achieved great notoriety, and the garden became a mecca for American men of science and foreign visitors as well. Benjamin Smith Barton, who accepted the post of professor of botany at the University of Pennsylvania when William Bartram declined the position, brought his students to study in the collections. The Reverend Manasseh Cutler, a Yale-educated clergyman, naturalist, and savant from Ipswich, Massachusetts, visited there in 1787 with a group of distinguished Americans who had gathered in Philadelphia for the Constitutional Convention.

Not a delegate to the convention himself, Cutler (1742–1823) was nonetheless well known to Benjamin Franklin and other American statesmen and men of science, many of whom were members of the long-established American Philosophical Society, which had been founded in Philadelphia in 1743. Cutler accepted membership in that group, and closer to home he numbered among the organizers of the Boston-based American Academy of Arts and Sciences, chartered by the General Court of Massachusetts toward the end of the Revolution in 1780. Moreover, a paper authored by Cutler appeared in the first volume of the Academy's memoirs. His contribution, "An account of the vegetable productions, naturally growing in this part of America, botanically arranged," came off the presses in Boston in 1783 and stands as the first botanical treatise published in America to use the Linnean classification system. Moreover, the treatise gave early evidence that botanical studies could be initiated by Americans and published in the young republic without recourse to the resources of

36. Benjamin Smith Barton (1766–1815), first professor of botany at the University of Pennsylvania and staunch promoter of a comprehensive flora of North America. Meriwether Lewis learned the rudiments of systematic botany from Barton, whose own training in medicine was completed in Edinburgh, Scotland. While Barton did not succeed in writing a flora of North America, in 1803 he did publish *Elements of Botany*, the first botanical textbook published in the United States.

European botanical centers or the approbation of European naturalists. Americans were clearly faced with the establishment not only of a new constitutional government but with the development of American-based science and American scientific institutions, ideals the Philosophical Society and the American Academy were to foster and promote in the years ahead.

André Michaux, Citizen of France

Of the foreign visitors welcomed in the botanical garden, one in particular was to continue the traditions of botanical and horticultural exploration pioneered by the Bartrams. André Michaux (1746–1803), shortly after his arrival at New York in 1785, only a year after Congress had ratified the peace treaty with England, traveled to Philadelphia and presented letters of introduction to William Bartram. Recently appointed botanist to Louis XVI of France, Michaux was to serve, like John Bartram before him, as a royally appointed botanist in North America. But unlike the elder Bartram, Michaux would not explore territory in his monarch's colonies. To the contrary, André Michaux traveled in the former English colonies as a diplomatic emissary, prepared to reciprocate for the privileges and courtesies extended to him in the new American republic. And before his return to France, Michaux would become Citizen Michaux. During his extended sojourn in the United States, revolutionary zeal swept across France three thousand miles away, and republican ideals and the rights of man were exercised in the storming of the Bastille and the dropping of the sharpened blade of the guillotine. The political turmoil at home handicapped Michaux in America, particularly in regard to his uncertain finances. Yet the Frenchman diligently undertook his mission while he simultaneously adopted and embraced republican views.

Having studied botany under the tutelage of Bernard de Jussieu near his childhood home in Versailles and then in the Jardin du Roi in Paris—where the Robins had established the collection during the Tradescant era—Michaux was the first trained botanist to come to the young republic in search of plants. De Jussieu, the foremost French botanist of the period, was dissatisfied with the Linnean sexual system of plant classification and sought to replace it with a natural, less artificial one that ensured the grouping of related forms based on overall similarity. A tour of England encouraged by de Jussieu had exposed Michaux to the English penchant for exotic plants. Returning to France, Michaux directed his energies to plant exploration. On embarking for American shores, he could claim to be a veteran of plant exploration in Europe and particularly in the Levant or Middle East, where he had spent three years collecting in Persia while serving as secretary to the French Consul.

Michaux's North American mission was a crucial one. As elsewhere in Europe, the forests of France had been devastated as a greater and greater demand for timber products had necessitated the felling of hundreds of forested acres. Chief among the reasons for this depletion was the enormous quantity of lumber required to build armada after armada during the long naval war that had been waged with England. Michaux's urgent assignment was to introduce into France American trees that might be acclimated to the French climate and might quickly replenish Gallic forest resources. Also sought were American species to enrich orchards and fields and ornamentals suitable for garden culture. Simultaneously, specimens would be prepared for addition to the rapidly

37. Bernard de Jussieu (1699–1777) was born at Lyons and studied medicine at Montpellier. In 1722 he was appointed subdemonstrator of plants at the Jardin du Roi in Paris, and he later became curator of the Trianon garden at Versailles. Bernard was a member of a family remarkable for the botanists it produced; Bernard's brother was Antoine de Jussieu (1686–1758), who was in charge of the Jardin du Roi, and their nephew, Antoine Laurent de Jussieu (1748–1836), became a noted botanist at the Muséum d'Histoire Naturelle in Paris.

38. Plan of the pleasure gardens at Versailles—sometimes referred to as "the queen of geometrical gardens." The layout was the product of the genius of André Le Nôtre, famous French landscape architect. The gardens are particularly famous for their fountains and waterworks, one of the most prominent being the Grand Canal, which measures 200 feet in width and extends for one mile in length. André Michaux grew up in the neighboring town of Versailles and his interest in plants was unquestionably enriched by the collections in the royal gardens, particularly those at the botanical garden at the Petit Trianon, where Bernard de Jussieu was curator.

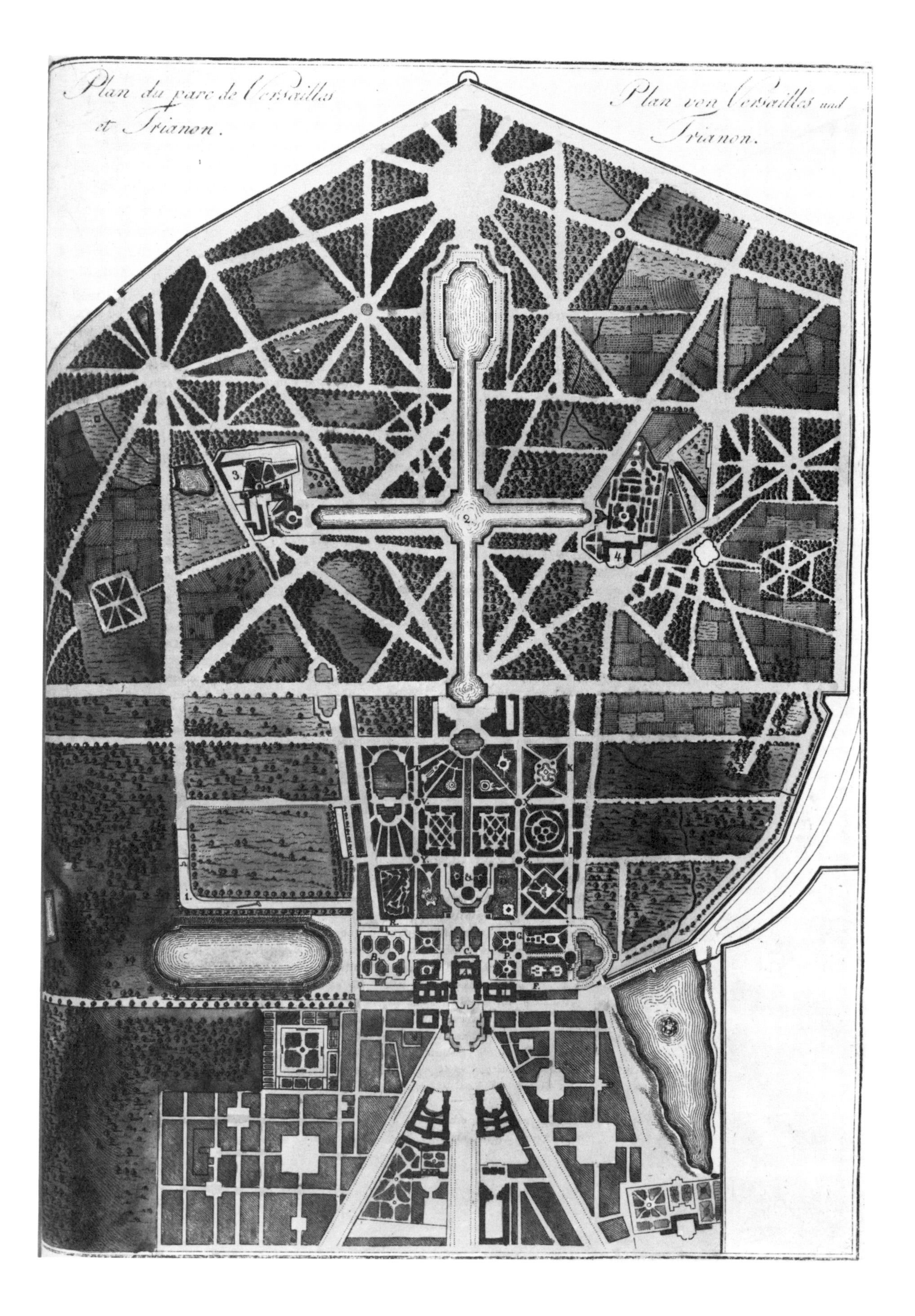
Plan du parc de Versailles et Trianon.
Plan von Versailles und Trianon.

39. An eighteenth-century allegorical engraving which, together with figure 40, reflects Europeans' growing awareness of the need for the establishment of sound forestry practices. Captioned "For Us," this engraving depicts the careful selection of a mature tree to be harvested for its timber.

growing herbarium in Paris, and wood samples would be cut from each tree species for empirical tests of strength, durability, and utility.

Realizing the importance of his task as well as the difficulties he would encounter in accumulating and maintaining as wide a range of living plants for shipment as possible, Michaux made his first priority the establishment of a nursery garden where seeds could be germinated and young plants could be grown until they attained a sufficient stature to endure and survive the long sea voyage to France. The tubs and pots in which the plants were grown would be placed on the open decks of ships, where the saplings would be exposed to the desiccating salt spray and parching sun of the open North Atlantic. Water for crew and passengers was always at a premium on board ship, and little could be spared for the plants, which in the sailors' view constituted unnecessary dunnage. In foul weather, the pots and tubs, their contents included, were frequently the first cargo to be jettisoned.

A parcel of about ten acres was purchased for the nursery garden in Bergen (now Hudson) County, New Jersey, and today much of the tract is included within the Hoboken Cemetery. The friendly relationships developed between the French and American governments during the Revolution and the presence of many francophiles in the new Republic combined to make the New Jersey purchase possible. Only American citizens were permitted to hold title to land in the state, but a special act passed

40. Captioned "For the Future," this engraving shows the planting of sapling trees to replenish timber supplies. Figures 39 and 40 are from a German treatise on forestry that dates from 1783. Similar pressures were being felt in England and France, and André Michaux's American undertaking was intended to address the problem faced by the French.

in 1786 by the New Jersey legislature allowed Michaux to make the purchase. The only stipulation tied to the sale required that the land be used solely for the purposes of a botanical garden.

A second, far larger tract of 111 acres was acquired, apparently with no legal entanglements, in the following year near the present location of Ten Mile Station on the outskirts of Charleston, South Carolina. The climates of New Jersey and South Carolina differed sufficiently to allow for the cultivation of a wide range of North American plants, and both gardens were near sea ports—New York in the north, Charleston in the south—where shipments could be dispatched to France. In fact, few French ships dropped anchor in Charleston harbor, and most of the botanical consignments from the South reached France after being exposed to salt air on board ships bound from Charleston to New York. Thus, transit was prolonged, and the success rate of shipping living plants was further reduced.

Throughout the eleven eventful years André Michaux spent in North America—years during which Americans began to prosper under their new experiment in government while the crowned heads of Europe shuddered at the events occurring in France—the New Jersey garden was tended by Pierre Paul Saunier, a seventeen-year-old journeyman gardener who accompanied Michaux to America. Initially with the assistance of his even younger fifteen-year-old son, François André, who returned to France in 1790 to further his education, Michaux personally took over responsibility for the development and maintenance of the South Carolina garden. This division of labor and Michaux's decision to leave the New Jersey establishment under Saunier's care was prompted by the greater diversity of the southern flora and the logistic advantage afforded by the South Carolina location. While Michaux explored widely throughout eastern North America from southern Florida north to Hudson Bay and west to the Mississippi River, he concentrated his efforts in the southern Appalachian region. In that area, at the edge of the frontier, he made repeated collecting forays westward over the Blue Ridge, into the Ridge and Valley Province, and up onto the broad but rugged and dissected expanse of the Cumberland plateau.

Like Catesby and the Bartrams, Michaux pushed ever onward in search of new plants. Collecting seeds and saplings in great numbers and always returning to his southern nursery—known as the French Garden to local inhabitants—Michaux filled the ever-expanding nursery rows with new and unusual plants, many never before the subjects of cultivation. Prior to his return to France in 1796, Michaux shipped upwards of sixty thousand living trees and thousands of seed collections, first to the Royal Nurseries, and, after the fall of the Bastille, to the National Nurseries. While the country had rejected its monarch, the Committee of Public Safety realized the necessity and urgency of continuing Michaux's work for the benefit of the Republic of France.

Many new plants, particularly rare ones of only local occurrence, were first located and documented with specimens by the wandering Frenchman. And of these, many were also brought into gardens and parks in Europe and America. The elusive oconee bells (*Shortia galacifolia* Torrey & Gray), not given a botanical name until years later when the American botanist Asa Gray found a fruiting specimen in Michaux's herbarium in Paris, numbers among Michaux's discoveries. This charming plant was not relocated in nature, despite repeated searches by Gray and other botanists he inspired to seek the plant, until a local youth, G. M. Hyams, stumbled upon a population in full flower near the town of Marion, North Carolina, in 1877. Nine years later in 1886, almost a century after Michaux's original discovery, C. S. Sargent discovered the locality were Michaux had made his collection near the headwaters of the Savannah River. Sargent, student of Gray and first director of the Arnold Arboretum, astutely consulted Michaux's journal in pinpointing the locality and retracing the explorer's footsteps.

The yellowwood or virgilia (*Cladrastis lutea* (Michaux f.) K. Koch, perhaps more correctly *Cladrastis kentukea* (Dumont de Courset) Rudd), another of Michaux's discoveries, now enlivens urban landscapes in both North America and Europe, and venerable old specimens shade the slope between Meadow and Bussey roads below the lilac collection in the Arnold Arboretum. Like oconee bells, the yellowwood is rare in nature, where the tree occurs on rich, forested slopes underlain by limestone in southwestern North Carolina and eastern Tennessee and in adjoining regions of Georgia, Alabama, and Kentucky. The tree also grows locally in south-central Indiana and southern Illinois. Records of its occurrence farther west in Missouri, Arkansas, and Oklahoma may stem from plants of cultivated origin.

When Michaux encountered the yellowwood along the Cumberland River near Fort Blount, some sixty miles east of Nashville, Tennessee, in March of 1796, the smooth, pewter-hued bark of the spreading tree must have signaled something novel in the perceptive mind of the wide-ranging collector. The ground was snow-

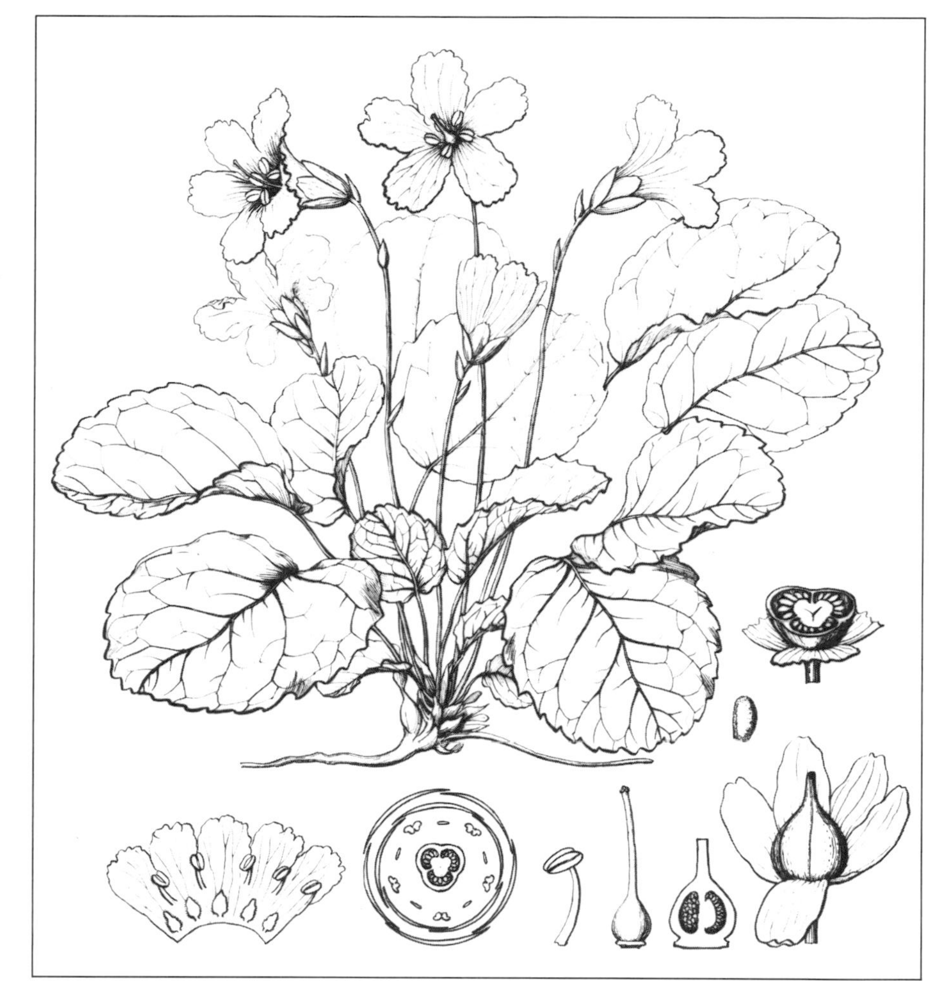

41. Pen and ink drawing of shortia (*Shortia galacifolia*) drawn by Charles E. Faxon, botanical artist and first curator of the herbarium at the Arnold Arboretum. This beautiful drawing, complete with details of floral morphology, was used to illustrate an article in the first volume of *Garden and Forest* published in 1888.

covered and the branches of the forest trees were leafless, but clusters of flattened, brownish seed pods hung from the naked branches of the yellowwood. Close inspection of the buds along the branches, each strangely encircled by the petiole scar of the previous season's leaf, gave further evidence of a new discovery. With the aid of an officer stationed at the fort, Michaux brought down some of the brittle branches of the tree and gathered seed pods for transport in his saddle bags. One branch, to serve as the obligatory wood sample, was cut, and the bright, clear, yellow color of its sawed ends provided proof of a tree not previously seen. Scattered saplings pried from the frozen ground were destined, at least temporarily, for a nursery row in the South Carolina way station, but they had commenced their long and unexpected journey to France.

Had chance dictated that Michaux cover the same terrain in the warm days that followed two or three months later, the period during which the yellowwood flowers, the collector may not have located the tree growing on the forested slopes along the Cumberland River. The fact that Michaux had gathered seed pods from the tree's naked branches indicates that the specimen had flowered the previous spring, and one of the idiosyncrasies of the yellowwood is its characteristic of producing flowers, at least abundantly, only in alternate years. Yet when the trees do flower, the sight is spectacular, and some observers claim the plant ranks as the foremost flowering tree of eastern North American forests. Long, pendent panicles of rich white flowers essentially similar in structure and size of those of the garden pea are produced at the ends of leafy branchlets and hang suspended against the backdrop of the warm, yellowish green canopy of compound leaves. The developing seed pods or legumes, which take the place of the flowers, apparently trigger the production of a hormone or chemical that inhibits the formation of flower buds, and this explains the two-year cycle of bloom. In fall, the leaves that form the tree's dense canopy turn butter yellow, providing a bright contrast with the more usual oranges and reds of the autumn landscape.

In late May and early June at the Arnold Arboretum, coincident with the flowering of the yellowwoods, strollers along Valley Road are captivated by an extensive planting of a shrub collected by Michaux. The flame azalea (*Rhododendron calendulaceum* (Michaux) Torrey) illuminates the lightly shaded hillside beneath the Arboretum's oak collection with a riot of colors reminiscent of an Indian summer mountainside. Produced in profusion on low, twiggy shrubs, which are frequently as wide as high, the large, broadly funnel-shaped flowers of the flame azalea range in color from yellow through yellowish orange to an intense scarlet. William Bartram observed this azalea in flower in northern Georgia as early as 1774 yet failed to procure seed for either Collinson or Fothergill. He nonetheless described its beauty in his *Travels through North & South Carolina, Georgia, East & West Florida,* which at long last took its place on the shelves of Philadelphia book dealers in 1791 and was quickly followed by a London edition in 1792. Bartram extolled the beauty of the plant, claiming that, seeing the plant in flower on Appalachian hillsides, "we are alarmed with the apprehension of the hills being set on fire," and noting further that "this is certainly the most gay and brilliant flowering shrub yet known" (quoted in Harper, 1958, p. 205). Michaux also noted the plant in his journal entry for May 11, 1795; he was probably writing by the flickering light of his campfire after a long but rewarding day in the Blue Ridge Mountains of North Carolina.

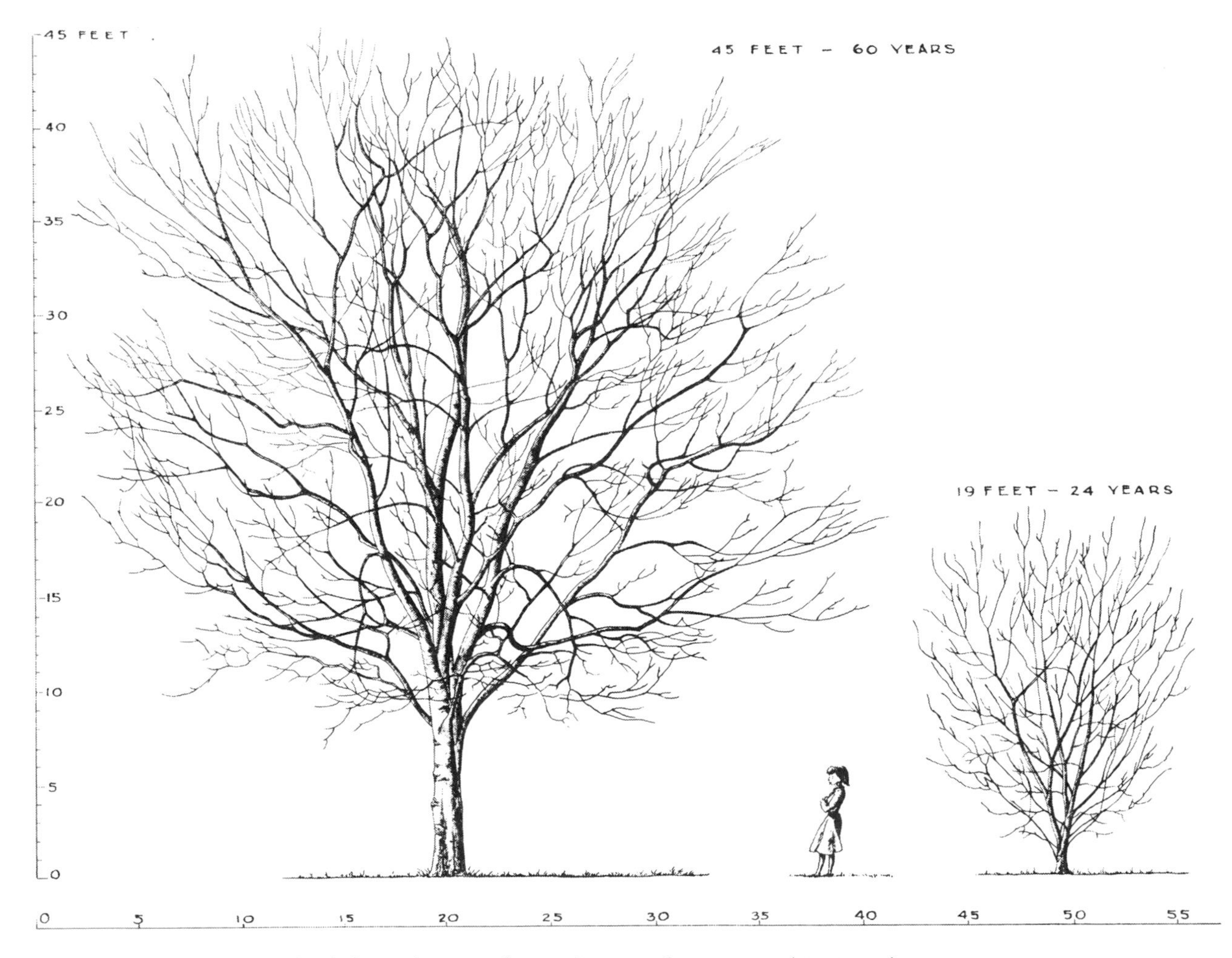

42. Habit of the yellowwood (*Cladrastis lutea*) with an indication of its size in cultivation after twenty-four and sixty years of growth.

Frontispiece.

Mico Chlucco the Long Warrior, or King of the Siminoles.

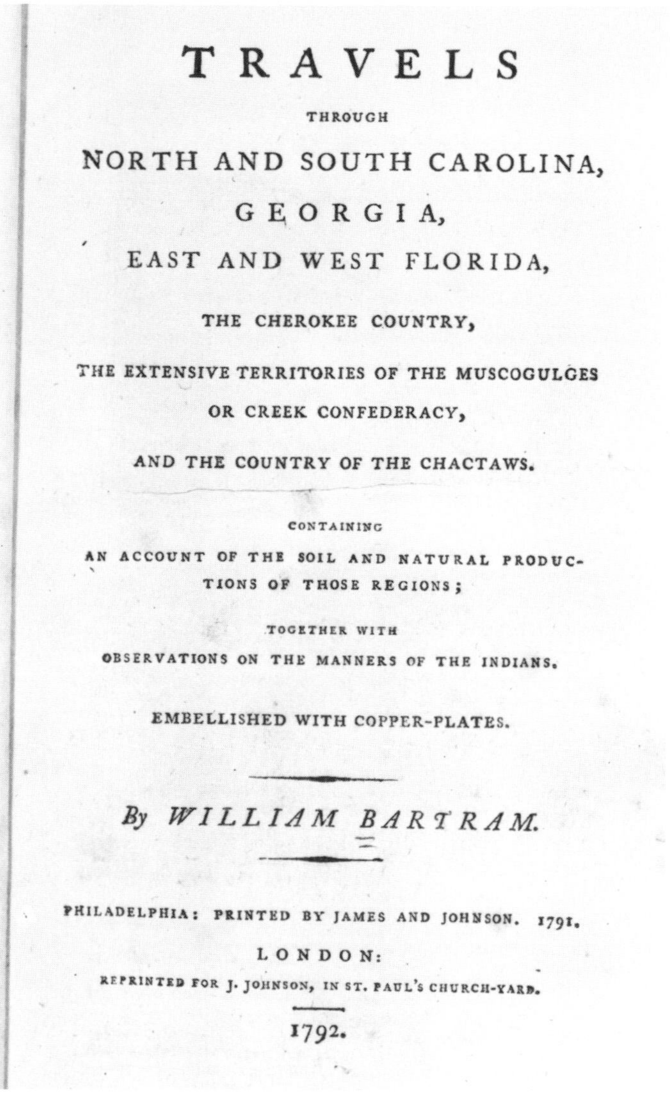

TRAVELS

THROUGH

NORTH AND SOUTH CAROLINA,

GEORGIA,

EAST AND WEST FLORIDA,

THE CHEROKEE COUNTRY,

THE EXTENSIVE TERRITORIES OF THE MUSCOGULGES OR CREEK CONFEDERACY,

AND THE COUNTRY OF THE CHACTAWS.

CONTAINING

AN ACCOUNT OF THE SOIL AND NATURAL PRODUCTIONS OF THOSE REGIONS;

TOGETHER WITH

OBSERVATIONS ON THE MANNERS OF THE INDIANS.

EMBELLISHED WITH COPPER-PLATES.

By WILLIAM BARTRAM.

PHILADELPHIA: PRINTED BY JAMES AND JOHNSON. 1791.

LONDON:

REPRINTED FOR J. JOHNSON, IN ST. PAUL'S CHURCH-YARD.

1792.

43. Title page of the first English edition of William Bartram's *Travels*, published the year after the first American edition appeared in Philadelphia. A Dutch edition was published in Haarlem in 1794. This book, which described the geography and natural history of the southeastern United States, ranks as one of the first international best sellers by an American author.

44. The mountain stewartia (*Stewartia ovata*) produces creamy white, camellia-like flowers. Its five-valued capsules contain two narrowly winged seeds per locule, and if all develop, a capsule can be expected to yield ten seeds.

Another shrub encountered in the high mountains of the Carolinas and mentioned in Bartram's *Travels* also came under Michaux's discerning eyes. The flowers of mountain stewartia (*Stewartia ovata* (Cavanilles) Weatherby) might be mistaken for those of the Franklin tree. They are similar in size and color, although the mountain stewartia's flowers expand from their initially tight, rounded buds in June, two months before those of the Franklin tree. A spreading shrub that occasionally attains the stature of a small tree, the mountain stewartia occurs in two separated regions in nature. While not a common plant in either area, it is most frequent on the upper Piedmont and in the mountains of the Carolinas, Georgia, Alabama, Kentucky, and Tennessee. Surprisingly, the so-called mountain stewartia is also known from two isolated stations near Williamsburg on the Virginia Coastal Plain.

The disjunct populations of the mountain stewartia on the coast of Virginia had not gone unnoticed, and the occurrence of the plant there added an element of confusion when the genus *Stewartia* was established by Linnaeus. For a second species, the silky stewartia (*Stewartia malacodendron* L.), also represents the genus on the Coastal Plain from Virginia to northern Florida and westward into eastern Texas. This latter species became known in 1742 when specimens and living plants were sent from Virginia by John Clayton (1686–1773) to Mark Catesby in England. On the basis of Clayton's specimens, Linnaeus founded the new genus—its name honoring the work at Kew of John Stuart, Earl of Bute—but John Mitchell (1680–1768), a Virginia physician, criticized the Swedish naturalist's characterization and placement of the genus in the sexual system. Mitchell's disapproval was doubtless unexpected and considered audacious but rested on his observations of the mountain stewartia from one of its two Virginia populations. Linnaeus worked with specimens of the more common plant of that region, the silky stewartia, thereby accounting for the confusion. The presence of two species was not sorted out for decades and only after the two species had been erroneously placed in two distinct genera, one conforming to Linnaeus's diagnosis, the other to Mitchell's.

Owing to the beauty of its white, camellia-like flowers, the silky stewartia enjoyed a wide popularity in English gardens, although it proved a difficult plant to maintain in the cool English climate. Of easier culture, the mountain stewartia quickly replaced the Coastal Plain species once it found its way into gardens. Despite repeated attempts to establish the silky stewartia in the Arnold Arboretum, it has not proven hardy in the Boston region. But its mountain cousin flourishes in the New England climate, and plants of the species Michaux located on the Blue Ridge of the Carolinas grow on Bussey Hill. These plants represent a form of the mountain stewartia (*Stewartia ovata* f. *grandiflora* (Bean) Kobuski), with flowers larger than the norm and with dark purple rather than whitish stamen filaments, which contrast elegantly with the pure white of the scallop-shaped petals.

During his extended sojourn in the young Republic, Michaux's activities continually faced the threat of curtailment because of financial difficulties wrought by political upheaval in France. Undaunted, the Frenchman continued his quest to explore ever farther westward to fulfill his mission. To help offset his financial setbacks, in 1793 members of the American Philosophical Society promised him backing if he would agree to mount a western expedition that would take the explorer across the Mississippi River, through the vast Louisiana Territory, and to the headwaters of rivers on the Pacific Slope. Thomas Jefferson, then Secretary of State, was instrumental in drawing up the proposal and securing subscriptions endorsing the undertaking from numerous fellow members of the Philosophical Society. Many of that group had met Michaux and enthusiastically supported the idea that essential scientific knowledge would be gained from the expedition. The list of subscribers included George Washington, Alexander Hamilton, John Adams, James Madison, and of course Jefferson himself, as well as many others.

Ironically, Michaux's plans to traverse the North American continent were dashed when the Republic of France dispatched its first diplomatic envoy, Edmond Charles Genet, to the United States in the spring of 1793. During the remaining years of Michaux's stay in North America his dream of reaching the Pacific coast was not to be realized, although he did reach the banks of the mighty Mississippi. It would remain for Lewis and Clark, during Jefferson's administration as president, to undertake the epic expedition that Jefferson and Michaux had planned. Collecting, however, continued to occupy Michaux's energies, although Citizen Genet, capitalizing on Michaux's urge to travel in new regions on the frontier, siphoned some of Michaux's time to help gain navigation rights on the Mississippi River from Spain. At long last, after serving his country in exemplary fashion as botanical collector and diplomat, Michaux packed his specimens, journal, and other belongings and embarked on the *Ophir* in the fall of 1796 to return to France.

Michaux's homecoming was fraught with disappointment and near disaster. Within sight of the Belgian coastline, a storm shipwrecked the *Ophir*, and the explorer's life was nearly lost. His unconscious body was heroically pulled ashore on a floating timber and his precious plant specimens, though thoroughly soaked, were salvaged. Lost, however, was the notebook containing the first portion of the journal recording Michaux's American experiences. After recovering from the shock of the ordeal and painstakingly redrying his large collection of botanical specimens, André traveled overland to Paris. On his arrival in the French capital, and to his overwhelming disappointment, Michaux learned that the vast majority of the living plants he had shipped to France had perished through neglect during the turbulent years of revolution and its aftermath.

In Paris, Michaux was reunited with his fully grown son, who had trained as a physician but shared his father's strong enthusiasm for botany. André took up residence near the newly named Jardin des Plantes. Disappointed at the meager impact his extended mission would apparently have for the forests, fields, and gardens of France, he nevertheless rejoined the circle of French botanists at the Jardin and set to work on his dried collections. His efforts resulted in two manuscripts, a monograph of the oaks of North America and the voluminous *Flora Boreali-Americana*, or Flora of North America. The *Flora* was the first botanical work that attempted to account for the spontaneous flora of the North American continent and drew on Michaux's intimate knowledge gained on his extensive travels in that region on behalf of his native country.

Wanderlust, however, still captivated Michaux. Given the opportunity to join an exploring expedition to the South Seas sponsored by the French government under the command of Captain Nicolas Baudin, the botanist eagerly joined the party as naturalist. Leaving François André in charge of final details and seeing his manuscripts through the presses, Michaux left France for the last time in 1800, on this occasion sailing for the Indian Ocean. Baudin intended to emulate the first exploratory voyage of Captain Cook, but severe difficulties developed between the captain and his scientific passengers. Because of the dissension, Michaux opted to leave the expedition when it dropped anchor off Madagascar. The unique flora of that mountainous isle fascinated the Frenchman, but before making extensive collections he determined, as in America, to establish a nursery garden where plants could be stockpiled for eventual shipment to France. Working strenuously in the steamy tropical heat, Michaux became exhausted and fell ill with fever. A victim of malaria, he died in November of 1802 in his remote nursery garden in Madagascar after a lifetime devoted to botany and plant exploration.

Michaux never held in his hands the volumes com-

Whereas Andrew Michaux, a native of France, and inhabitant of the United States has undertaken to explore the interior country of North America from the Missisipi along the Missouri, and Westwardly to the Pacific ocean, or in such other direction as shall be advised by the American Philosophical society, & on his return to communicate to the said society the information he shall have acquired of the geography of the said country, it's inhabitants, soil, climate, animals, vegetables, minerals & other circumstances of note: We the Subscribers, desirous of obtaining for ourselves relative to the land we live on, and of communicating to the world, information so interesting to curiosity, to science, and to the future prospects of mankind, promise for ourselves, our heirs executors & administrators, that we will pay the said Andrew Michaux, or his assigns, the sums herein affixed to our names respectively, one fourth part thereof on demand, the remaining three fourths whenever, after his return, the said Philosophical society shall declare themselves satisfied that he has performed the sd journey & that he has communicated to them freely all the information which he shall have acquired & they demanded of him: or if the sd Andrew Michaux shall not proceed to the Pacific ocean, and shall reach the sources of the waters running into it, then we will pay him such part only of the remaining three fourths, as the said Philosophical society shall deem duly proportioned to the extent of unknown country explored by him, in the direction prescribed, as compared with that omitted to be so explored. And we consent that the bills of exchange of the sd Andrew Michaux, for monies said to be due to him in France, shall be recieved to the amount of two hundred Louis, & shall be negociated by the sd Philosophical society, and the proceeds thereof retained in their hands, to be delivered to the sd Andrew Michaux, on his return, after having performed the journey to their satisfaction, or, if not to their satisfaction, then to be applied towards reimbursing the subscribers the fourth of their subscription advanced to the said Andrew Michaux. We consent also that the said Andrew Michaux shall take to himself all benefit arising from the publication of the discoveries he shall make in the three departments of Natural history, Animal, Vegetable and mineral, he concerting with the said Philosophical society such measures for securing to himself the said benefit, as shall be consistent with the due publication of the said discoveries. In witness whereof we have hereto subscribed our names and affixed the sums we engage respectively to contribute.

Go. Washington one hundred Dollars
John Adams Twenty Dollars
Benjamin Hawkins Twenty Dollars.
Ra. Izard — Twenty Dollars
Saml Johnston Twenty Dollars
Robt Morris Eighty Dollars
Jno. Henry ten dollars
G Cabot Ten Dollars
John Rutherfurd Twenty dollars

H Knox fifty dollars
Th: Jefferson fifty dollars
Alexander Hamilton Fifty Dollars
Rufus King — Twenty Dollars
John Langdon Twenty Dollars
John Edwards Sixteen Dollars
John Brown Twenty Dollars
Tho Mifflin Twenty Dollars

Jona Trumbull Twenty Dollars
James Madison Jr. Twenty dollars
J Parker — Twenty Dollars
Alex White Twenty Dollars
John Page twenty Dollars
John Beckley 10 Drs
Wm Smith — Twenty Drs
Jere Wadsworth Thirty
Richard Bland Lee ten dollars
Thos Fitzsimons ten dollars
Saml Griffin Ten dollars
Wm B. Giles Ten Dollars
Jn W Kittera Ten Dollars.

prising his two significant contributions to American botany, yet his work is clearly evident in the many epithets of North American plants that bear his name as author. Less well known are the numerous plants he successfully introduced into cultivation despite the prolonged transatlantic voyages and the turbulent conditions in France that combined to greatly reduce their numbers. The list of cultivated subjects associated with his name would be far longer had his consignments received the care and attention Michaux anticipated.

A Penchant for Americans

During the period André Michaux reaped harvests of seeds and dug consignments of young saplings for shipment to France, other European plantsmen actively foraged for horticultural rarities in the young American Republic. The English hunger for American plants, and the vogue of growing American species together in areas referred to as American gardens, continued to create great demand for exotics of American origin. Bartram's *Travels*, moreover, whetted the appetites of experienced plantsmen for species described and figured in its pages that were not yet known in cultivation. Additionally, talented gardeners as well as novices became inspired to attempt the cultivation of novel plants as a result of new publications aimed at describing and illustrating recently introduced species. Public taste for gardening increased dramatically, as evidenced by the circulation of *Botanical Magazine, or Flower Garden Displayed* that soared to three thousand copies per issue shortly after the periodical first appeared in the book stalls of London. Launched in 1787 by William Curtis (1746–1799), who had resigned his post as Demonstrator in Botany in the Chelsea Physic Garden to devote his time and energy to his own garden and to popularize horticulture through the press, *Botanical Magazine* found a ready and appreciative audience. Initially published in twelve parts annually, every issue contained three hand-colored plates, each accompanied by a page or two of text that provided botanical details and horticultural requirements for successful culture.

In response to growing demands for plants, new commercial nurseries were founded, long-established ones flourished, and competition between firms, particularly for the introduction of choice plants, intensified. A small group of enterprising collectors, viewing the United States as a botanical treasure trove and seeing opportunities for financial gain coupled with the excitement of exploration in nature's garden, undertook the continued horticultural exploitation of the former colonies. John Fraser, a Scotsman, had been encouraged in his growing interest in plants at the Chelsea Physic Garden by William Forsyth, who had succeeded Philip Miller as curator. Initially Fraser intended to become a mercer, but his career took an unexpected turn when Forsyth, with assistance from William Aiton, head gardener at Kew, and Sir James Smith, president of the London-based Linnean Society, backed Fraser's plans for a collecting trip to Newfoundland. Between 1780 and 1784 Fraser traveled in this northern region still solidly within the British domain, probably recognizing that he would not be particularly welcomed in the young republic to the south.

After biding his time in the northern clime of Newfoundland, surrounded by a boreal flora that provided relatively few novelties, the Scotsman set his sights on the far richer flora to the south. By 1785 political attitudes in the former colonies had moderated to the extent that Englishmen—or their Scottish agents—could travel freely. In the same year that André Michaux sailed to America from France, Fraser headed for the southeastern United States, where he, too, would undertake the arduous exploration of the southern Appalachian region. The two plantsmen's paths were to cross, but despite their mutual goals and interests, their collecting days together were few. Each had developed his own method, and whether or not they sensed competition and rivalry, each preferred to tramp in solitude.

Fraser, laden with living plants expertly packed for the

45. Subscription list for André Michaux's proposed trans-continental journey to the Pacific Ocean under the auspices of the American Philosophical Society. This contract, drawn up in 1793, was never realized, as plans for Michaux's proposed journey fell through. It was not until Lewis and Clark mounted their expedition that the goals outlined for Michaux were accomplished. Familiar names in the list of subscribers include those of George Washington, John Adams, Robert Morris, Henry Knox, Thomas Jefferson, Alexander Hamilton, and John Trumbull.

46. John Fraser (1750–1811) was born in Scotland but moved to London as a young man, where he established a nursery business in Sloane Square, Chelsea, in the 1780s. Smitten with an urge to travel and undertake botanical and horticultural exploration, Fraser traveled in Newfoundland between 1780 and 1784, and undertook extensive exploration in the southeastern United States between 1785 and 1807.

voyage, returned to England in 1788. His success in transporting living plants became legendary and may have involved the inspired use of sphagnum moss, the moisture-holding properties of which he was sure to have become familiar with in boreal bogs. Fraser returned to the enchanting forests and mountains of the southeastern United States on three additional sojourns between 1788 and 1796. Again burdened with a precious cargo of choice plants, Fraser returned to England in 1796 and later in the year traveled to St. Petersburg (now Leningrad) in Russia, where he was able to name his own price for a collection of "Americans" sold to the Empress Catherine. The czarina, amassing treasures to fill the Hermitage, her vast St. Petersburg palace, displayed extraordinary activities as a collector, and her interests included not only works of art but a natural-history cabinet and living exotics as well. On Catherine's death in 1797, the new czar and czarina, Paul I and Maria, enthralled with the horticultural novelties from North America, appointed the Scotsman their botanical collector. Once again Fraser trod the forests and woodlands of the Southeast, this time with a royal commission from the crowned heads of Russia. The czar's far-flung imperial domain stretched from St. Petersburg in the western part of Russia to the Indian villages and trading posts in the uncharted wilderness of Russian America (Alaska) on the Western side of the North American continent, just across the Bering Strait.

Following what seems to have become established tradition, Fraser was accompanied on his southeastern forays by his son, also named John. Together they tramped across the flat but watery expanse of the Carolina Coastal Plain, suffering the intense summer heat and facing the dangers of cooler but snake-infested swamps. In the thicketlike growth of pocosins and at the edges of tranquil bays the Frasers located a shrub that had been illustrated but not described by William Bartram in the pages of his celebrated *Travels.* The name Bartram provided for the drawing placed the plant in the genus *Andromeda* L., a group that served at the time as a catch-all for numerous newly discovered plants of the heath family, or Ericaceae. Many of the species temporarily placed there were subsequently recognized as generically distinct, and such was the ultimate disposition of the plant the Frasers shipped back to Europe in 1800. It is now placed in the genus *Zenobia*—the name of the independent-minded queen of ancient Palmyra, which also serves as the plant's common name.

Individuals of the solitary species in this genus (*Zenobia pulverulenta* (Bartram) Pollard) are variable in

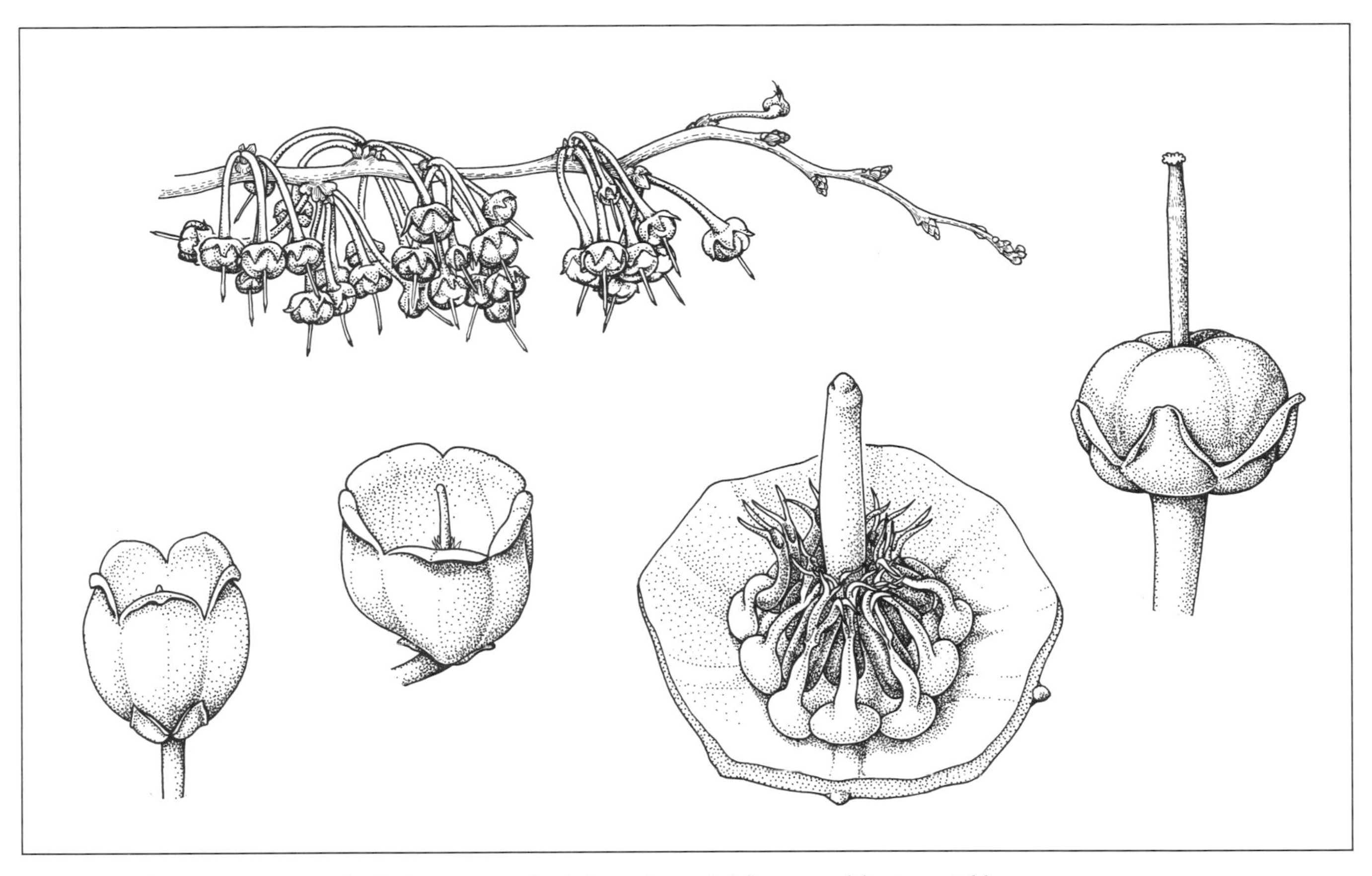

47. The zenobia is an attractive shrub that is grown both for its beautiful flowers and for its variable foliage. Shown here are details of the flowers and immature to mature fruits. The corolla of one flower has been partially removed to show the stamens and the protruding style.

respect to a character of their leaves and stems. Some plants, which are small shrubs spreading or arching to four or five feet in height, produce leaves and stems that are covered with a waxy whitish or silvery bloom. This superficial coating can be rubbed from the surfaces of the elliptical blades to expose the intrinsic green of the chlorophyll-packed tissues of the leaves. Stems and leaves of other individuals lack the waxy bloom, and the shining, deep-green foliage gives these plants an entirely different aspect in the landscape than their glaucous counterparts.

In the Arnold Arboretum, plants of both types of zenobia grow intermixed in the moist ground at the base of the seepage slope abutting the flood plain of Spring Brook, just above its confluence with Bussey Brook. Visitors to the Arboretum in June have the opportunity to view these plants in flower and also to experience their faint but delicious fragrance. In the bell-like structure of their united petals, the pure white to pinkish nectar-producing flowers resemble those of lily-of-the-valley, but they are larger and hang suspended in groups of four or five together along the elongated inflorescence axes, which terminate the leafy branches. As the growing season progresses, curious capsular fruits take the place of the nodding flowers and at maturity turn upright, a position that must promote widespread wind dispersal of the hundreds of small seeds held within when the capsules dehisce at the end of the growing season.

The Frasers returned to Europe in 1802 after venturing to explore the mountainous districts of Cuba, initially disguised as American citizens. Hostilities had developed between Spain and Great Britain, and the Scotsmen anticipated that their English passports would prevent their planned activities on the Spanish island. Learning the intent of their mission, however, the Spanish governor allowed them to travel freely, explaining, "My country, it is true, is at war with England, but not so with the pursuits of these travellers" (quoted by W. J. Hooker, 1831, p. 301). After enduring further hardships on the voyage home on board a leaky, rotting ship, the elder Fraser was distraught to learn of the untimely death of the czar. Only after a period of years—during which time the father attempted to have his claims settled by Alexander, the new czar—did the Frasers return to the southeast in 1807, this time to further their own financial interests, without royal patronage.

On this trip—destined to be the elder Fraser's last expedition in North America—the father-and-son team ascended the Blue Ridge, where they collected the mountain rosebay, or catawba rhododendron (*Rhododendron catawbiense* Michaux), on the vast expanse of Roan Mountain. A magnificent broadleaved evergreen shrub, the mountain rosebay forms dense, mounded thickets and is characteristic of the unique, high elevation and otherwise grassy, parklike "bald" that covers twelve hundred acres on the rounded peak that straddles the boundary between North Carolina and Tennessee. While the mountain rosebay is seen to best advantage on Roan Mountain—and today thousands of visitors annually make the pilgrimage during June, when the shrubs are in flower—the species occurs at high elevations in the Appalachians from Virginia and West Virginia south into Georgia and Alabama. The plant occurs on the balds of other peaks in the region of Roan Mountain and along ridge tops and on steep mountain slopes, where their impenetrable growth has given rise to the descriptive terms "rhododendron slicks" or "rhododendron hells." So dense is their growth that some claim bear paths must be followed to penetrate these thickets.

The flowers of the mountain rosebay are lilac purple in color and are produced in large rounded clusters or

48. Of all the shrubs introduced into European gardens by John Fraser and his son, the mountain rosebay or catawba rhododendron (*Rhododendron catawbiense*) must surely have been the most important from a horticultural standpoint because of its subsequent use in rhododendron hybridization.

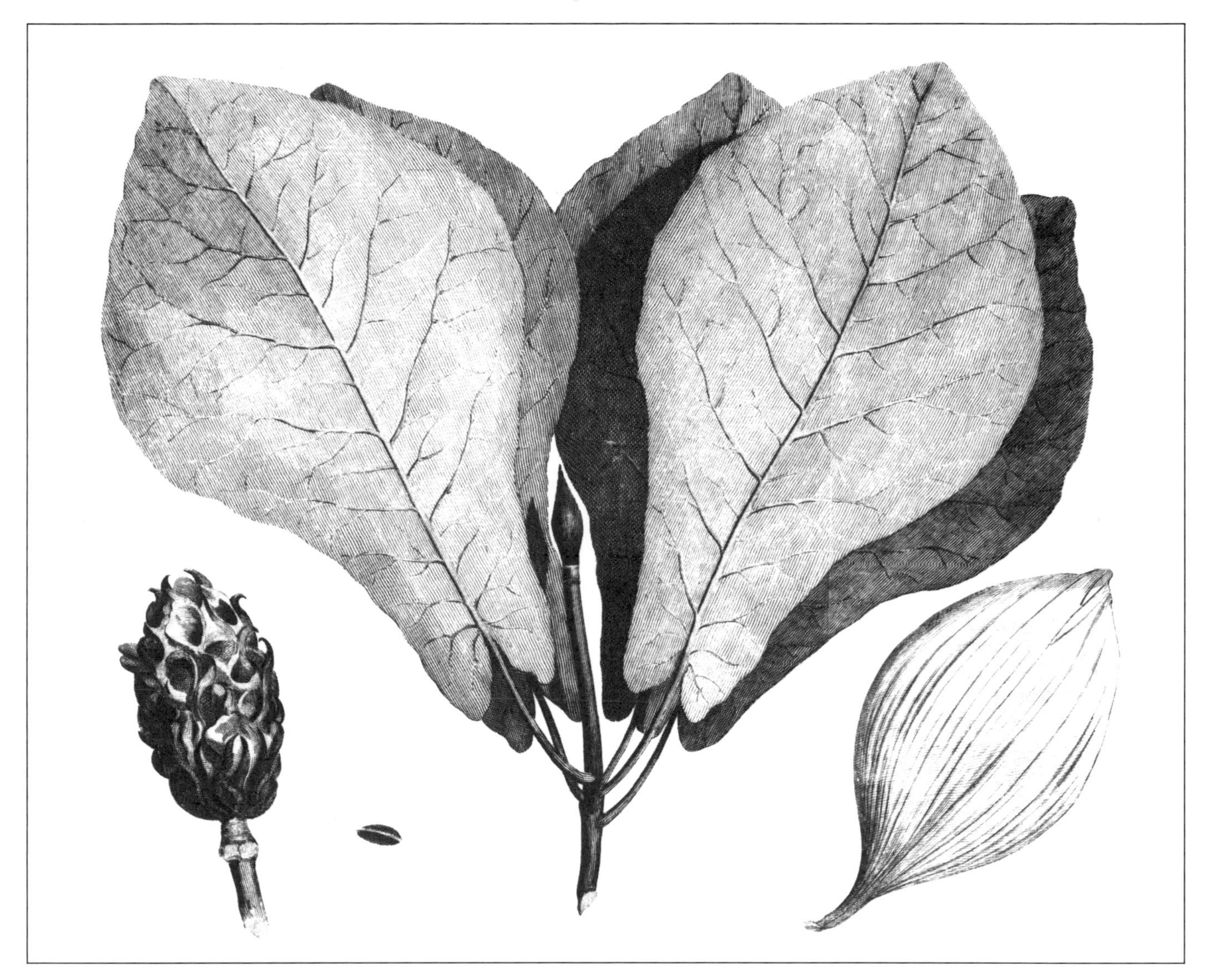

49. *Magnolia fraseri* was named by Thomas Walter (1740–1789) to honor John Fraser, who introduced the new species into English gardens. The costs involved with printing this plate of the "New Auriculated Magnolia" for inclusion in Walter's *Flora Caroliniana*, however, were borne by Fraser, who probably reckoned that sales of the plant at his Chelsea nursery would be greater if the gardening public had become somewhat familiar with the plant through this illustration. Thus, its cost could be justified. The *Flora*, issued in 1788, utilized binomial nomenclature and soon replaced Catesby's *Natural History* as the primary reference on the flora of the Southeast.

trusses five to six inches across. Individual flowers, each with five petals joined together at the base, have olive-green spots on their inner surfaces. The clusters terminate the numerous branches of the shrubs, and each is backed by the closely spaced, dark green, glossy leaves. Because of the large number of inflorescences produced, the shrubs offer a spectacular sight when in flower, yet the beautiful foliage, which usually robes the plants to their base at ground level, is reason enough for the plant to be a much desired garden subject.

André Michaux first discovered this horticultural prize, yet the Frasers are credited with making living plants of the mountain rosebay known to nurserymen and gardeners in England in 1809. Once in the hands of English horticulturists, this hardy Appalachian species provided a genetic matrix that was quickly appreciated by plant breeders, and a raft of garden hybrids, many of uncertain parentage but all involving the mountain rosebay, were produced. This group, frequently referred to simply as "catawbiense hybrids"—or "iron clads," owing to their hardiness—has provided cold-tolerant cultivars with flowers that range in color from creamy white and pink through crimson to deep, bluish purple. Many of these iron clads, which originated in English nurseries, especially at the Knaphill Nursery of the Waterer firm, grow along Bussey Brook at the base of Hemlock Hill in the Arnold Arboretum. In this favorable location plants of the mountain rosebay grow alongside hybrid progeny which combine its genetic background with that of Eurasian rhododendrons too tender to withstand the New England climate.

The elder Fraser's collecting activities were curtailed when he suffered broken ribs and other injuries in an accident with his horse early in 1810. Returning to London in the spring of that year, he suffered further medical complications, and his physical condition worsened. After spending the better part of a year in his sick bed, Fraser died in April of 1811. His son, who had shared many of his father's plant-collecting adventures, continued to introduce new plants from the southern Appalachians and established a commercial outlet for these plants in England—a nursery located in Kent and reminiscently named The Hermitage. The botanical and common names of two of the Frasers' introductions, Fraser fir (*Abies fraseri* (Pursh) Poiret) and Fraser magnolia (*Magnolia fraseri* Walter), both from the high mountains of the southern Appalachians, commemorate the father's, and by association, the son's efforts to make North American plants known in Europe as exotic garden subjects.

Initiating Two-way Traffic

In the infant American republic, garden plants were becoming increasingly recognized both as subjects for study and as objects of intrinsic beauty. The Bartram garden was well known to all scientifically minded citizens, botanists as well as the emerging class of horticulturists. But the renown of The Woodlands, the three hundred-acre estate of William Hamilton, extended the length and breadth of the new nation and across the Atlantic to Europe as well. Located on the banks of the Schuylkill River close to the botanical garden and nursery of the Bartrams, The Woodlands provided a decided contrast to the former establishment. The Bartram collection was of botanical interest, but little thought had been given to the placement of the plants in the garden. Hamilton's estate, by contrast, had been landscaped, following the European mode, as a pleasure ground, which nonetheless incorporated a wide array of exotic and native plants. Before and after the Revolution, Hamilton aggressively imported plants from Europe, and The Woodlands—now, like Michaux's New Jersey nursery, part of an urban cemetery—included the largest collection of foreign shade and fruit trees in the young nation.

Among the trees Hamilton introduced into North America via his estate gardens was the Lombardy poplar (*Populus nigra* L. 'Italica'), a rapid-growing fastigiate form of the black poplar from northern Italy. Appearing something like a fanciful, arboreal exclamation point in the landscape, especially when planted among lower-growing, round-headed trees, the Lombardy poplar enjoyed an extremely wide popularity as a street tree, an ornamental subject, and a tree that could be closely spaced to form effective screens and windbreaks. By the early eighteen hundreds, it was already widespread; in reminiscences of New York City at the beginning of the nineteenth century, John W. Francis wrote in 1858, "In 1800-'4 and '5, they infested the whole island [of Manhattan], if not most of the middle, northern, and many southern States" (J. W. Francis, 1858, p. 23).

Because the Lombardy poplar produces only staminate flowers—and as a consequence seeds are never produced—propagation is dependent on successfully rooting cuttings from a parent plant, the original American parents being the trees at The Woodlands. The Lombardy poplar's rapid spread up and down the Atlantic seaboard from Philadelphia attested to Hamilton's generosity in making cuttings available to friends and horticultural enthusiasts. The great number of trees

50. This view of the mansion at The Woodlands was drawn during the middle of the nineteenth century, when many of the trees Hamilton had planted seventy years earlier had reached their maturity. The property eventually formed the nucleus of a Philadelphia cemetery.

planted, moreover, gave evidence of the abilities of skilled American plantsmen at propagation, and of the willingness of the public to incorporate a new tree into the American landscape.

While the Lombardy poplar is of rapid growth, often attaining fifty feet in half as many years, the tree is susceptible to a canker-forming fungus disease, as well as an array of other fungal ailments, one of which, like the tree itself, is Italian in origin. As a result, many individuals are short-lived. Susceptibility to infection, moreover, appears to increase when the tree is grown in often adverse urban conditions. Consequently, the popularity of Lombardy poplars has diminished, and the tree is only infrequently incorporated into American landscapes today.

Another European tree introduced by Hamilton is not only long-lived but essentially disease-free and capable of withstanding the poorest soils and polluted city conditions. Because of these attributes, as well as its rapid rate of growth and its pleasing, symmetrically rounded crown, the Norway maple (*Acer platanoides* L.) was destined to become an extraordinarily popular street and landscape tree in North America. Since its importation by Hamilton it has become so ubiquitous in eastern North America that some claim this denizen of cities and towns has been overplanted. The abundant seeds of the Norway maple frequently produce offspring that form groves in undisturbed sites, and occasional individuals have joined native American species in woodlots and forests on the fringes of urban areas. Under these

51. William Hamilton (1745–1813) and his niece, Anna Hamilton Lyle, in a portrait attributed to Benjamin West. This painting was probably executed when Hamilton visited England soon after the Revolutionary War. A man of great wealth, Hamilton was in a position to emulate in America the natural style of landscape gardening that was coming into vogue in England. Before returning to his estate, Hamilton wrote from England, "I shall, if God grants me a safe return to my own country, endeavor to make it smile in the same useful and beautiful manner." The Woodlands became a horticultural showplace, which was toured by many of the prominent people who visited Philadelphia during the late eighteenth and early nineteenth centuries.

52. The fastigiate Lombardy poplar (*Populus nigra* 'Italica') was one of the many new trees introduced to North America by William Hamilton that attained great popularity and was widely planted as an ornamental in the New Republic. Unfortunately, the tree is not long-lived, and its use as a landscape tree has diminished significantly. Today, relatively few nurseries offer the tree, and it is rarely recommended for planting.

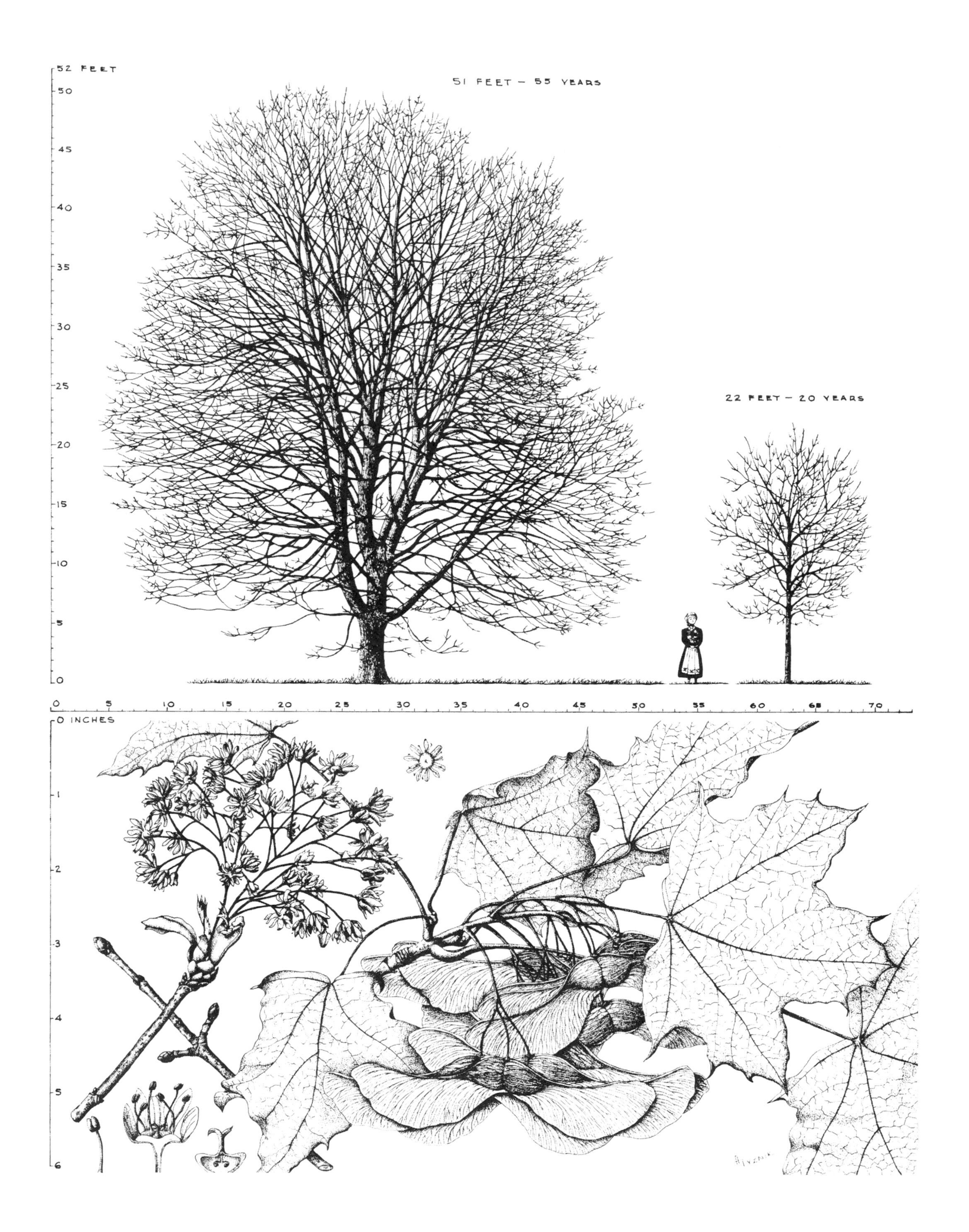
52 FEET
51 FEET — 55 YEARS
22 FEET — 20 YEARS
0 INCHES

circumstances this European species appears as a native North American tree, fully adapted to the North American forest community.

Like the foliage of many trees native to continental Europe, the yellow fall leaves of the Norway maple lack the brilliance exhibited by the native sugar and red maples and other American species. Yet in April and May, its abundantly produced, rounded clusters of chartreuse flowers cover the branches and brighten the spring landscape. Once seen, this seemingly unvarying color alone will enable recognition of the species in cities and towns across eastern North America.

Seeds or saplings of another maple, this one native to mountainous regions of central and southern Europe, were also growing in American soil for the first time during the early years of the nineteenth century. The sycamore maple or false plane tree (*Acer pseudoplatanus* L.) has become almost as ubiquitous in American cities and towns as the Norway maple, its more northern European cousin. Sycamore maple derives both its common name and specific epithet from the resemblance of its large, five-lobed, palmately-veined, deep green leaves to those of the true sycamore or plane tree. On old trees, which can attain seventy to one hundred feet in height, the bark sometimes flakes and peels away, creating a mottled appearance, which is also vaguely reminiscent of the more beautifully patterned trunks of majestic sycamores.

Despite its superficial similarities to the plane tree, the sycamore maple more closely resembles related species of maples. In its habit of growth it may be easily confused with the Norway maple. Its flowers, however, while green, are not chartreuse and are not produced in clusters as the leaves emerge in spring. Sycamore maple flowers are disposed in pendant racemes two to four inches in length and appear in May and June, after the leaves of the tree have fully expanded. Partially hidden by the large, long-stalked leaves, the inflorescences may go unnoticed, but the elongated clusters of samaras or "keys"—the familiar, wishbone-shaped, winged fruits characteristic of all maples—become noticeable in summer and particularly in fall when the leaves drop to the ground.

Not a tree planted for its flowers or its fall color—the leaves turn a dingy brown with only a hint of yellow before falling—sycamore maples nonetheless became extremely popular as a stately shade tree in North America. Like the Norway maple, the tree grows rapidly and tolerates a wide spectrum of urban conditions, as well as the salt spray of the ocean in exposed coastal situations.

Christopher Gore, son of a successful Boston merchant and himself a veteran of the War of Independence, appears to have introduced the sycamore maple into the United States from Europe early in the nineteenth century, perhaps when he served in Thomas Jefferson's administration as *chargé d'affaires* in London in 1803. Earlier, Gore served as first district attorney of Massachusetts and was a member of the Massachusetts convention that ratified the federal constitution. In 1796 President Washington appointed the Harvard-educated lawyer chief commissioner to settle American claims under the terms of the Jay peace treaty with England, an appointment requiring an extended eight-year transfer of the Gore household to London.

With diplomatic service in Britain came the opportunity to visit and observe the gardens and estates of Gore's wealthy English counterparts. The diplomat sensed the dominant role the refined arts of horticulture and landscaping played in English society and witnessed the beginnings of the transformation of London into a city of parks, squares, and gardens. In partnership with his wife, Gore devoted considerable effort while in Europe to planning for the construction of their new suburban home in Waltham, Massachusetts, which commenced upon their return to Boston in 1804. The landscaping of Gore Place, the three-hundred acre estate situated along the Charles River where their imposing

53. By contrast with the Lombardy poplar, the Norway maple (*Acer platanoides*) has increased in popularity as a symmetrical shade tree of rounded habit since it was introduced into North America from Europe by William Hamilton toward the end of the eighteenth century. Tolerant of city conditions, it is frequently planted as a street tree, although the shade it casts is often so heavy that few plants will grow at its base.

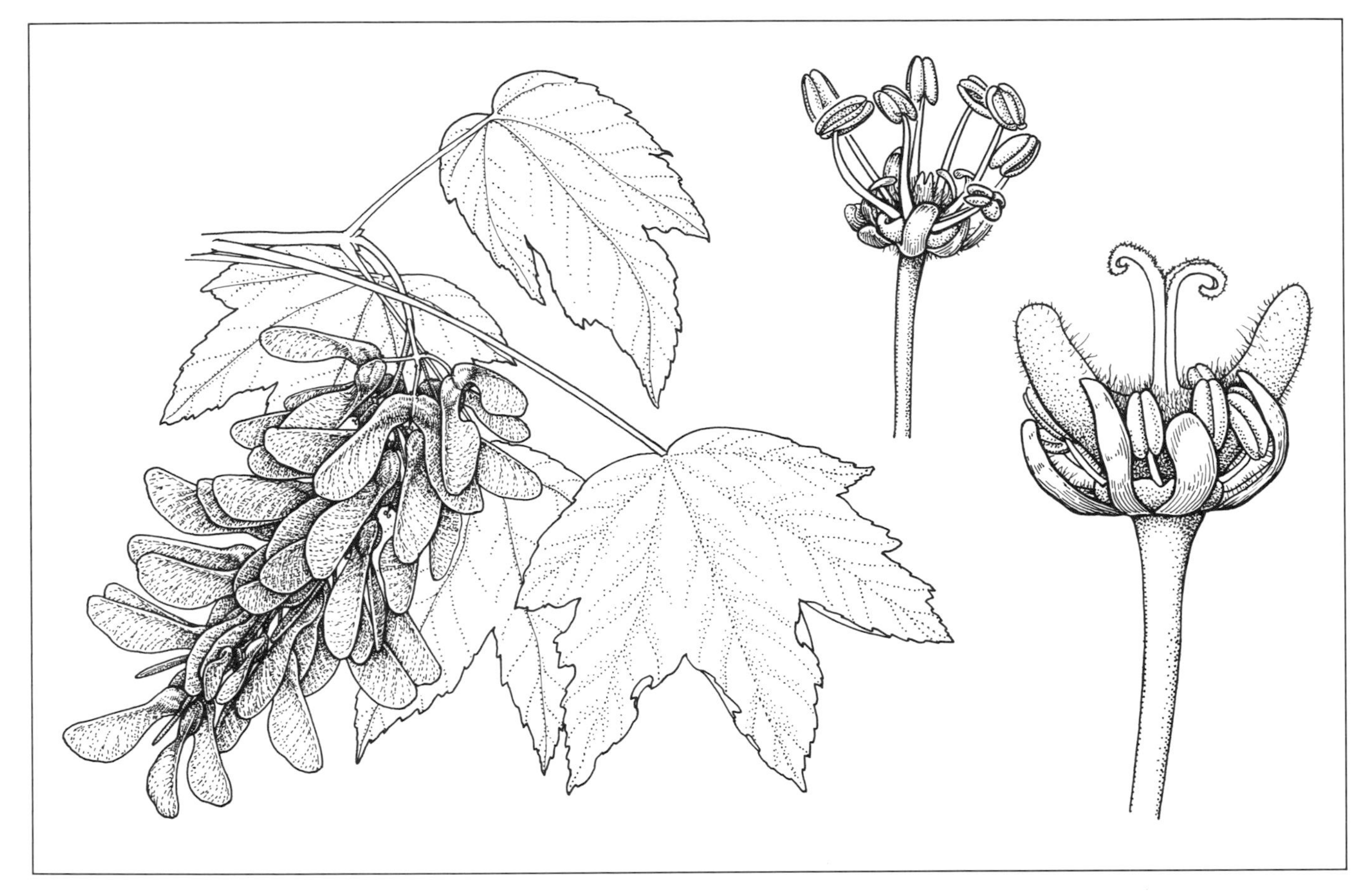

54. Christopher Gore was probably the first American to cultivate the sycamore maple (*Acer pseudoplatanus*) as an ornamental shade tree in New England. As a consequence of its wide cultivation in Europe over many generations, its spontaneous occurrence has become obscured. It is probably indigenous to the mountainous districts of Europe in the Pyrenees, Alps, and the Carpathians. In North America sycamore maples run a close second to Norway maples as frequently planted street trees.

55. In New England, Christopher Gore (1758–1827) was one of a group of wealthy citizens who developed an abiding interest in horticulture and began to practice the art of gardening on a grand scale. Gore, shown here in a portrait by John Trumbull, had become familiar with English examples of the natural style of landscape gardening then in vogue and had the means to implement similar landscape improvements at his Waltham, Massachusetts, estate, Gore Place.

56. View of the south front of Gore Place. Built of red brick as a summer residence, the federal style mansion is situated on an estate that comprised 300 acres in Gore's day and afforded ample space for agricultural as well as horticultural experimentation. As chief American negotiator for claims resulting from the Jay Treaty with Great Britain, Gore and his wife, Rebecca, lived in England and traveled in Europe. The plans for Gore Place were developed by the Gores while they were in Europe and reflect the dual influence of English country houses visited by them and that of the French architect Jacque Guillaume Legrand, who drew the architectural plans.

57. Painted sketch of The Vale, the country estate of the Lyman family, close Waltham neighbors of the Christopher Gores. Here, in then-rural Waltham, Theodore Lyman (1753–1839) experimented with many horticultural innovations, including glass- or stove-houses in which tender exotics could be cultivated throughout the year. In this painting the Lyman greenhouses can be seen to the left of the mansion. Also noteworthy is the row of four Lombardy poplars in the lower right-hand corner of the sketch. While most of the land that once comprised the estate has been developed as Waltham became urbanized, the core property of The Vale survives to this day and is a property of the Society for the Preservation of New England Antiquities. Today, the house and grounds—including the greenhouses, which are considered the oldest structures of their kind in North America and which support a fine collection of exotics fashionable in early nineteenth-century America—are open to the public.

federal residence still stands, also occupied the Gores' attentions. Sycamore maples, then fashionable and widely planted in Britain, seemed appropriate and necessary subjects for trial across the Atlantic in Waltham. Once established, propagation material of this imposing shade tree could be shared with neighbors like Theodore Lyman, whose country seat, The Vale, was nearby.

To achieve the splendor of Gore Place and The Woodlands required great wealth, and to judge from their success in attaining horticultural perfection in addition to amassing collections of exotic plants, both Hamilton in Philadelphia and Gore in Boston were surely wealthy men. Their collections, moreover, were not limited to flowering trees and shrubs on the grounds of their estates but filled glasshouses and conservatories, where tender exotics from the South Seas grew and flowered in tropical luxuriance. Some glasshouses were given over exclusively to oranges, grapes, peaches, and other fruits that were ripened to perfection for eating out of hand and for decorative center pieces and the dessert tray at the dinner table. The Lymans shared these enthusiasms at The Vale, and the glasshouses or "stoves" they constructed remain in use to the present day.

On his return to Boston in 1804, Gore did not seek retreat and a retiring life on his country estate. To the contrary, he resumed his law practice in Boston, was elected to the Massachusetts legislature, and in 1809 commenced one term as Governor of the Commonwealth. Active as a leading citizen of the region, he involved himself in the affairs of the Massachusetts Society for Promoting Agriculture, a group he had joined in 1792 when the organization had been formed as an off-

58. David Hosack (1769–1835), New York physician-botanist, Columbia college professor, and founder of the short-lived Elgin Botanic Garden. By his own admission, Hosack wanted to model his scientific career on that of his English colleague Sir Joseph Banks.

shoot of the American Academy. Other notable Bostonians belonged to this group, including his Waltham neighbor Theodore Lyman, the merchant prince Joseph Barrell, the architect Charles Bulfinch, and other well-known men such as John Hancock and John Adams. Manasseh Cutler, from outlying Ipswich, also joined their ranks, as did Benjamin Bussey, whose Roxbury estate, Woodland Hill, was to form, later in the century, the nucleus of the Arnold Arboretum.

By 1805, many members of the Society for Promoting Agriculture, Gore and Bussey among them, felt the pressing need for a botanical garden in the Boston region and recognized the necessity for instruction in botany and natural-history topics in general within an academic setting. If progress was to be made in the science of botany, if scientific agriculture was to be developed for the benefit of the agrarian-based American economy, and if horticulture was to continue to attract growing numbers of the populace, the time to act had come. Quite clearly, if a professorship of natural history could be funded and a botanical garden created, the dependency of Americans on the long-established European centers of study would diminish.

A precedent had already been set in New York in 1801. Largely through the energies and foresight of one man, Dr. David Hosack, the Elgin Botanical Garden was under development on the northern edge of the growing metropolis of New York City. After studying medicine in Scotland and receiving additional medical as well as botanical instruction in London, Hosack was convinced that a botanical garden was essential for training students for the medical profession. In London he pursued

59. A view of the Elgin Botanic Garden in New York City, which was established by Dr. David Hosack in 1801. Today Rockefeller Center in midtown Manhattan occupies the site on which Hosack developed the garden.

botanical subjects with William Curtis of *Botanical Magazine* fame and under the tutelage of Sir James Edward Smith, president of the Linnean Society. Sir James introduced Hosack to the Royal Society, where the young physician presented a paper that was later published in the Society's *Transactions*. Smith also arranged for the young New Yorker to have a personal interview with the Society's president, Sir Joseph Banks, a man Hosack was to emulate throughout his distinguished career.

Returning to New York to establish his medical practice and to accept a professorship in the medical school at Columbia College, the young doctor was hard pressed to provide his students with the caliber of instruction he had received in Edinburgh and London. Although he had spared no expense while in Europe procuring books and scientific instruments, there was no substitute for living plants in demonstrations of plant structure; and to elucidate the Linnean classification system required a broad spectrum of living examples. Carefully prepared herbarium specimens went far toward satisfying Hosack's teaching needs; yet students could not dissect and critically examine the dried plants lest they be damaged and rendered useless for future lectures and demonstrations. The obvious solution was to establish a botanical garden to satisfy teaching requirements as well as to promote increased botanical and horticultural knowledge in general.

In Boston, similar frustration had been experienced by Benjamin Waterhouse (1754–1846), a professor in the competing medical school at Harvard. While scientific advancement in Boston was largely centered in the Academy of Arts and Sciences, that organization was integrally linked with the college and its faculty. Many members of the Academy were also faculty members and in the ranks of the Agricultural Society. It was not surprising that when the idea of a professorship of natural history backed by members of the Massachusetts Society for Promoting Agriculture was proposed, it was assumed that the position would be established within

Harvard College. And despite the fact that the Massachusetts legislature would provide some of the necessary funding for a botanic garden, it was assumed that this institution would be located near the Harvard College campus in Cambridge as well.

The increasing claim horticulture and gardening held for the American public during the Jeffersonian era is reflected not only by the establishment of botanical gardens in New York and Cambridge but also by the numerous American seedhouses and nurseries that flourished and expanded or were established during the period. These included commercial firms, among which was the Bartram establishment, that could provide the American public with native and exotic trees, shrubs, and bedding plants in addition to essential fruit trees and vegetable and crop seeds. Bernard M'Mahon (ca. 1775–1816), Irish by birth, established a nursery in Philadelphia in 1802 and was soon issuing catalogues of seeds of American plants he had for sale. M'Mahon also wrote and published *The American Gardener's Calendar; Adapted to the Climates and Seasons of the United States,* the first edition of which appeared in 1806. The *Calendar,* like *Botanical Magazine* in England, was an immediate success and went through eleven consecutive editions between 1806 and 1857.

As a knowledgeable and skilled nurseryman, M'Mahon held a respected place within the American natural-history circle, and he could count President Jefferson among his clients and frequent correspondents. In 1803, during his first term as President of the United States, Jefferson finally received appropriations from Congress for an expedition up the Missouri River and on to the Pacific coast; it goes without saying that botanical and horticultural as well as other natural-history concerns were uppermost in the minds of both Jefferson and M'Mahon. Meriwether Lewis and William Clark would lead the expedition, which André Michaux had hoped to mount years earlier. The expedition would cross the vast Louisiana Territory recently purchased from France in 1803, thereby bringing firsthand observations of the little-known region back to American scientists in cities on the Atlantic coast. In preparation for the journey, which left St. Louis in May of 1804, Lewis spent nine months in Philadelphia studying botany under the instruction of Benjamin Smith Barton. During his stay in Philadelphia he was also in frequent communication with M'Mahon concerning horticultural matters.

On the return of the expedition in September of 1806, some confusion ensued with regard to the disposition of the natural-history collections gathered on the 6,000-

60. William Dandredge Peck (1763–1822), first Massachusetts Professor of Natural History and first director of the Harvard Botanic Garden in Cambridge, which was established in 1801 with the support of the Massachusetts Society for Promoting Agriculture. A catalogue of the garden's collections published by Peck in 1818 gives valuable insights into the range of plants then in cultivation in Boston and the New England region. For example, only 0.02 percent of the listed plants originated in China and Japan; today, the majority of garden exotics in North America are of Asiatic origin.

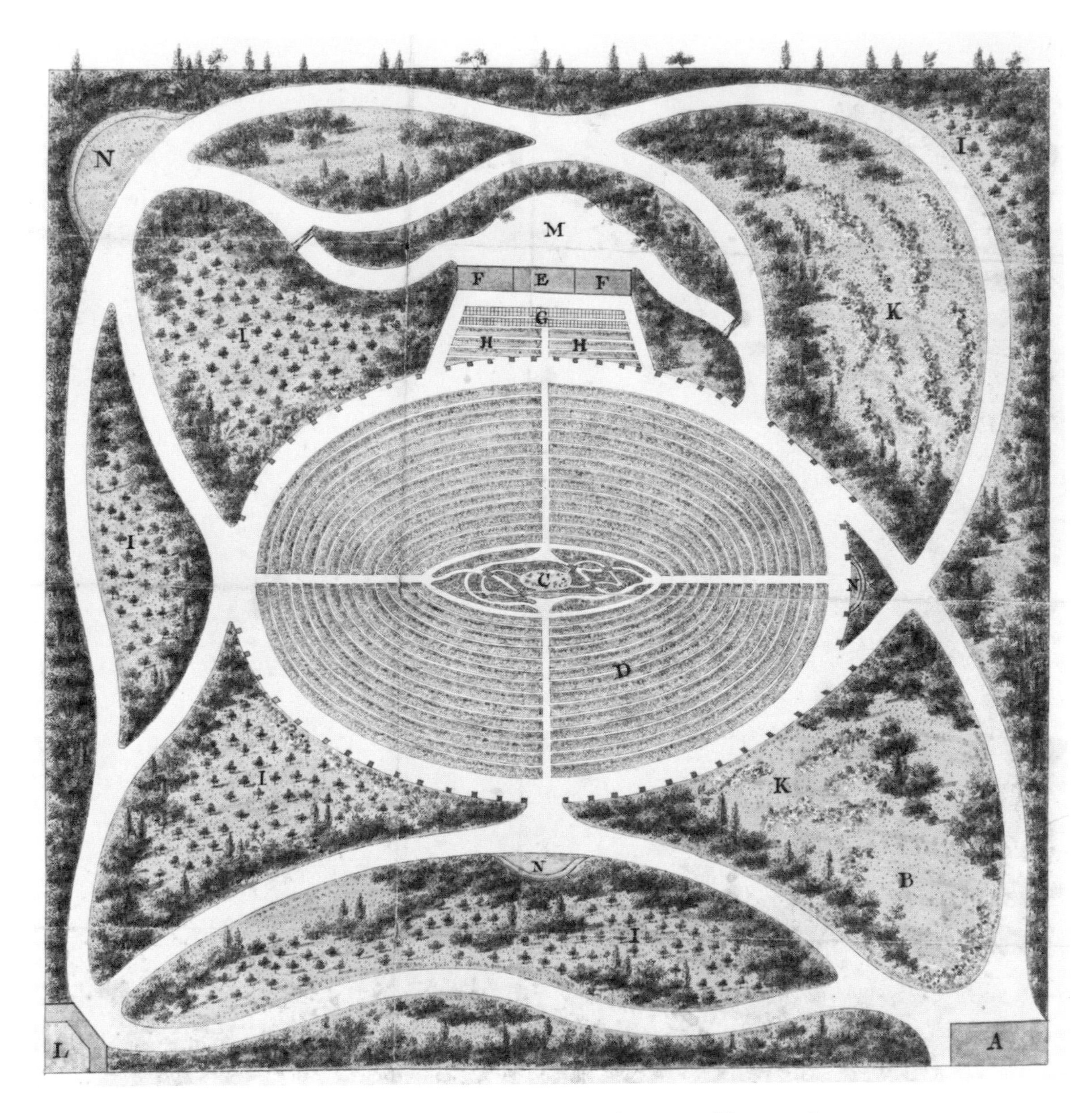

61. Early plan of the Harvard Botanic Garden. "A. House. B. Plots or small lawns with flowering shrubs. C. Basin or reservoir with running fragrant waters. D. Garden of arrangement or botanic school. E. Hot house. F. Green houses. G. Frames. H. Hot beds. K. Portions of land for various trees and flowering shrubs. L. House for the gardener. M. Yard for the stercorary [manure pile], and the back of the hot and green house for sheds to receive ladders, tools, garden pots, etc. N. Green seats or turf banks."

According to the garden's early regulations, "Members of the Board of Visitors, original Subscribers to the foundation of this Institution, the Governour [sic] and Legislature of the State, the Corporation, Board of Overseers, Professors and Instructors of Harvard College, strangers of distinction, and clergymen are admitted to the Garden, gratis. Other visitors are admitted by tickets." After World War II the Garden was disbanded and housing for married Harvard students was erected on the site.

mile journey. While it was readily apparent that the American government was capable of mounting and successfully undertaking an epoch-making expedition, the scientific community was not so well organized when it came to determining by whom and where the materials brought home would be processed and studied. Finally, it was decided that the seeds collected by Lewis and Clark would be shared between Hamilton at The Woodlands and M'Mahon, each testing their skill and luck in attempts to coax the seeds to germinate.

A new evergreen shrub with beautiful, pinnately compound leaves grew from the seeds included in one of the carefully sealed packets brought back by Lewis and Clark from the Pacific Slope of the Oregon country. Plants of this species now flourish in a protected location below the wall a few steps from the Center Street gate of the Arnold Arboretum. In the spring, clusters of chrome-yellow flowers are produced by these shrubs in bracteate racemes, and these in turn are clustered together near the ends of the slender but sturdy branches. Six outer sepals and an equal number of petals comprise the floral display and surround the relatively large ovary at the center of each flower. Inserted between the petals and the ovary are six stamens, each with two small, inwardly-facing valves or flaps hinged at the top, which flip upward to expose the golden pollen to foraging insects.

By early summer the cheerful flowers give way to the developing fruits, each ellipsoid in shape and coated with a waxy bloom. At maturity the berries assume a deep blue color, and their clusters appear like small bunches of grapes. And like grapes, the berries can be made into a pleasing jelly. These resemblances gave rise to the common name of the plant, Oregon grape (*Berberis aquifolium* Pursh), yet here the comparison with the true grape ends. In its shrubby habit and particularly in the structure of its compound leaves, the Oregon grape bears no resemblance to the celebrated vine.

Based on the foliage of the plant, a more apt common name might be Oregon holly, a similarity alluded to by the specific epithet, *aquifolium*, of its scientific name. For each of the five or seven leaflets, which together comprise a single leaf, is furnished with spine-tipped teeth along its margin, and in their heavy texture and green lustre the leaflets resemble the leaves of the holly tree, branches of which are often used for decoration during the Christmas season. At that festive time and throughout the winter months, however, the leaflets of the Oregon grape assume a rich purplish cast or bronzy hue.

62. Thomas Jefferson's name was commemorated in the generic name of the twin-leaf, *Jeffersonia diphylla*, a genus of two species of woodland herbs. Named by Benjamin Smith Barton in 1793, specimens of the plant had been collected by André Michaux and given to the American botanist for identification.

63. The settlement at Astoria, located on the south bank of the Columbia River near its mouth, replaced Fort Clatsop as a fortification and trading post on the Oregon coast. Lewis and Clark established Fort Clatsop in 1805, and Astoria, the oldest settlement in Oregon, was founded in 1811 by John Jacob Astor, who required a depot in the region to handle the Russian, Chinese, and Alaskan trade for his Pacific Fur Company.

Bernard M'Mahon succeeded in growing the Oregon grape in his nursery in Philadelphia, and plants soon passed into the hands of enthusiastic horticulturists and receptive nurserymen. The demand for these charming evergreen shrubs was staggering, and the prices they commanded were in proportion to their novelty and rarity. By 1825, when the plant had become widely known up and down the Atlantic seaboard, the Prince Nursery firm of Flushing, New York, listed plants in their catalogue at twenty-five dollars each, in today's currency doubtless equivalent to several hundreds of dollars!

While M'Mahon received accolades for growing the Oregon grape and other horticultural novelties from the Pacific coast, Frederick Pursh made the plants garnered by Lewis and Clark known to the scientific world. Of German birth, Pursh (1774–1820) had immigrated to the United States in 1799, where he sought employment as a gardener. With experience and training at the Dresden Botanical Garden in his native Germany, he soon located a position in Maryland, and in 1803 William Hamilton counted Pursh among his employees at The Woodlands. Once in the Philadelphia region, Pursh took every opportunity to consult the living collections in the neighborhood and to make the acquaintance of men with botanical interests. By 1805 Pursh had left Hamilton's employ and was working as curator and collector for Benjamin Smith Barton at the University of Pennsylvania.

Barton, inspired by André Michaux's earlier example, intended to embark on an American-based project to

64. Among the horticultural novelties brought back from the Pacific coast by the Lewis and Clark Expedition was the so-called Oregon grape (*Berberis aquifolium*), which was shared widely by plantsmen along the Atlantic seaboard and in Europe. These drawings illustrate the habit of a flowering branch as well as an individual flower and unopened buds.

produce a North American flora, and President Jefferson, eager to promote the undertaking, make Lewis and Clark's dried plant specimens available to the Philadelphia physician-botanist for study. Support for the undertaking came from Cutler in New England, Hosack in New York, Stephen Elliott in North Carolina, and Henry Muhlenburg in Lancaster, Pennsylvania. Muhlenburg, a Lutheran minister and college president, had previously proposed a collaborative effort whereby local botanists would be responsible for their respective regions in preparing a national flora. Owing to the pressures of other obligations, however, Barton never turned his attentions to the Lewis and Clark discoveries, and his dream of a continental flora authored by Americans and published in the United States never materialized. But at M'Mahon's urgings, Meriwether Lewis provided materials for Pursh to name, describe, and illustrate. After collection forays throughout much of eastern North America, ostensibly to further Barton's plans for a comprehensive flora, Pursh was in a position to act on his own.

Leaving the employ of Barton, Pursh next gardened for David Hosack at the Elgin Botanical Garden for a brief period in 1809 but returned to Europe in 1811 with specimens, illustrations, and plans for his own flora of North America. Basing his operations in London in the herbarium of the wealthy savant Aylmer Bourke Lambert, Pursh—despite his difficulties in leaving alcohol in the cask or bottle—also availed himself of the rich library and herbarium collections of Joseph Banks and James Smith. With access to previous collections of North American plants, coupled with his own specimens and his firsthand knowledge of numerous North American plants he had inspected in gardens, he managed to produce a manuscript. Pursh's *Flora Americae Septentrionalis,* published in late 1814, proved to be the comprehensive study Barton had planned. And in its pages the discoveries of Lewis and Clark, the Oregon grape included, were illustrated, described, and named by the German botanist. Pursh, following Michaux's earlier example, had succeeded where his American counterparts—still in the process of establishing traditions and institutions to accommodate botanical study and horticulture—had not.

Last Days in the Southeast

Many North American horticultural novelties discovered by John Lyon, another active plant hunter, also first received botanical names and were described in the pages of Pursh's *Flora.* From a horticultural perspective, chief among these was another choice evergreen shrub, this one from the high mountains of the Virginias, North Carolina, and Tennessee. Fetterbush (*Pieris floribunda* (Pursh) Bentham & Hooker), sometimes known by the misnomer mountain andromeda, was initially grouped along with *Zenobia* in the catch-all genus *Andromeda* by the German author. Subsequently, the shrub was recognized as constituting a distinct genus, *Pieris,* its generic name based on the region in ancient Macedonia where the nine Muses of Greek mythology were first revered.

The landscape value of the fetterbush has won it, like the mythical Muses, much esteem and veneration, particularly by keen plantsmen. And while its common name has a derogatory ring, it alludes to the plant's neat, compact and dense, low-growing habit, which stands as one of its primary garden attributes. To the contrary, a traveler in the high mountains—where the plants form low, dense thickets—may have difficulty extricating himself or his horse once the growth has been penetrated. Under those circumstances, fetterbush may seem an appropriate name.

Rarely exceeding four or five feet in height, and usually about as widespread as high, the stiff, twiggy branches of the evergreen shrub are densely clothed by elliptical leaves one to three inches in length. Of thin texture, the leaves are nevertheless leathery and a warm, lustrous, dark green. A prominent midvein courses the blade from its base to the apex, but the secondary veins are so finely reticulated that their presence is nearly obscured. The abundant flowers, which share a structure similar to those of the zenobia and sourwood, appear in April and are produced in erect panicles that terminate the branches. Their pure white color appears even whiter against the backdrop of the persistent foliage.

By early summer, while seeds are slowly maturing in the developing fruits, the annual increment of spring growth of the fetterbush is completed. At the ends of the still-flexible green stems, which harden and become woody before the end of the growing season, can be seen the small, almost embryonic inflorescences that will enlarge and mature in the following spring. On careful inspection, the flower buds can be identified in the axils of linear bracts along the branches of the miniature panicles. Noting this precocious development of flower buds

65. The fetterbush (*Pieris floribunda*) was discovered by John Lyon growing on Pilot Mountain near Winston-Salem, North Carolina. This handsome, low-growing evergreen shrub, has a limited and scattered distribution along the Blue Ridge in the southern Appalachians.

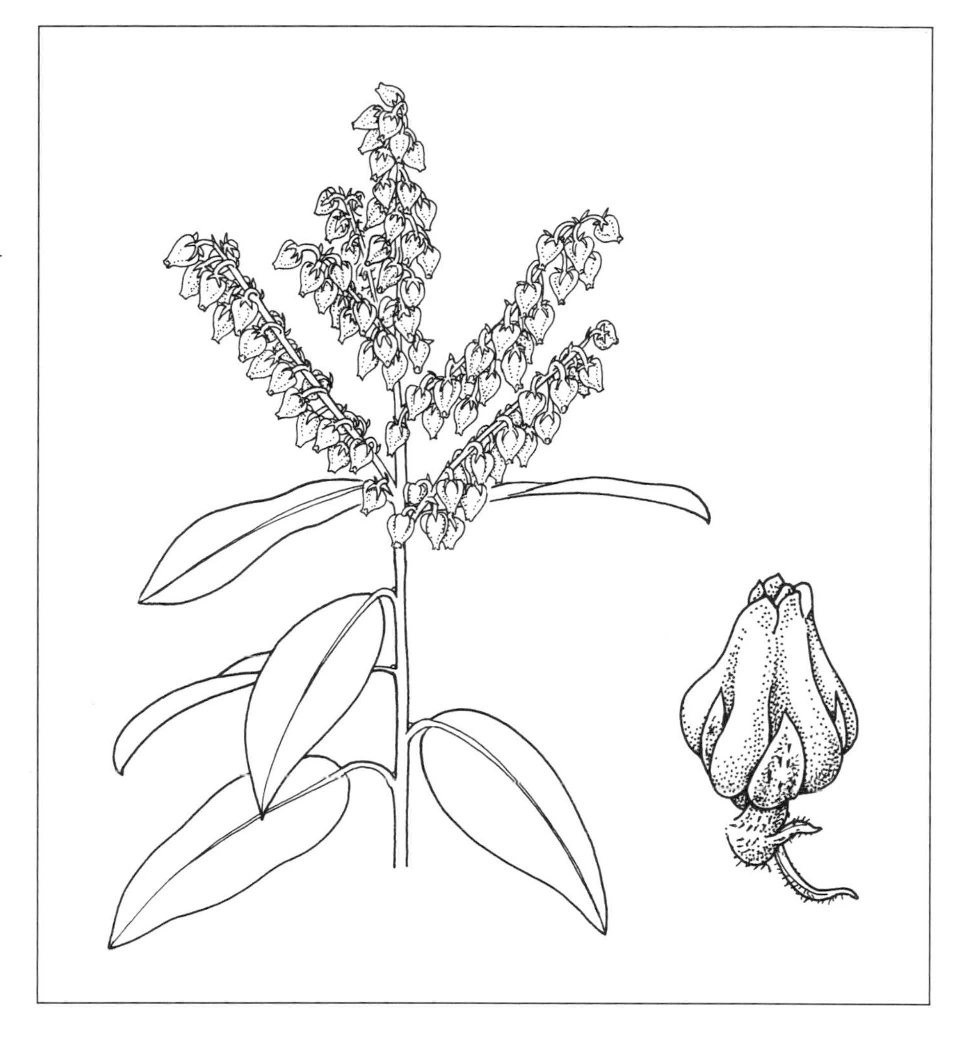

almost a year before anthesis, horticulturists in England initially relegated the shrub to the class of tender plants suitable only for greenhouse culture.

Now cultivated without winter protection in England, the fetterbush has also proven to be a hardy shrub in southern New England as far north as Boston. Because of its handsome habit and beautiful floral display, plants of this delightful shrub have been incorporated into the landscape of the Arnold Arboretum in several locations. One individual can be seen to advantage in the rockery along Valley Road, while a group of a dozen or more plants grows at the edge of Valley Road at the base of Hemlock Hill. In this location—across the road from a sourwood tree and not far from the zenobias—the plants form an impenetrable shrubbery not unlike the thickets of fetterbushes encountered in the southern Appalachians.

John Lyon (1765–1814) discovered the fetterbush on September 16, 1807, when he climbed Pilot Mountain, an isolated monadnock on the upper Piedmont northwest of Winston-Salem in Surrey County, North Carolina. Thousands of miles from his birthplace in Scotland, Lyon had immigrated to the United States about the same time André Michaux and John Fraser began their exploration of the Southeast. But unlike Fraser and Michaux, Lyon had no governmental backing or royal patronage. He came, as did thousands of other immigrants, seeking gainful employment in the young republic. And while he may have met William Hamilton before traveling to America, he was at work at The Woodlands as gardener during the closing years of the eighteenth century. It was not until 1799 that this Scotsman undertook botanical and horticultural exploration of the new nation, at first for Hamilton and then, in 1803, to further his own personal interests.

Determined to capture a share of the bull market for plants in England, Lyon retraced the footsteps of Catesby, the Bartrams, Michaux, and the Frasers. His journal entry for June 1, 1803, records that he was in Savannah, Georgia, after five days of collecting in the vicinity of Fort Barrington, where he encountered the Franklin tree. His was the last sighting of the plant in its native habitat, and he recorded for posterity what he saw and his thinking concerning its obvious rarity: "It is sufficiently remarkable that this plant has never been found growing naturally in any other part of the United States as far as I can learn, and here there is not more then 6 or 8 full grown trees of it which does not spread over more than half an acre of ground, the seed has most probably been brought there originally from a great distance by a Bird of passage" (Ewan and Ewan, 1963, pp. 22, 23).

Demand by nurserymen in England for plants of the Franklin tree may have been largely responsible for its extinction in nature, and Lyon himself may have contributed. The Scotsman was certainly active in collecting and surpassed the Frasers' success in transporting living plants to England. One account of a London sale of Lyon's plants in 1806 recounted,

> He brought an extensive collection to England; the plants composing which were partly disposed of by private account, but were chiefly sold by auction in a garden at Parsons' Green, Fulham. The catalogue of these plants fills 34 closely printed pages, it enumerates 550 lots, and the sale occupied four days. Several of the lots were composed of large quantities of one-year-old seedlings in pots; and ten lots at the end of the sale consisted each of 50 different sorts of seeds. This, it is believed, was by far the greatest collection of American trees and shrubs ever brought to England at one time, by one individual. (Loudon, 1838, 1:122)

Financially lucrative, Lyon's system of plant collection followed by their maintenance in Philadelphia-area nursery gardens and their eventual transport to London for sale on the auction block was repeated on several occasions during the early years of the nineteenth century. But the risks and privations encountered in the field were substantial, as earlier collectors and naturalists had found, and Lyon's lot was no exception. On one foray a mad dog bit the collector on the leg, forcing Lyon to sear the three punctures he sustained with a burning-hot iron and to depend on self-administered folk remedies. When his horse went astray he was sometimes forced to travel on foot, and poor roads and the lack of maps or adequate directions often resulted in lost bearings and restless nights spent without an evening meal and the comfort of a bed. The ultimate disaster than can befall the collector-naturalist in the field is to meet an untimely death, thousands of miles from home, family, and friends. This was the fate of John Lyon, who became stricken with a fever in the North Carolina mountains in 1814 and died in Asheville in September of that year.

Many of the plants John Lyon successfully brought to the auction block in London were already known in cultivation. Yet the fetterbush was totally new, and the yellowwood was imported directly into England for the first time by Lyon. Another shrub, which had been figured by William Bartram in the pages of his *Travels*, undoubtedly stimulated active bidding at the auction held in 1806. The oak-leaved hydrangea (*Hydrangea quercifolia* Bartram), first sent to England in 1803 by William Hamilton, Lyon's former employer, was not imported in any quantity until Lyon carried many of these shrubs to England.

This novel hydrangea was figured in Curtis's *Botanical Magazine* in 1806, and in the accompanying text caution was expressed concerning the potential hardiness of this denizen of the "deep South," suggesting that it "may probably require the protection of a greenhouse" (Sims, 1806, *t.* 975). Lyon had apparently collected the oak-leaved hydrangea in Florida, and it is now known to occur eastward from the Mississippi River in Louisiana, Mississippi, and southwestern Tennessee through south-central Tennessee and Alabama and into western Georgia and the panhandle of Florida. In New England the plant survives in open ground in the Boston region, although three or four years may be required before individuals become established in Yankee soils; during that period they usually suffer some dieback from winter cold.

Despite the time required for its establishment, the oak-leaved hydrangea is worthy of greater recognition as a garden plant. An arching, irregularly shaped shrub frequently attaining heights up to ten or twelve feet in its native habitats, it is more characteristically a lower shrub of four or five feet in height when cultivated in New England. Its stout branches, wrapped with a flaking, cinnamon-brown bark, bear large, inverted triangular leaf scars of previous seasons' leaves. The leaves themselves occur opposite one another along the branches, and their size, shape, and summer and fall coloration combine to give the shrub a bold luxuriance in the landscape.

Each heavily textured leaf blade, between three and eight inches in length, is broadly oval or rounded in outline, but the margins are deeply cut to form five or seven lobes, which give the blades an appearance similar to those of a red oak. In color they are a deep, usually dull but rich green during the growing season; yet as the cool days of fall advance, they assume a variety of shades ranging from purplish to crimson or orangish-brown.

The warm fall coloration and the distinctive lobed shapes of the leaves of the oak-leaved hydrangea are hallmarks that aid in distinguishing the species from others of the genus. Even lacking its foliage, the freely produced, large, pyramidal inflorescences that terminate the branches also allow for easy recognition of the plant wherever it is growing. The inflorescences, like those of many other species of *Hydrangea*, are comprised of flowers of two distinct types, yet only an Asian taxon shares its characteristic pyramidal shape. By far the most numerously produced flowers in an inflorescence are small, whitish ones, which on close inspection are seen to consist, as expected, of sepals and petals, stamens and carpels. Around these complete and fertile flowers occur large, showy, neuter flowers—flowers that lack reproductive capabilities—which consist of three to five enlarged, petal-like sepals.

The attraction of these inflorescences, which are produced throughout the summer months until frost in fall, consists of the combination of the delicate, almost ethereal quality of the fertile flowers surrounded and offset by the bold, rounded contours of the neuter ones. As the central flowers develop into fruits—small, urnlike capsules—the neuter flowers persist, become dry and papery, and turn tannish-brown. And the dried infructescences persist on the branches of the shrub throughout the fall and winter months and into the following growing season, adding year-round interest to the oak-leaved hydrangea.

66. As ornamental as its leaves, the large, pyramidal inflorescences of the oak-leaved hydrangea are comprised of large "neuter" flowers, which lack reproductive parts, and many small functional ones, which lack showy bracts. Several horticultural selections featuring inflorescences with an abundance of neuter flowers are available in the horticultural trade.

By 1814, the year Pursh's flora appeared in London and the year of John Lyon's death, the great majority of the plants William Bartram had depicted, noted, and described from the American Southeast—many tantalizing the European horticultural public—had been rediscovered, collected, and introduced into European gardens. Many of the same species were also in cultivation outside of their native ranges in American gardens further north—in Philadelphia, New York, and Boston—on estates and in fledgling botanical gardens. The latter, only two in number, gave witness to initial American attempts to establish local centers for the study of plants—the perceived basis of the natural wealth of the new nation—for the promotion of horticulture and the advancement of botany and medicine.

The establishment of nurseries and seedhouses near the growing urban centers of America followed hand in hand with increasing horticultural awareness and the obvious requirement for providing supplies of essential fruit trees and vegetable seeds. Bernard M'Mahon in Philadelphia provided necessary plant materials to his clientele, yet actively promoted interests in ornamental plants in governmental circles. In New England, John Kenrick became skilled in grafting and budding—essential techniques for the asexual propagation of large numbers of woody plants. Sometime around 1797 Kenrick opened a commercial nursery of ornamental trees in the Nonantum section of Newton, Massachusetts, and devoted two acres of the fledgling establishment to saplings of the Lombardy poplar alone, its popularity in the Boston area greater than that of any other single ornamental tree.

André Michaux, the Frasers, and John Lyon all served to make plants available from the southeastern region of the United States at a time when novelties suited to garden culture gained in appreciation. European plantsmen, particularly the English, found their appetites for eastern American plants more nearly satisfied, while their American cousins found their own thirst increasing and unquenched. If eastern North America had been largely explored and its floral wealth exploited—even to the extent that the Franklin tree had been seen in nature for the last time—where would new novelties be found in the future?

The Bartram era, during which the southeastern region of the United States was the mecca for horticultural and botanical collectors, was coming to a close. The young nation, firmly established on the North American continent, was turning its eyes westward beyond the ever-expanding frontier, and a new era in botanical exploration of North America was dawning. Simultaneously, with whaling vessels entering the Pacific, American businessmen and merchants were becoming increasingly interested in the prospects of trade in the Far East.

Plate 7. The umbrella tree (*Magnolia tripetala*) was the subject of this beautiful botanical portrait by Georg Dionys Ehret, perhaps the foremost botanical artist of the eighteenth century. This hand-colored engraving appeared in Christoph Jakob Trew's *Plantae Selectae*, and although this lavishly illustrated book was published in Nuremberg between 1750–1773, it illustrated recently introduced exotic plants growing in London gardens.

Plate 8. The rhodora (*Rhododendron canadense*), a charming small shrub of northern climes, was introduced into cultivation in England by Joseph Banks. Native to New England, the rhodora has wide ecological tolerances and grows in a variety of habitats ranging from bogs and damp thickets to exposed, rocky mountain summits.

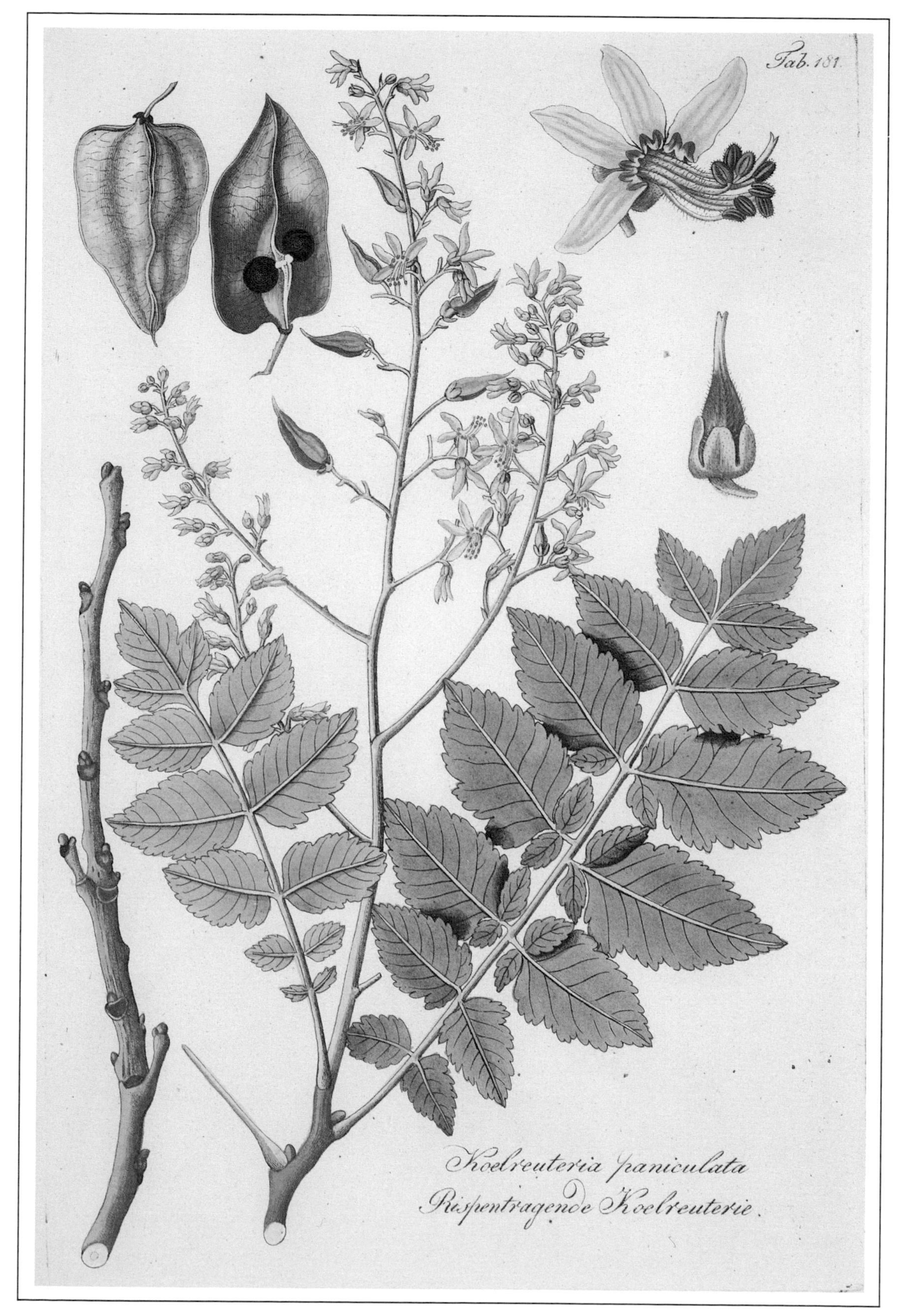

Plate 9. This plate of drawings illustrates the morphological details of the golden-rain tree (*Koelreuteria paniculata*), including an individual flower and two of its papery, seed-bearing capsules.

Plate 10. The salt-spray rose (*Rosa rugosa*) has become naturalized on sand dunes in the Great Lakes region and along the Atlantic seaboard from Quebec to New Jersey. In situations where it has become naturalized the plant frequently appears to be a native North American species.

CHAPTER THREE

On the Rim of the North Pacific

Deshima

During 1634, the fourteenth year of the Pilgrim settlement at Plymouth in the Massachusetts Bay Colony, a small, man-made, harborside island was constructed on the opposite side of the world. Composed of rubble and debris laboriously hauled to the harbor's edge, the tiny island took shape and came to measure 557 feet along its northern shore and 706 feet along its southern edge, which faced the harbor itself. Broadly curved and fan-shaped in outline, the completed island was 197 feet wide and added a total of 32 acres of dry ground to the earth's land surface. Before construction was complete, a high wooden fence enclosed its perimeter, a village of frame houses constructed in Japanese fashion was erected on either side of a centrally intersecting T-shaped thoroughfare, and a guardhouse stood at the main entrance. An arched stone bridge served as a short umbilical cord linking the island to the mainland, a bridge over which all supplies and communications with the outside world would be funneled, but not before each item had been carefully scrutinized and inspected and information sifted and digested by a multitude of guards and interpreters. The construction site lay in Nagasaki harbor off the southern Japanese island of Kyushu, and the man-made island was called Deshima (Dejima).

Xenophobia had gradually gripped Japan over the years since 1542, when the Portuguese had been the first Europeans to penetrate Japanese waters and establish contact with the peoples of the island nation. In addition to a few English, Russian, and Dutch adventurers interested in establishing trade, a wave of Portuguese Jesuit missionaries, followed by Spanish Franciscan fathers, had come to the North Pacific islands over the next few decades, intent on establishing the Christian faith in the virgin territory of Japan. But the growing negativism of the Japanese toward foreigners and their missionary activities led most Europeans to leave the country by 1634; the Portuguese who remained were forced to take up residence under heavy guard on prisonlike Deshima.

The troubled and suspicious Japanese attitude toward the proselytizing foreigners rose to new heights in 1636. The Portuguese were forced to leave Deshima and were banished from Japanese waters altogether; trading with English and Russian merchants was also abolished, and travel and trade in foreign countries by the Japanese themselves was outlawed.

Retreating southward, the Jesuit fathers joined other Portuguese members of the Society of Jesus at Macau, a natural island positioned off the southern Chinese mainland at the head of the estuary of the Pearl River near the entrance to the harbor at Canton (now Guangzhou). As the only port where commerce with European nations was allowed by the equally reclusive Chinese government, Canton was the entry point into an immense kingdom also ripe, in the minds of the Jesuits, for their missionary zeal.

With the expulsion of the Portuguese from Deshima, Japan silently closed its doors to the West and to European influence. Nonetheless, vessels of the Dutch East India Company were still allowed to call at the island port of Hirado near Nagasaki harbor for trading purposes, and the Japanese found this limited interchange lucrative and attractive. Restricted entirely to trade and without meddling clerics interfering with the traditional Japanese social structure and religious beliefs, the relations between the Dutch and the Japanese matured to the point that the Hollanders, or red hairs, were offered the use of unoccupied Deshima as a base for conducting business. To the advantage of the Dutch, East India Company ships could be resupplied from the island base for the long voyage to Java—rechristened Batavia—where the company had firmly established another Pacific outpost. In return for lucrative trade agreements and an outpost at Deshima, the Dutch East India Company paid a high price. The company agreed to strict

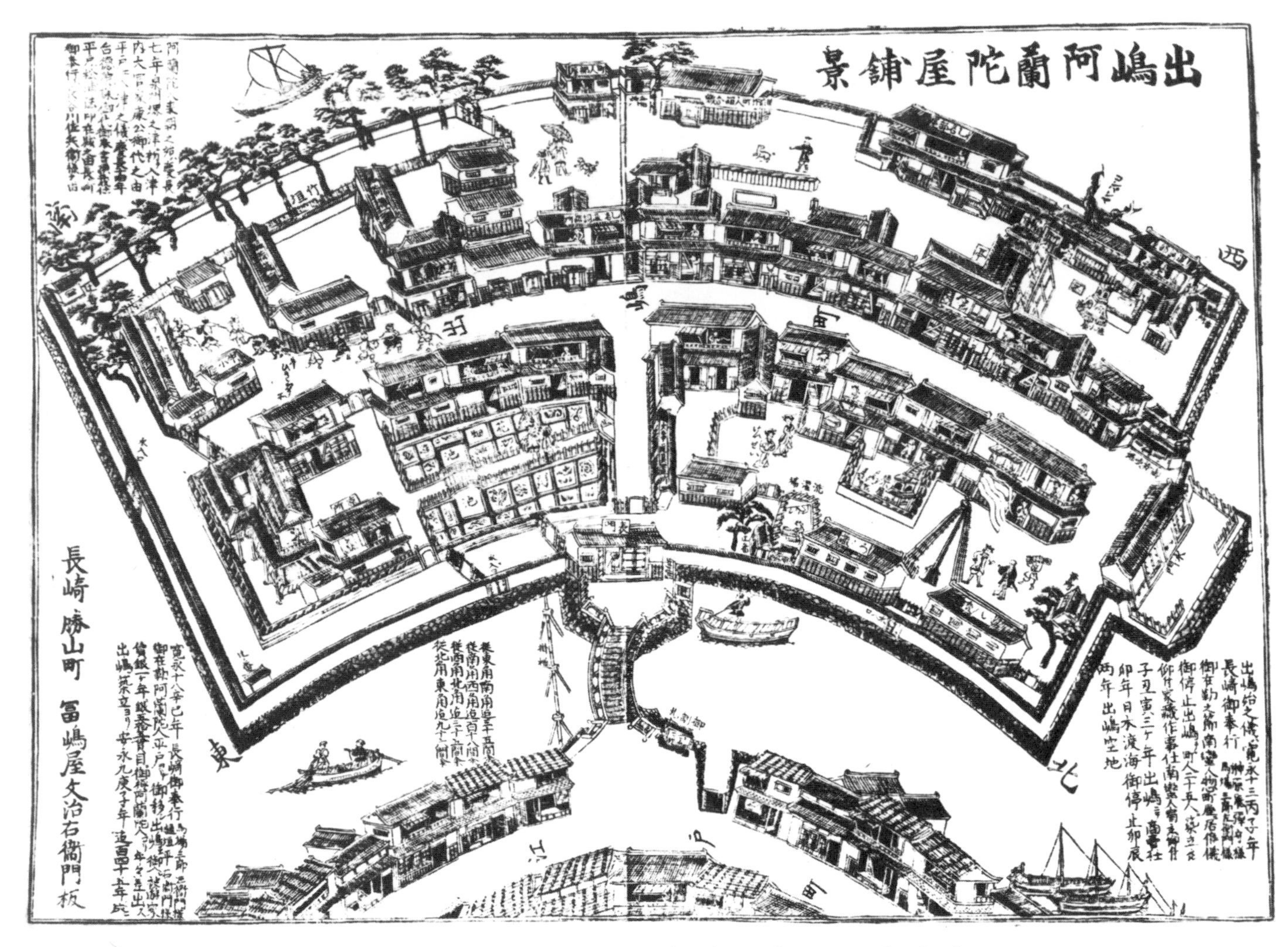

67. This Japanese woodblock print provides a detailed bird's eye view of Deshima, the man-made island in Nagasaki harbor constructed for occupation by Portuguese traders in 1634 and later by Dutch traders. The botanical garden established on the island can be identified just above and to the left of the main gate.

limitations of freedom of movement for their officers and employees while in Japan, and as a nation the Dutch promised not to enter into trade negotiations or treaties with the detested Portuguese.

Life on the tiny island proved to be existence under house arrest. The only exception to their restricted movement on Deshima was the yearly pilgrimage (or *hofreis*) by the Dutch contingent to Edo (now Tokyo), where they brought gifts and paid respect to the Shogun, stopping en route at Kyoto to pay homage to and bestow additional gifts on the powerless but nonetheless revered Mikado. The journey commenced when the Dutch party crossed the short, arched bridge that linked Deshima with the mainland and was joined by their Japanese escorts. On each sojourn the red hairs were outnumbered by Japanese guards, interpreters, and officials as many as three hundred to one, yet all traveling expenses for the entourage came from Dutch coffers! And after the two- to three-month round trip was completed, the Hollanders were again incarcerated on Deshima, where the usual term of service under East India Company regulations was two interminable years.

The Dutch East India Company settlement on Deshima was established in 1641, the same year that the Tradescants were gardening in South Lambeth, England. For the next two hundred years the only Europeans who ventured to Japan came aboard East India Company ships, and the only view of the Western World that the

68. A Dutch dinner party on Deshima as captured by a Japanese artist. The servants, shown here serving the meal, were Javanese, brought to Japan by the Dutch from their outpost at Batavia.

69. On their yearly pilgrimage to Edo, Dutch officials were frequently transported in palanquins like the one shown here, which gave meager opportunity for sight-seeing, much less collecting botanical and other natural-history specimens.

Japanese had was through the Hollanders on Deshima. Nevertheless, Western knowledge of the Japanese and of the natural world of Japan—its geography, natural resources, belching volcanoes, fauna, flora, and frequent earthquakes—slowly escaped without Japanese awareness from Deshima to Europe, as generation after generation of Dutchmen and other Europeans signed on with the Dutch East India Company and were assigned two-year stints at Nagasaki.

One of the first accounts to be published in Europe that gave an inkling of the diversity and richness of the Japanese flora came off the presses of a local printing establishment in the small German town of Lemgo in 1712. This curious volume, entitled *Amoenitates Exoticae,* included descriptions and illustrations of Japanese plants in the last of its five fascicles. Other sections of the book were devoted to a variety of topics, including descriptions of the Japanese practices of acupuncture and moxibustion (a cauterizing remedy that involved burning the dried young leaves of wormwood, *Artemisia moxa* De Candolle, directly on the skin of the patient), life in the Persian court, and geographic and natural phenomena witnessed in the Middle East. Its German author, Engelbert Kaempfer, based the book on his personal experiences as an astute and scholarly world traveler. His detailed accounts of Japanese life and precise botanical information concerning Japanese plants had been gleaned during his tenure as physician for the Dutch East India Company while stationed at Deshima.

Kaempfer (1651–1716) arrived in Japan in 1690 for a two-year tour of duty. Intellectually brilliant and with a warm and inquisitive personality, the Germany physician quickly befriended the Japanese guards and the numerous translators stationed at Deshima. He soon gained their deep respect and confidence through his superior medical expertise and willingness to share his knowledge and offer instruction. To his advantage, Kaempfer could tolerate alcoholic beverages well and had little difficulty in winning drinking bouts with the Japanese. Ever hospitable, Kaempfer found that as the Japanese responded to his cordiality, their tongues began to wag. Constantly alert for information, Kaempfer soaked up any tidbit of interest about the island nation, and all information was meticulously recorded in his journal and notebooks.

Plants from the mainland in the vicinity of Nagasaki were brought to Deshima at Kaempfer's request, and a botanical garden—large, considering the size of Deshima—was established inside the main gate. The plants, the physician claimed, were required for medicinal concoctions; yet one suspects that the unfamiliar plants were wanted to satisfy his insatiable curiosity about everything Japanese, not for their immediate value as simples. More intimate knowledge of Japanese plants as well as the geography and natural resources of Kyushu, the Inland Sea, and the large island of Honshu were the rewards of the otherwise long, tedious, and costly annual journey to Edo.

Returning to Germany when his term with the East India Company ended, Kaempfer began to write about his life of travels and, in particular, about Japan. Although his *Amoenitates Exoticae* appeared in 1712, Kaempfer's most significant contribution to Western knowledge of Japan and the Japanese was left unpublished when the sixty-five-year-old explorer-physician died in 1716. His manuscript, written in German, might not have been published at all had it not fallen into the hands of Sir Hans Sloane sometime after Kaempfer's death. Eager to make its contents widely known, Sloane arranged for his librarian to make an English translation, and Kaempfer's two-volume *History of Japan* was published in London under Sloane's auspices in 1727. The English edition—already once removed from the original German—was soon translated, and French editions, published in 1729 and 1732, were followed by a Dutch one in 1733.

The *History of Japan,* coupled with the accounts and descriptions, supplemented by illustrations, of plants included in the *Amoenitates Exoticae,* fueled a growing European interest and curiosity about Japan. Horticulturally, this interest became manifest as the nuts or seeds of one of the novel trees illustrated by Kaempfer were brought back to Europe on a Dutch East India Company ship that weighed anchor in Nagasaki harbor, destined for its home berth in Holland. By the middle of the 1730s the maidenhair tree, or ginkgo (*Ginkgo biloba* L.), was growing in the botanical garden at Utrecht, and shortly thereafter, as plantsmen learned of its existence, the ginkgo was planted and became established throughout Europe. Its spread was possible because the tree could be propagated by layering, whereby portions of the lowest branches of the tree were buried in soil and allowed to take root. Once the root system of the branch or layer had taken firm hold, the branch was severed from the parent tree, and the new individual could be lifted from the ground and planted elsewhere.

With its unique, fan-shaped leaves, which in outline suggest the shape of Deshima, the ginkgo stands as the epitome of Oriental trees, and it was probably the first Asian tree to become widely cultivated as an ornamental shade tree in Europe and America. From Holland it was

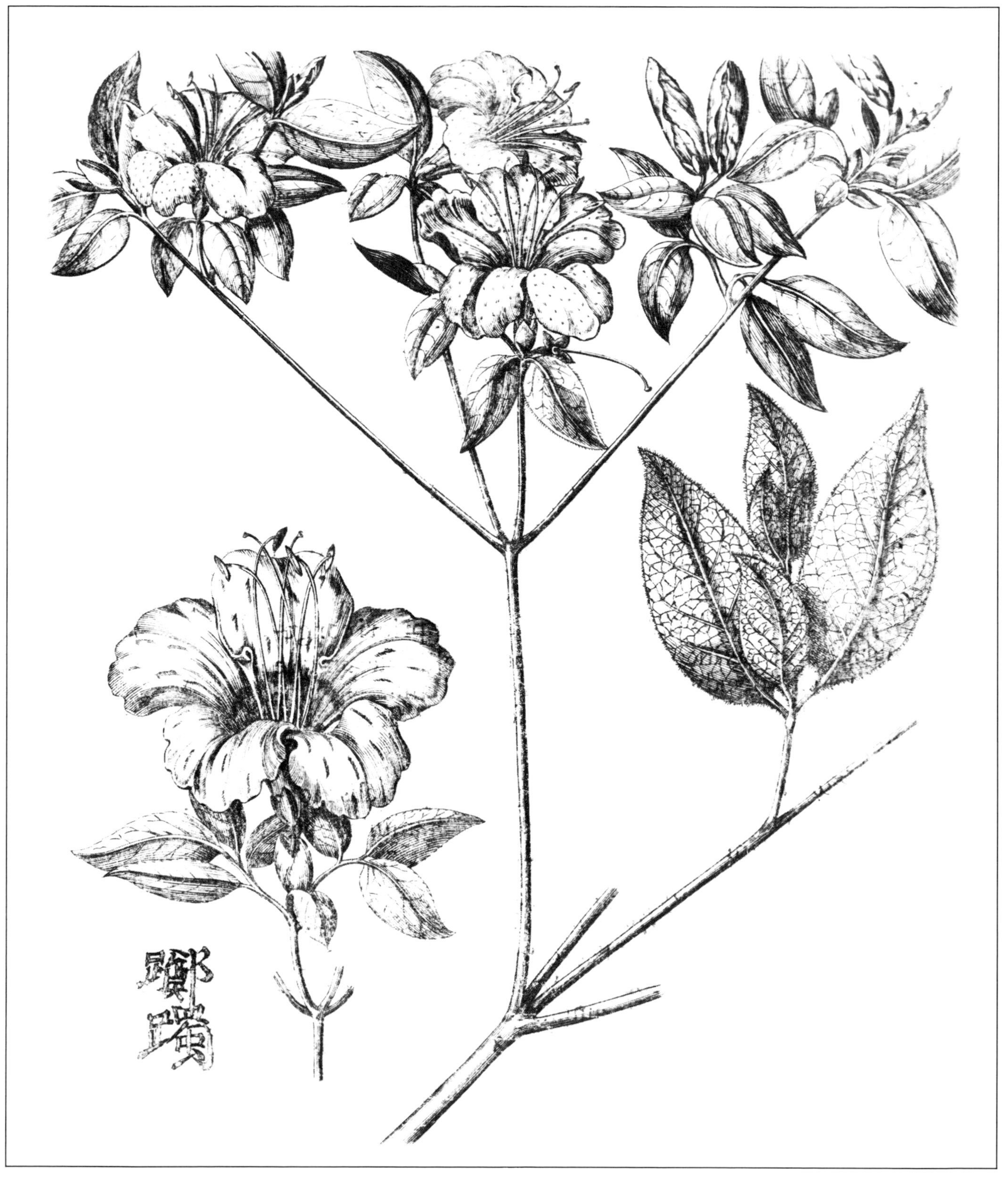

70. This illustration, reputedly representing *Rhododendron kaempferi*, appeared in Kaempfer's *Amoenitatum Exoticarum* and constituted the first illustration of an Asian azalea to be published in Europe.

71. These ginkgo nuts, known as silver apricots in China, have had the outer, fleshy seed coat removed. This outer layer contains high concentrations of a combination of fatty acids, including butyric acid, that give the seeds a rank, fetid odor. Because of their offensive smell, nurserymen suggest planting only male trees that have been propagated by asexual means. In some people the fleshy outer seed coat can also cause severe contact dermatitis.

introduced into England about 1754, and sometime after 1768 it was growing in Vienna. In 1771 Linnaeus received the plant in Sweden from an English correspondent; also from England it was taken to France in 1780.

When trade resumed between England and the United States after the War of Independence, the tree crossed the Atlantic in 1784, to be given space in the expanding collection of exotic trees on William Hamilton's estate in then-suburban Philadelphia. The year after Hamilton first imported the tree from England, André Michaux arrived in the United States and brought the ginkgo as a gesture of good will on behalf of the government of France. In Hyde Park, New York, David Hosack planted a young sapling on his estate above the Hudson River shortly after the beginning of the nineteenth century, and in Boston Gardiner Greene grew a prized specimen in his garden on Pemberton Hill.

Greene's ginkgo was already in the garden when he purchased the property in 1798, but it was not destined to remain rooted in that location. In 1835, after Greene's death, his cherished specimen—then estimated to be about thirty feet in height—was carefully dug and moved to the Common as a gift from Greene's children to the citizens of Boston. The transfer of the tree, which involved the construction of a special low-slung, four-wheeled cart built specifically for the purpose, occasioned much comment in the Boston press, particularly since some of the expenses involved were borne by frugal, taxpaying Bostonians. Years later, it was recalled by one eyewitness that Greene's ginkgo was known as "a very rare tree from Japan, a region almost as little known to us then as the moon" (H. W. S. Cleveland, 1888, p. 227). As botanical knowledge of the ginkgo increased, it was realized that it was indeed a unique tree, one that might as well have come from the moon as from Japan or anywhere else on earth.

By the middle of the nineteenth century, as the science of paleobotany was emerging, giving factual evidence of the gradual change that had taken place in the earth's flora over the course of geologic time, the ginkgo was recognized as a "living fossil," a term Charles Darwin coined for long-enduring forms that have survived to the present day (C. Darwin, 1859, p. 107). This characterization was correct and appropriate, as the ginkgo, or maidenhair tree, represents a living legacy of an ancient lineage known primarily from the fossil record. Its early origins are also evidenced by its motile, swimming sperm cells, which characterize primitive plant groups such as the ferns and cycads. This exceptional feature distinguishes the tree from the gymnosperms, with which it does share unprotected, "naked" seeds and the curious growth pattern of its branches, which are differentiated into two types: distinct long-shoots and curious, stubby, short-shoots.

Evidence from the fossil record indicates that the first ancestors of the ginkgo appeared in the Triassic period of the Mesozoic era; fossil remains of these forms are preserved in rock formations laid down some 225 million

72. This botanical illustration of the ginkgo clearly shows the differentiation between long- and short-shoots and its fan-shaped leaves. Also depicted are details of male and female flowers. Ginkgolides, complex compounds extracted from the leaves, have been used for medicinal purposes in China for generations to treat a variety of diseases.

years ago. By the Jurassic period, which commenced 195 million years ago, a great variety of ginkgo-like forms had become abundant and flourished in now widely separated areas across both the Northern and Southern Hemispheres. Fossil remains of leaves identical to those of the living ginkgo predominate in deposits from the Tertiary period, which began 65 million years ago, but from that time forward the abundance of ginkgo-like forms began to diminish. With the single exception of the maidenhair tree, all of the multitude of types eventually fell victim to extinction. Surprisingly, this sole surviving representative of the most ancient lineage of extant trees was among the very first to be introduced into Europe and North America from the mysterious Orient, where it had survived as a cultivated relict of the distant geological past.

Almost totally ignorant of the Japanese flora, the first European and American horticulturists to grow the ginkgo did not realize that the tree was cultivated in Japan and was technically not a native tree of the heavily forested island nation. When, and under what circumstances, it had been brought to Japan will remain forever unknown; yet it is certain that the tree was taken to Japanese shores from China and probably also from Korea. For more than two hundred years after the ginkgo became firmly established as an element of garden landscapes in Europe and America, it was also thought to be a cultigen in China and Korea, a species known only from solitary, cultivated, often gigantic individuals towering above temple grounds and gracing palace gardens. Nowhere was it known to occur in natural populations, taking its place with other trees on undisturbed, wooded slopes with its lofty crown occupying portions of the highest levels of the forest canopy.

Occasional claims of naturally occurring ginkgos have been reported, the most recent assertion published in the 1950s. All of these contentions have been discredited, but the hypothesis has been advanced that the region where the tree occurred and may still occur naturally comprises a mountainous area south of the Yangtze River (now the Chang Jiang) in eastern China, where Anhwei (Anhui), Kiangsu (Jiangsu), and Chekiang (Zhejiang) provinces abut. This argument seems logical and convincing, yet the debate is likely to continue, particularly since further field reconnaissance will probably leave the question unresolved. Regardless of whether or not the ginkgo still survives outside of cultivation, its close association with Chinese culture down through the annals of that ancient civilization has ensured that the tree was cultivated, given protection, and saved from extinction. Today, owing to the continued intervention of man, it once again grows and flourishes around the globe where climatic conditions allow, as its ancestors once did when dinosaurs roamed the earth.

Trees from the Celestial Empire

Growing from cracks in pavement, rooted in the narrow confines of strips of rubble between the asphalt of parking lots and the foundations of apartment houses, thriving on the cinder- and trash-strewn embankments paralleling urban rail systems, and invading forested acres far from city centers, the tree of heaven (*Ailanthus altissima* (Miller) Swingle) has claimed the North American landscape as its own. From the Atlantic to the Pacific coast and from Canada southward, this malodorous tree ranks as the most ubiquitous arboreal alien in the New World, extending southward in the Southern Hemisphere to Argentina. But nowhere is it more common than in the urban centers of the Atlantic seaboard megalopolis stretching from Boston to the District of Columbia. Synonymous with city, the tree of heaven heaves pavements, cracks foundations, grows from eave troughs and gutters, and up through chainlink fences and the fire escapes of tenements. Frequently, this lofty tree provides shade and offers the relief of summer green where no other tree could grow.

Of rapid growth, attaining heights of from fifty to sixty feet in fewer years, the tree of heaven has become a marker of derelict neighborhoods, vacant city lots, and neglected parklands. But the connotations surrounding the tree were not always these negative ones. In 1784, when William Hamilton willingly planted the very first saplings of the species to grow in American soil—plants that had been brought from England, possibly in the same shipment that included the ginkgo—they were viewed as great novelties from a far distant land. Philip Miller at the Chelsea Physic Garden shared Hamilton's emotions when, in 1751, he planted the first seeds of the tree to be had in all of Europe. Entrusted with these biwinged, easily wind-dispersed kernels, Miller hoped for results that would add to his justifiable reputation as a skilled plantsman. The seeds did not fail him; germination was successful, and the seedlings grew rapidly, increasing in height by between three and five feet in each succeeding year. As the plants attained greater stature, they also began to proliferate by sucker shoots, thereby providing additional plants that could be severed from the parent plants and shared with other plant

73. A fast-growing tree of tropical luxuriance, the tree of heaven was first illustrated in an article authored by the French botanist Réné Louiche Desfontaines in 1788; Desfontaines's illustration is shown here. Little did the first European and American plantsmen who grew the tree expect that two centuries later it would have escaped from gardens to become a commonplace tree outside of cultivation.

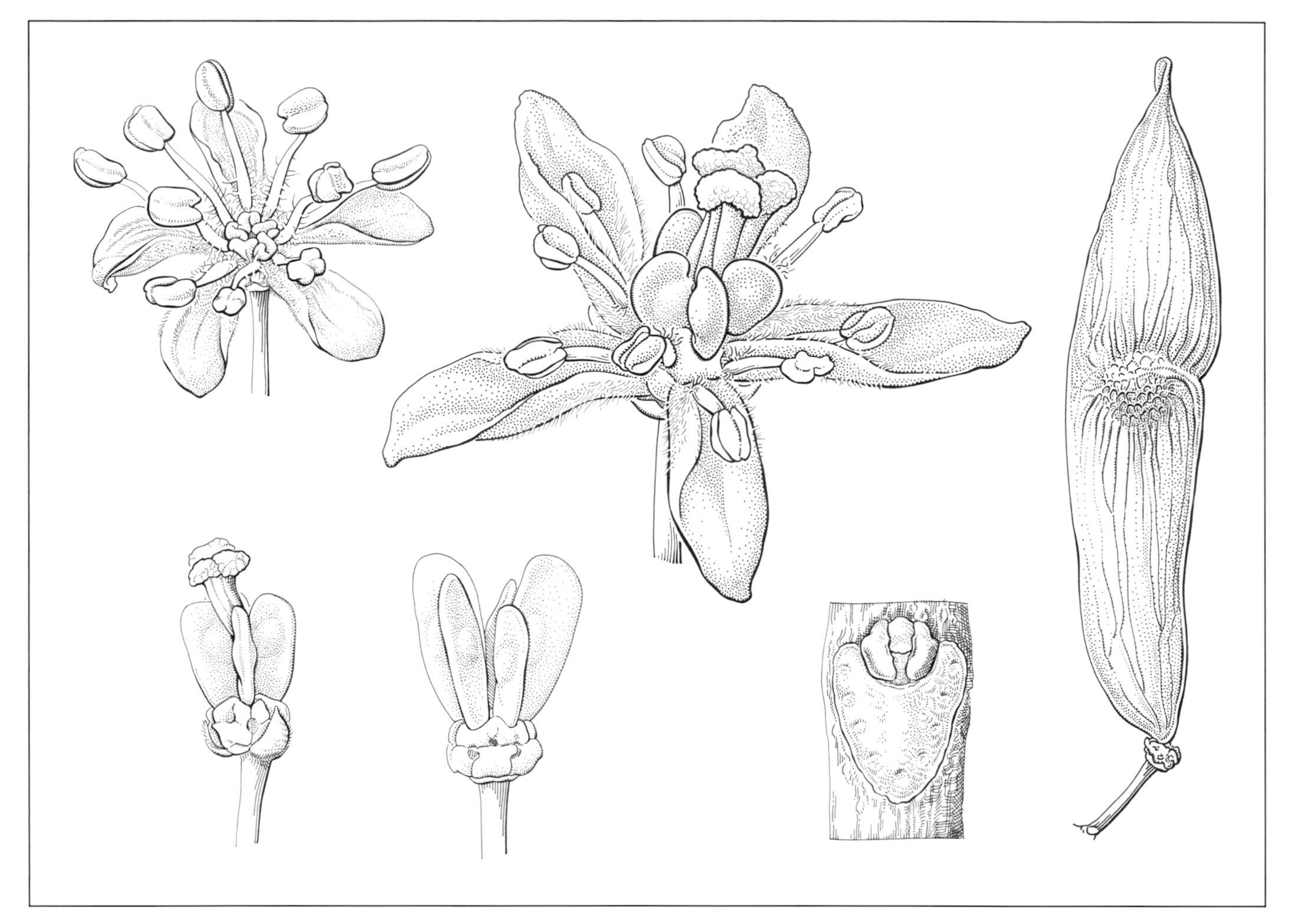

74. The flowers of the tree of heaven are of two types, staminate ones with functional stamens and an abortive gynoecium (left), and hermaphroditic ones (right) in which both the male and female structures are functional. After pollination, the winged seeds begin development and potentially five can be produced by each hermaphroditic flower. Two stages in the development of the seeds are shown here, one before, and one after the stigma and style have severed their attachment with the developing seeds. The seeds of the tree of heaven are frequently held on the trees until late into the winter and early spring months. Their germination rate is high, and urban and suburban gardeners in eastern North America are well acquainted with both the pale, silvery-tan seeds and the seedling trees. The unique leaf scars on the thick and stout, cane-like branches of the tree of heaven as well as its distinctive, winged seeds aid in its easy identification during the winter months.

enthusiasts like Peter Collinson and eventually, in America, William Hamilton.

It is incredible today to consider that the tree of heaven was once a pampered garden subject and revered as a botanical novelty on the estates of a few wealthy individuals. But the distance separating the European and American gardens, where the tree was infrequently cultivated in the eighteenth century, from the region of northern China where it is indigenous and commonplace was alone sufficient to inspire awe. The seeds that Philip Miller planted had been received in England by the Royal Society from a French Jesuit Father, Pierre Nicholas le Chéron d'Incarville, who was stationed in the Jesuit mission in Peking (now Beijing), the capital of the far distant and mysterious Celestial Empire. It would seem that the common name applied to the plant, tree of heaven, denotes both its rapid growth and habit as well as its native land, a country populated by celestials, a term frequently used in the eighteenth and nineteenth centuries to refer to the Chinese.

As a member of a religious order that had managed to maintain a toehold in China since the time of Marco Polo, d'Incarville was in a unique position to become familiar with the flora in the vicinity of Peking, where other Europeans were normally excluded. While the Chinese conducted limited trade with European merchants through the ports at Canton and Macau, the French missionaries had gained a tenuous foothold further inland. To the surprise of the Portuguese Jesuits, who were banished from Japan, the French Jesuits in China had miraculously succeeded in establishing Catholic missions in Macau, Nanking (now Nanjing), and the imperial city of Peking. The Chinese bureaucracy had allowed these intrusions into their otherwise closed soci-

75. An early view of Peking (now Beijing) with the western hills in the background. Pierre Nicholas le Chéron d'Incarville (1706–1757) explored this region for plants and sent seeds of several species to the Jardin du Roi in Paris.

76. An allegorical representation of the vast region of the Chinese empire and the plans the Jesuits had for the conversion of the Chinese to Christianity with the sanction and help of a few high-ranking Chinese mandarins. Matteo Ricci, one of the first Jesuits to gain an influential position in China, is figured on the right, while a mandarin helps to hold the map on the left. Ignatius Loyola, the founder of the Jesuit order, is probably portrayed at the upper left.

77. The observatory in Peking was founded using astronomical instruments that were gifts from Western nations to the Chinese emperor in the hopes of gaining favors and trade agreements in return. The well-trained Jesuits brought much practical scientific knowledge to the emperor's court, which responded by allowing the presence of the religious order in China.

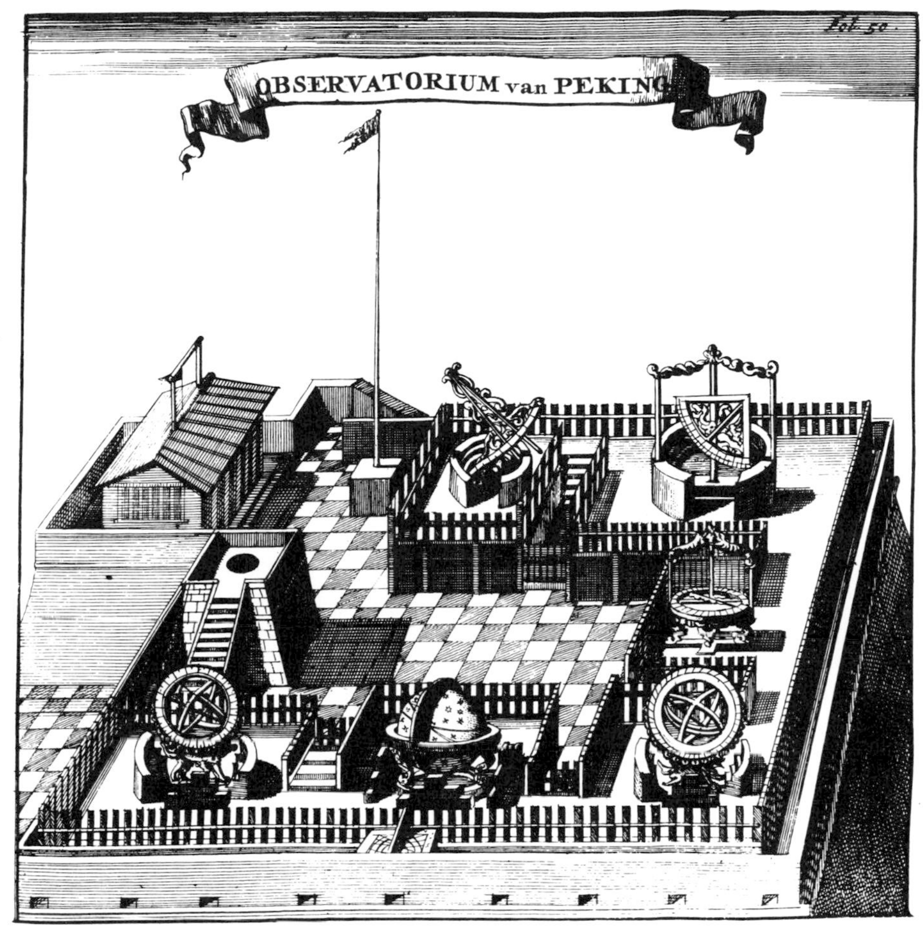

ety not because of concern for the salvation of Chinese souls but because the educated and scholarly Jesuits brought with them highly valued scientific knowledge and technological skills from the West.

European merchants attempting to inaugurate trade in China were viewed by the hierarchy of mandarins as inferior beings. And the Middle Kingdom, the self-proclaimed center of the world, considered all foreign nations vassal states. The restrictions that applied to foreigners in China were, if anything, more restrictive than those that applied to the Dutch in Japan. Despite an embassy sent to the emperor in Peking—reminiscent of the annual pilgrimage to Edo—the Dutch East India Company had had little success in establishing trade agreements with the Chinese. Furthermore, they risked jeopardizing their relationships with the Japanese, lest they be perceived as currying favor with the Portuguese at Macau.

More success came to agents of the English East India Company, which had been established at the turn of the seventeenth century by a group of enterprising London merchants. While limited trade between the English and Chinese commenced about 1600, for 240 years—until the Opium War of 1840—this trade relationship was under the strict control of the mandarins. For the English as well as other European nations, and after 1784 the Americans, it was a prolonged period punctuated by fits and starts and recurrent gains and losses as the seemingly mercurial and inscrutable Chinese opened and then closed ports and relaxed restrictions, only to reinforce them with renewed vigor.

As a consequence of China's strictly enforced isolationist policies, initial Western knowledge of China in general, and of the Chinese flora in particular, hinged on occasional reports of privileged travelers and infrequent consignments of plant specimens and packets of seeds sent to Europe by the proselytizing but scientifically minded French Jesuits. Among them, d'Incarville was the first with botanical training to serve in China.

In 1747, before he had the opportunity to send seeds

of the tree of heaven to the Royal Society, d'Incarville had entrusted seeds of another tree to a member of a Russian caravan that managed to visit Peking every three years. Carried across the vast expanse of Mongolia and Siberia on the dusty route back to Moscow, seeds collected from pagoda trees or scholar's trees (*Sophora japonica* L.) growing in Peking were eventually sent from Russia to the Jardin du Roi in Paris. Dispatched to Bernard de Jussieu, who had tutored d'Incarville and nurtured his interest in botany before the cleric had left his homeland for China, the well-traveled seeds germinated in French soil despite their prolonged overland journey. Young trees were carefully tended in the Paris garden, and excess seedlings were shared with other plantsmen. By 1753 the pagoda tree made its debut in English gardens, and fifty-seven years later, in 1811, plants of this Chinese tree—erroneously listed as a Japanese species—were growing in the greenhouse at the Elgin Botanical Garden in New York City.

It comes as no surprise that the pagoda tree was given the protection of a greenhouse when it was brought to New York from Europe. Essentially nothing was known of its cultural requirements. Commercial contacts, which American merchants tentatively inaugurated with the Chinese when the sailing vessel *Empress of China* left New York harbor destined for the Orient in 1784, had—like the efforts of the English—been restricted to trading and bartering at the port of Canton, a city that enjoys a subtropical climate. The climatic conditions prevailing in the environs of Peking further north, like most everything else concerning the Celestial Empire, were a closely guarded Chinese secret. In fact, the pagoda tree has a wide environmental tolerance and is cultivated as a street and shade tree in Canton as well as Peking. And it has proven to be a hardy plant preadapted to the climatic conditions of a wide area of eastern North America, from the southeastern states northward into southern New England. A grove of venerable old specimens shade the slope across Bussey Hill Road below the lilac collection in the Arnold Arboretum and annually reward visitors with their floral display in August and early September.

Like the native American yellowwoods that grow near by, the pagoda tree is classified as a member of the pea or legume family. And like those of the yellowwood, the petioles of the alternate, pinnately compound leaves of the pagoda tree are swollen at their base, encircling and enclosing the vegetative buds. Its abundantly produced creamy-white flowers have a close resemblance to those of yellowwoods and share a structure unique to one large subgroup of the legume family. Each flower consists of an outer and uppermost banner or standard petal, two lateral wing petals, and two somewhat united lower petals termed the keel, because their combined shape is similar to the shape of a boat's hull. Borne in large, terminal panicles, thousands of flowers cover the rounded contours of the pagoda tree. If pollination and fertilization are successful, each flower develops into a narrowly cylindrical green legume. These long-persistent pods are regularly constricted along their lengths between adjacent seeds and in areas where ovules have not developed. Splitting a pod lengthwise with a pocket knife will disclose the mature or maturing seeds and the small, unripened, abortive ovules attached alone one side of the fruit wall.

In all likelihood seeds of a third Chinese tree frequently planted in and around Peking were collected by d'Incarville and given to a trusted confidant of the Russian caravan to be carried on the long westward trek to Europe. The black, pea-sized seeds of the golden-rain tree (*Koelreuteria paniculata* Laxmann; Plate 9) are produced in papery, three-valved capsules that assume greenish to yellowish hues and shades of pinkish-tan before finally turning brown at maturity. In China these seeds were used as beads, and in former times the trees, known in Chinese as *luan,* were commonly planted at the graves of high governmental ministers. By contrast, pagoda trees frequently shaded the graves of lower officials and scholars, while easily propagated willow trees were grown at the burial sites of peasants.

Like pagoda trees, golden-rain trees flower during the summer months, normally from mid-June into July, although some individuals flower later in the season into September. The bright yellow flowers are produced in large, diffuse inflorescences, and close inspection reveals that the flowers are of two types. Initially, staminate flowers with pollen-bearing anthers borne on long filaments outnumber carpellate flowers, which produce abortive anthers on very short filaments. As greater numbers of flowers open, carpellate flowers—which will develop into papery capsules if pollination is successful—begin to outnumber the staminate ones. Another dramatic change in the flowers will reward careful scrutiny. As the four petals open, they are uniformly yellow, but by the time the anthers of staminate flowers have dehisced to expose the pollen held within, glandlike appendages at the base of the petals turn orange to brilliant red. It is likely that these bright, contrasting spots aid in attracting and guiding pollinating insects to the rich source of nectar hidden at the base of the petals.

78. These pen and ink drawings illustrate a portion of one of the large inflorescences of the pagoda or scholar's tree as well as two views of a flower and two of its legumes or fruits.

While in flower and after the bladderlike capsules have developed, golden-rain trees rarely fail to attract the interest and attention of visitors to the Arnold Arboretum. From the middle of June into September these middling-sized trees add color to the Arboretum landscape with their large inflorescences of yellow flowers, while the long-persistent capsules extend the trees' interest into the winter months. The large, dark green, compound leaves of the tree, each with its eleven to eighteen leaflets irregularly toothed along the margins, are an added trademark of the species. In China, the leaves are used in the preparation of a black or blue dye, but in Europe and North America golden-rain trees are grown solely for the aesthetic impact they create in the landscape.

In eastern North America, golden-rain trees grown from seed will produce flowers and fruits at a young age. In the climate of St. Petersburg, the former Russian capital, where the seeds d'Incarville sent from China germinated at the Academy of Sciences, no flowers or fruits were produced until 1771—some twenty-odd years after they had been received—despite the protection afforded by a greenhouse. When flowering finally occurred, the event did not pass unnoticed but allowed the Finnish naturalist Erik Laxmann, then resident in St. Petersburg, to classify the plant according to the Linnean sexual system. A year later Laxmann published a description of the tree, supplemented by a carefully drawn illustration, naming the new genus to note the pioneering accomplishments in plant hybridization of Joseph Gottlieb Koelreuter (1733–1806), professor of natural history at the German University of Karlsruhe.

Since d'Incarville had also forwarded seeds to Bernard de Jussieu in Paris, the golden-rain tree was established as an exotic in France at about the same time Laxmann published the first botanical description of the plant. By 1788 the tree numbered among the few of Chinese origin grown at Kew in England, and German plantsmen were introduced to the new plant by 1796. In 1809 Thomas Jefferson received seeds of this species from Madame de Tessé of France for cultivation at Monticello in the Virginia Blue Ridge. And the noted American naturalist and entomologist Thomas Say received plants at his home on the Wabash River in New Harmony, Indiana, from Robert Carr of Philadelphia in 1832. Say's plants had been transported to New Harmony by water from Philadelphia via New Orleans and subsequently up the Mississippi and Ohio rivers. The man-assisted peregrinations of golden-rain trees initiated by Père d'Incarville in Peking in the middle of the eighteenth century have continued to the present day because of the interest of plant-conscious men and women around the Northern Hemisphere.

A Traveler in Disguise

The debut of the ginkgo, the tree of heaven, the pagoda tree, and the golden-rain tree as cultivated trees in Europe and North America served to introduce horticulturally inclined citizens of the middle and late eighteenth century to living examples of four commonly cultivated trees of China and Japan. But the circumstances under which these arboreal aliens had been introduced were passive and, with the exception of the ginkgo, centered on the activities of a single individual. Decades would pass before the extent, diversity, and relationships of the flora of eastern Asia would become more fully known and understood. Philip Miller, Bernard de Jussieu, William Hamilton, and David Hosack, all pleased to experiment with a few new species from the Orient, could not begin to imagine that, beginning around the middle of the nineteenth century, succeeding generations of horticulturists would witness the introduction of a veritable multitude of eastern Asian species into the West. These exotic plants were destined to join the ranks of native European and American trees and shrubs in our gardens and parklands and line the streets of our cities and towns.

Few plants were introduced into Europe from China after the initial successes of d'Incarville, despite the efforts of the English to make inroads to improve their relationships with the Chinese. In Japan, the Dutch continued to man their outpost at Deshima, and it was through that relationship that another sample of the rich diversity of the Japanese flora came to the attention of botanists and cultivators in Europe and America.

Sixty-three years after Englebert Kaempfer returned to Germany from Japan, Carl Pieter Thunberg reached Nagasaki harbor on a Dutch East India Company ship in August of 1775, three years after embarking from Holland. Swedish by birth, Thunberg was a young, recently graduated physician; and, like that of his German predecessor on Deshima, his interest in plants was avid. Of necessity, Thunberg's medical education in Sweden had included botanical training, and he was fortunate to have numbered among the students at Uppsala who had studied under and been inspired by the great Linnaeus. Upon

79. Johannes Burman (1706–1779), an Amsterdam physician and botanist who was a friend and correspondent of Linnaeus, befriended Carl Peter Thunberg when the Swedish physician was continuing his studies in Holland. At Burman's urging, Thunberg accepted the position of physician in the Dutch East India Company and thereby had the opportunity to travel to Japan to make a collection of plants for exhibition in Holland.

completion of his formal academic course in Sweden, Thunberg was awarded a traveling scholarship that enabled him to further his studies in Holland and in France. Attending medical lectures and demonstrations in Paris was a distinct privilege and also allowed opportunities for visits at the flourishing Jardin du Roi. There, the young Swede made the acquaintance of leading French botanists, among them the eminent Bernard de Jussieu. Furthermore, through botanical acquaintances in Holland, Johannes and Nicolaas Burman, Thunberg was given the opportunity to continue and extend his travels. Backed by wealthy and influential merchants in Amsterdam and Haarlem, and with an appointment as surgeon extraordinary to the Dutch East India Company, Thunberg accepted an offer to travel to Japan with the express purpose of collecting botanical specimens for exhibit in Holland.

Dutch East India Company authorities welcomed the young medic for service at Deshima, particularly since the skills of a talented surgeon could be put to good use

80. While a wide diversity of Japanese maples can be grown in New England, the great majority are cultivars of *Acer palmatum*, and most have been selected for variations in the color or shape of their leaves. In others, growth habit and ultimate plant size have been the features selected.

by company officials. The ban on translations of European books had recently been lifted by the Shogun, and Western science was being studied by increasing numbers of Japanese. In particular, interest in European medicine—methods of dissection and surgery—and astronomy was increasing, yet the practical expertise of the Japanese in these fields fell behind their gains in theoretical knowledge. Aware of these developments, the Dutch East India Company wagered that greater trade concessions might be granted if Japanese physicians and scholars could be impressed by the skill, scientific knowledge, and teaching ability of the company's surgeon. Thunberg's intelligence and out-going, almost naive personality suggested he was the man for the job. Moreover, the young physician was eager to join the Dutch at Deshima. By doing so he would enter the ranks of Linnaeus's globetrotting disciples as an apostle, exploring the natural wonders of the world and applying Linnaeus's system of classification in unraveling the complexities in the miraculous scheme of nature.

One hazard had to be overcome, but Dutch East India officials were willing to accept the risk. Since only Dutchmen were permitted in Japan, Thunberg must be thoroughly enough versed in Dutch to delude the Japanese and not cause suspicion concerning his nationality. To perfect his Dutch, Thunberg interrupted his voyage to Japan at the Cape of Good Hope, where the Dutch had an established colony. This layover providentially provided a period for recuperating from a severe case of lead poisoning that the physically sturdy physician-naturalist suffered on the passage from Holland. On the long voyage southward to the Cape, the ship's cook had inadvertently substituted white lead for flour, and Thunberg, the captain, and the first mate fell seriously ill.

Thunberg spent three years among the colonists at the Cape, perfecting his knowledge of Dutch and filling his time practicing medicine and gathering specimens of the unique South African flora. His voyage to Japan was finally recommenced, and after an additional six-month layover in Batavia he at last reached Nagasaki harbor and Deshima on August 17, 1775.

Thunberg's sojourn in Japan was condensed into a fourteen-month period, not the usual two-year stint of Kaempfer's time. But during that period, successfully passing as a Dutchman, the surgeon-naturalist was extraordinarily active in instructing Japanese students in Western medicine and botany and in collecting specimens of Japanese plants. He began his botanical investigations by carefully inspecting the hay and fodder brought to Deshima for the European livestock the

81. Perhaps the most unusual shrub Thunberg encountered in the Japanese flora was the *Hana-ikada* of the Japanese or the helwingia of Western botanists (*Helwingia japonica*), which bears its insignificant flowers and shiny black fruits on the surfaces of its leaves. The species is dioecious, and an old staminate plant has been cultivated along Chinese path in the Arnold Arboretum for many, many years. Unfortunately, repeated attempts to obtain a carpellate or female plant have been to no avail, although an effort will continue to be made.

Dutch had shipped and established there for the milk and red meat not otherwise available in Japan. Although trade restrictions had been tightened since Kaempfer's day, the restraints on the Dutch encamped on Deshima had been relaxed. To his surprise, Thunberg was able to make botanical excursions to the mainland in the neighborhood of Nagasaki. Additional specimens were prepared and observations were made on the annual spring pilgrimage to Edo. Surrounded on the foray by Japanese translators, who had become his apt, attentive, and eager students, Thunberg was able to keep his plant presses filled. In return for their assistance, the Japanese students learned the rudiments of Linnaeus's revolutionary classification system and augmented their firsthand knowledge of the methods and doctrines of Western science.

In the months immediately following Thunberg's arrival in Japan, trees on the steep, forested mountainsides blazed with fall color and must have stunned the Swede. The spectacle was unlike anything he had witnessed in Europe or South Africa. In Japan, as in New England, the brilliance of fall color is largely a consequence of the abundance of maples that grow in the forests. Generations of Japanese horticulturists had cultivated and selected the extremes of variation displayed by these trees for garden and home ornament, and many plants became the objects for bonsai, the age-old Japanese tradition of dwarfing trees for culture in clay pots and ceramic containers.

A multitude of other Japanese trees and shrubs, many of great beauty or novelty—magnolias, roses, lilies, honeysuckles, an evergreen shrub classified as an andromeda (*Pieris japonica* (Thunberg) D. Don), the raisin tree (*Hovenia dulcis* Thunberg), which has thickened pedicels with a sweet, raisinlike taste on which the fruits are borne, and a bizarre shrub with the flowers and fruits produced on the surfaces of its leaves (*Helwingia japonica* (Thunberg) F. G. Dietrich)—as well as a host of other plants attracted the attention of the astute collector. One of the commonplace roses (*Rosa rugosa* Thunberg), to single out a solitary example, was destined to be commonly planted in North American gardens during the nineteenth century. It has also become widely naturalized on the sand dunes along the Atlantic coast of New England, and this robust, summer-flowering Japanese species—first described by Thunberg and occasionally known as salt-spray rose (Plate 10)—is an element of the American flora that, for many people, conjures up memories of summer months spent vacationing on Cape Cod, on Martha's Vineyard, or in Maine.

By the time Thunberg returned to Europe in 1776 he carried with him a rich harvest of dried botanical specimens representing a portion of the Japanese flora, particularly plants native to coastal regions. The Burmans and other European naturalists with botanical interests must have been overjoyed and overwhelmed with Thunberg's booty, particularly since their removal

82. Carl Peter Thunberg (1743–1828) was featured on this Swedish commemorative postage stamp in 1973, one of a set of four honoring Swedish explorers. Here, Thunberg appears with a tree peony and a Japanese dwelling, recalling his Japanese sojourn during 1775 and 1776.

from Japan involved smuggling them on board the Dutch ship in full view of the guards at Deshima. As in Kaempfer's day, an embargo still prevented the exportation of almost everything from Japan except the few items sanctioned for trade.

Horticulturists and horticulturally minded botanists surely felt disappointment and even frustration, however. While the Swede had managed to push open the door on Japanese natural history to a far greater extent than Kaempfer, Thunberg by and large had lacked the opportunity—and perhaps the desire—to bring back seeds or living plants for cultivation in Europe. Botanical data for descriptive and classification purposes could be obtained by simple recourse to his easily transported dried specimens. The science of botany—focusing on the classification of plants following the Linnaean method—could proceed outside of the garden, as Thunberg's *Flora Japonica,* published in 1784, attested.

Philipp Franz Balthasar von Siebold

In 1835, fifty-one years after Thunberg's *Flora Japonica* had incorporated more than seven hundred Japanese plants into the twenty-four classes of Linnaeus's sexual system, the first fascicle of another flora with an identical title was delivered to the libraries of learned academies and universities throughout Europe. A few copies crossed the Atlantic, and in 1836 Asa Gray, newly appointed librarian at the New York Lyceum of Natural History, probably obtained a copy for that institution's growing library.

Of large format, the new *Flora Japonica* roughly followed the outlines of the new natural system of classification that had been proposed by Antoine Laurent de Jussieu, Bernard de Jussieu's nephew, and refined by Augustin Pyramus de Candolle, the leading Swiss botanist of the era. Unlike Thunberg's earlier treatise, the new flora was sumptuously illustrated with elegant, hand-colored engravings that graphically depicted ornamental Japanese plants in page-sized plates. The engravings alone made the *Flora* worth purchasing by plantsmen outside the inner circle of the botanical literati and gave Western horticulturists, nurserymen, and gardeners their first view of a wide array of the highly ornamental plants of Japan.

The title page of the initial fascicle boldly indicated joint authorship of the *Flora* by "Dr. Ph. Fr. de Siebold" and "Dr. J. G. Zuccarini." While Professor Zuccarini had contributed many of the technical plant descriptions, responsibility for the *Flora* rested firmly with the first-named author, Philipp Franz Balthasar von Siebold. This multi-talented genius, then resident in a suburb of Leiden, was frequently seen walking in the streets of that Dutch city wearing a traditional Japanese kimono. On these strolls Siebold was usually accompanied by his Chinese amanuensis costumed in mandarin garb. To the Dutch burghers and the students of the university town, the pair constituted a minor spectacle, and German-born Siebold, although known to be famous, was considered eccentric. This prevalent opinion was strengthened, moreover, when it came to be known that the German removed his shoes before entering his house, ate using long sticks of wood as utensils, and frequently slept not in his bed but on tatami mats on the floor of his chamber.

In the garden surrounding his Dutch home Siebold grew an amazing variety of plants. Among the choicest specimens, which never failed to attract the attention of his numerous distinguished, often noble, visitors, were chrysanthemums, peonies, and roses, plants that were for Siebold tangible reminders of the six years between 1823 and 1830 he had lived in Japan. His house was also crowded with mementos and keepsakes of his prolonged residence in the island nation. It was widely rumored among his friends and acquaintances that it was with bittersweet regret that Siebold had returned to Europe from

Japan, the country he had hoped to make his adopted nation.

Born on February 16, 1796, in the Bavarian town of Würzburg, Siebold entered into an intellectual family that had produced several generations of distinguished physicians. Both his father and his grandfather were medical men, and one of his aunts, trained in obstetrics, attended the Dutchess of Kent at the birth of her daughter, who was destined to become Queen Victoria. Following the established family pattern, Siebold entered the Würzburg medical school at the age of nineteen. But during his schooling he migrated to a circle of naturalists, made up of both students and faculty, and found stimulation and expression for his youthful interests in plants, animals, geology, and mineralogy. Upon receiving his medical degree, Siebold established a small practice in his native Würzburg, but after two years dissatisfaction and wanderlust had gripped him completely. Determined to leave Germany, he set out for Holland, intent on launching a new career as a scientific explorer. Fortunately for Siebold, he was successful in obtaining a post as surgeon major in the Dutch East Indies Army, and the young physician, anxious to explore the world, set sail from Rotterdam for the Dutch outpost in Java in September of 1822.

Surprisingly, the governor general requested an interview with the German physician shortly after his arrival in Java, and Siebold left the meeting with instructions to proceed to Japan. Recurrent reports of Russian, English, and American ships entering Japanese waters had troubled the Dutch and suggested that one of these nations would soon demand the opening of Japan, breaking the 200-year monopoly on trade held by the Dutch. While acknowledging this inevitability, the Dutch hoped to solidify their relationships with the Japanese in order to maintain favored-nation status when the door to foreigners was finally opened. After all, their 200-year investment hung in the balance.

Based on their previous experience, the Dutch reasoned that strengthening their influence would require the skills of another scientifically minded physician—like either Kaempfer or Thunberg—and his presence at Deshima was required immediately. To his advantage, Siebold, who on the surface appeared ripe for the job, was already in Java and prepared and willing to undertake the mission.

Possibly no other plant explorer to enrich European gardens—and secondarily American gardens—with plants from foreign lands displayed such extraordinary intellectual brilliance as Philipp von Siebold. And yet his

83. A Japanese cartoon representing Philipp Franz Balthasar von Siebold (1796–1866), Bavarian physician-botanist and student of Japanese geography, society, and culture. Siebold gained the respect of the myopic Japanese by skillfully performing the first cataract operations in Japan; thus, the eye in this drawing brought attention to this fact.

Japanese sojourn nearly ended before it began. Siebold's Dutch was spoken with a heavy and thick German accent that was quickly noted by both the Japanese translators and guards at Deshima. Immediate action followed to prevent Siebold from joining the Dutch community in Japan, and it was only after the chief Dutch official succeeded in convincing the Japanese that Siebold came from a remote, mountainous district of Holland—thereby explaining his accent—that the German was permitted to join the Dutch on Deshima.

Eager for success in his duties, Siebold was also personally ambitious and captured every opportunity to learn, absorb, and discover as much about Japan and the Japanese as possible so that he might become the leading European authority on every aspect of the culture. Through medicine and his skill as a surgeon, Siebold achieved fame and notoriety throughout Japan. He performed the first cataract operations witnessed by the Japanese, and his ability to restore sight to the near-blind increased his reputation and the reverence with which the Japanese came to regard him. Because he refused payments for his medical services and advice, the Japanese responded in their time-honored tradition of sending gifts in appreciation, and to his delight Siebold's collections of Japanese objects expanded. Anything and everything, including what might have seemed to be trivia to anyone less curious, was examined and recorded with interest and intellectual acumen.

In his frenzied daily routine, which included teaching medical students and tending the sick, many of whom came great distances for consultations, plant and animal specimens were gathered, purchased, and also brought for classification from distant localities by intrigued Japanese students. Japanese artists were trained to prepare illustrations from nature, and in response to Siebold's request to the governor general in Java, Karl Hubert de Villeneuve, a highly skilled draftsman and artist, arrived at Deshima in 1825 to aid in documenting plants and animals with drawings. Siebold's Japanese students were an obvious source of valuable information, and as a part of their training they were required to prepare theses on a wide variety of topics relating to their homeland. Clearly, Siebold viewed his stay as an all-encompassing fact-finding mission, and it became increasingly a labor of love. In the process, Siebold became infatuated with the natural beauty of Japan, its ancient civilization, and the customs and traditions of its people. He also fell deeply in love with an eighteen-year-old girl whom he was allowed to marry. Siebold's wife went to live with him on Deshima, and in 1827 a daughter, Ine, was born.

The botanical garden inside the gates at Deshima—neglected since Thunberg's time—was put back into use, and when Siebold was allowed to move his base of operations to the mainland and occupy a house on a hillside above Nagasaki, the surrounding grounds were transformed into a botanical garden and arboretum.

The pilgrimage to Edo, while still required, had been limited to the three ranking Dutch officials at Deshima, and instead of being an annual event, by Siebold's time it was undertaken only every four years. Consequently, Siebold looked forward with great expectations to participating on the trip scheduled for 1826. At first enormously successful, the trip was ultimately a disaster for Siebold.

In his eagerness to expand his knowledge of Japan and gather important documents pertinent to scholarly understanding of the country, the physician-naturalist overreached himself. Siebold accepted as a gift a copy of the highly guarded imperial map of Japan from the court astronomer at Edo. Of all contraband, the map was the most highly guarded because of its potential military value, and when it was discovered that Siebold was in possession of a copy, he was immediately ordered to return to Deshima, where he was imprisoned for over a year. In October of 1829 he was pardoned, but a worse fate was handed him. He was banished from Japan and forced to return to Europe. On January 2, 1830, after six years and three months spent in the country where he had hoped to spend the remainder of his life, Siebold left his wife and daughter and set sail for Holland.

Thirty years passed before Siebold once again returned to his beloved Japan. And by August of 1859, when he sailed into Nagasaki harbor, extraordinary changes had occurred. Six years earlier, a squadron of smoke-belching American ships under the command of Commodore Matthew Perry had entered Edo Bay. Japan was opened to foreign trade, as the Dutch had predicted thirty years before, and Japan was forever changed.

Siebold once again was forced to leave Japan, this time as a result of intrigue surrounding the establishment of relations with the foreign powers that had come to leave their calling cards in the wake of Perry's success. The Dutch, anxious and concerned that Siebold would create more problems than he claimed he could help solve, tricked the veteran physician into leaving Japan in 1862 on the pretext that he would ultimately return as the Dutch advisor in Japan.

Siebold returned to Europe for the last time in 1862. As in 1830, the deck of the vessel on which he sailed was a veritable nursery. Returning to Leiden, where in the

84. One of Siebold's introductions from Japan, *Hydrangea anomala* subsp. *petiolaris* is a vigorous, high-climbing, deciduous shrub that produces an abundance of large corymbose inflorescences during the summer months. Like other species of hydrangea, the inflorescences are comprised of small functional flowers and larger, bracteate "neuter" flowers that add significantly to the ornamental aspects of the plants. During the winter months the naked stems of the climbing hydrangea can be viewed against the substrate on which it attaches itself by small aerial rootlets. The bark on the larger stems exfoliates in small, longitudinal strips.

85. The so-called panicled hydrangea (*Hydrangea paniculata* 'Grandiflora'), frequently referred to as the "p-g hydrangea" in the nursery trade, is one of the most common late summer flowering shrubs in urban areas of New England. Its large inflorescences are comprised of a great many bracteate, sterile or neuter flowers, and the weight of an individual inflorescence often causes it to hang from the contours of the shrub.

1830s he had established a nursery specializing in Japanese plants, he added a considerable number of new introductions to the stock already in cultivation. His catalogue for 1863 listed an astounding 838 species and varieties available to the gardening public. Essentially all of the plants illustrated in Siebold and Zuccarini's two-volume *Flora Japonica* were included in the catalogue, and Siebold had been directly involved with making them available to cultivators in Europe.

Siebold forsythia (*Forsythia suspensa* (Thunberg) Vahl var. *sieboldii* Zabel; Plate 11), with long, slender, pendulous branches and abundantly produced brilliant yellow flowers, which bloom in early spring, was the first of the genus to be introduced into European gardens. From Europe Siebold forsythia was introduced into North America and, along with other members of its genus subsequently introduced, has become one of the most familiar harbingers of spring. Almost everywhere forsythias brighten dooryards during the first warm days; and when the shrubs are not pruned but allowed to assume their natural form, they become almost fountainlike cascading mounds of yellow flowers.

The Japanese wisteria (*Wisteria floribunda* (Willdenow) De Candolle), another of Siebold's most popular plant introductions, is a high-climbing, twining vine with compound leaves that produces masses of lavender or white pea-like flowers in pendulous, long racemes in April and May. This wisteria is known to practically everyone; yet another climbing plant, the so-called climbing hydrangea (*Hydrangea anomala* D. Don subsp. *petiolaris* (Siebold & Zuccarini) McClintock), is far less well known. Small adventitious, aerial rootlets fasten the stems of the plant on the tree trunks, brick walls, or other substrates over which it grows.

A far more familiar hydrangea (*Hydrangea paniculata* Siebold 'Grandiflora'), sometimes known as panicled hydrangea, also numbers among Siebold's introductions. Widely available in the nursery trade, this is the common hydrangea that begins to flower toward the end of summer and into the fall. Its large, terminal, pyramidal clusters of flowers are nearly all sterile and are initially white. As they age, they become purplish-pink and eventually turn brown. The weight of each inflorescence is often sufficient to cause them to hang somewhat suspended on the compact shrubs, and they normally persist on the plants into the winter months and sometimes into the following growing season. In the pyramidal shape of its inflorescences, this Japanese species is thought to be closely related to the oak-leaved

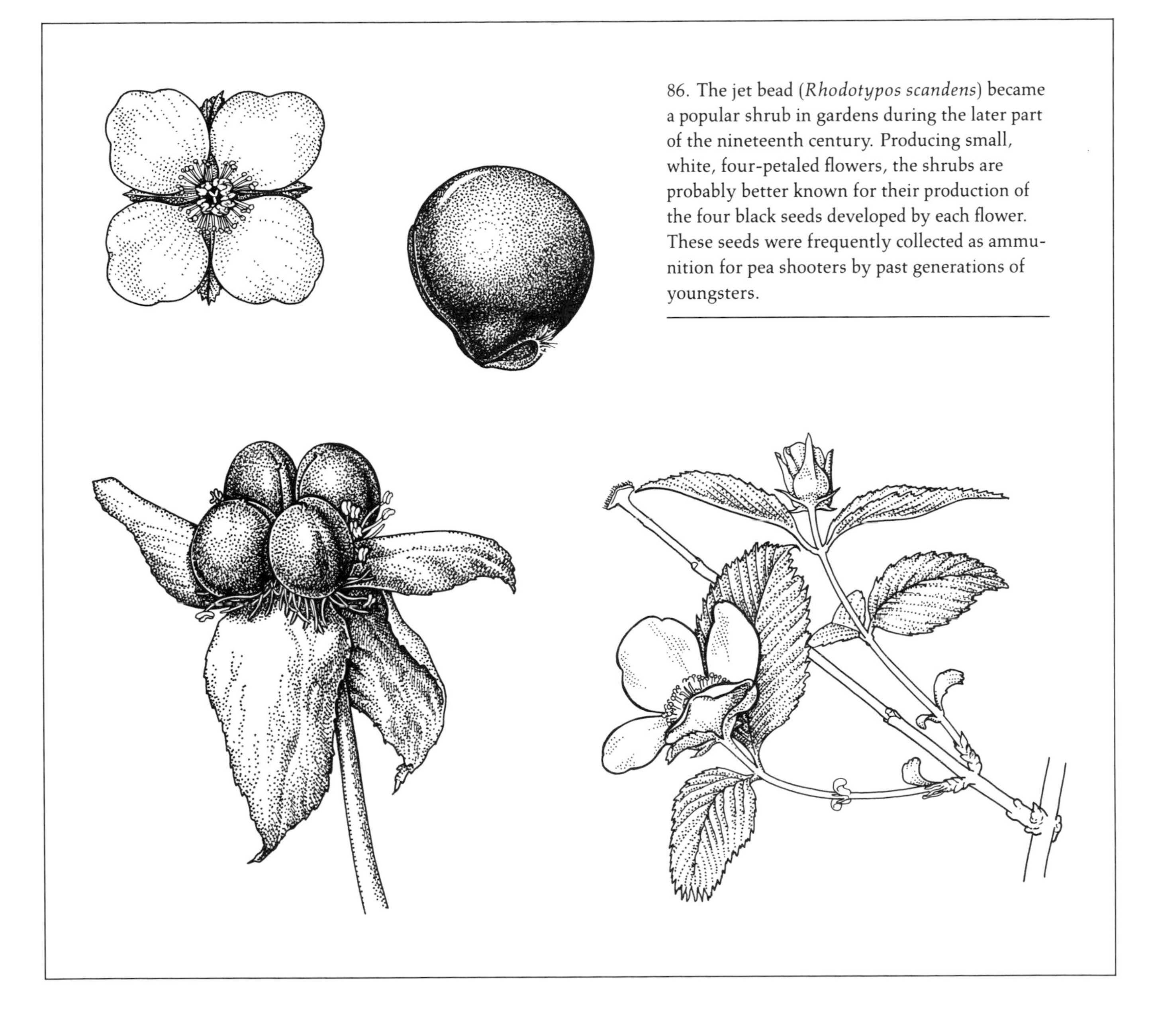

86. The jet bead (*Rhodotypos scandens*) became a popular shrub in gardens during the later part of the nineteenth century. Producing small, white, four-petaled flowers, the shrubs are probably better known for their production of the four black seeds developed by each flower. These seeds were frequently collected as ammunition for pea shooters by past generations of youngsters.

hydrangea discovered by William Bartram in the southeastern United States.

Another shrub favored during the Victorian era and frequently planted in gardens of the last century has generally lost its place in today's gardens. It nonetheless began its life as a European garden plant in Siebold's Jardin d'Acclimatation in Leiden. Originally described by Thunberg, jet bead (*Rhodotypos scandens* (Thunberg) Makino) is still cultivated in the comprehensive Bradley collection of rosaceous plants in the Arnold Arboretum. While Thunberg based his original description of this species on specimens he collected in Japan, and Siebold introduced the plant from that country, it may not be truly native to Japan. It does occur in the indigenous flora of China, and like many other plants, the ginkgo included, jet bead may have been introduced to Japan from China.

Because of their beauty, two other plants merit mention in conjunction with the horticultural and botanical exploits of Siebold in Japan. Perhaps the most beautiful of all ornamental crabapples when in flower, the Japanese flowering crab (*Malus floribunda* Siebold ex Van Houtte) has become the standard against which the ornamental virtues of other crabapples are compared. A small tree to

87. While the Korean mountain ash (*Sorbus alnifolia*) presents a pleasing sight in spring, when it is virtually covered by corymbs of white flowers, it is in the late summer and fall, when the beautiful multicolored fruits are produced against the background of golden foliage, that the tree attains its greatest ornamental appeal.

twenty-five feet in height with a densely branched, broadly rounded crown, this crabapple produces flower buds that are red to rich rose in color. As the flowers open in late April and May, the petals gradually fade to white. Literally obscuring the branches of the tree in their abundance, the flowers transform each branch into a garland.

Like the Japanese flowering crabapple, the second tree that requires mention is also classified as a member of the Rosaceae or rose family, and a splendid example of this species grows adjacent to the Arborway wall where the Bradley collection merges with the maple plantation in the Arnold Arboretum. Siebold placed this species in the genus *Crataegus*, a group that includes the Washington thorn (*Crataegus phaenopyrum* L.), which had been introduced into European horticulture from North America in Philip Miller's era. Like the Washington thorn, Siebold's introduction is extremely handsome in both flower and fruit. Now placed in the genus *Sorbus*, the tree is commonly known as the Korean mountain ash (*Sorbus alnifolia* (Siebold & Zuccarini) K. Koch) because of its natural occurrence on the Korean peninsula; however, it is also indigenous to Japan as well as to central China.

A tree of forty or fifty feet in height generally with a rounded to oval crown, the Korean mountain ash branches from near the base of its thick, sinuous trunk, and its pewter-hued bark suggests that of the European beech (*Fagus sylvatica* L.). Its leaves, too, resemble those of the beech—or those of an alder, hence its specific epithet—inasmuch as they are simple and shallowly toothed at the margin. In these respects this species differs from the great majority of mountain ashes, which are well known for their pinnately compound leaves, which resemble those of the true ashes (species of *Fraxinus*). Like other species of the genus, however, the Korean mountain ash produces white flowers in broad, rounded nosegays in spring, followed by clusters of small, applelike fruits in the fall.

The aesthetic aspects of the Korean mountain ash are best appreciated in the fall, when the bright red fruits cover the tree and contrast with the varying hues of golden yellow and apricot of its leaves. The fruits remain long after the leaves have fallen; and, seen from middle distance, the leafless trees present an ethereal reddish haze in the landscape. Had Philipp Franz Balthasar von Siebold failed to bring or send back any tree but the Korean mountain ash from the Land of the Rising Sun, our debt to him would still be great.

88. A portrait of Philipp von Siebold in his later years, his chest resplendent with the many medals awarded him by European governments and learned societies for his numerous accomplishments.

Adventure and Distress in the Service of the Society

Little news from Europe reached Philipp von Siebold during his first prolonged residence on Deshima in Nagasaki harbor. Totally imbued with Japan, the physician-naturalist would probably have paid little attention had he learned that across the vast expanse of the north Pacific a Scot had been sent on what would prove to be a difficult mission to ferret out new plants, prepare botanical specimens, and carefully pack collections of seed for shipment to London. The vegetation of coastal regions of the Pacific Northwest where these activities were to be focused had been poorly investigated before David Douglas arrived at the mouth of the Columbia River aboard the Hudson's Bay Company's ship *William & Ann*, on April 7, 1825. Political unrest in China, where Douglas had anticipated and hoped to be sent, and the willingness of the Hudson's Bay Company to cooperate in the Pacific Northwest combined to fix the plant collector's destination.

Prior to Douglas's mission, only a few dried plant specimens from the region known as New Albion—now California—and the Oregon Country had been received in Europe. The majority had been collected as opportunity allowed by Archibald Menzies, surgeon-naturalist on Captain George Vancouver's expedition to the region

89. Archibald Menzies (1754–1842), first collector to bring specimens from the Pacific Northwest directly to the botanists of England. Menzies was able to sample the flora of the Pacific Coast of North American between 1791 and 1795 while serving as surgeon-naturalist on the voyage of exploration under the command of Captain Vancouver. Many of Menzies's discoveries were subsequently introduced into cultivation by a fellow Scot, David Douglas.

between 1790 and 1795. A celebrated navigator and explorer, Vancouver had sailed with Captain James Cook on Cook's second and third voyages of discovery. With a squadron under his own command, he was ordered by the British Admiralty to sail to the Pacific Northwest coast of America to reestablish claim over territory the Spanish had seized but subsequently relinquished.

Ironically, in May of 1792 Robert Gray, a Boston sea captain in the employ of two influential Bostonians, Joseph Barrell and Charles Bulfinch, successfully sailed the merchant ship *Columbia* through the hazard-ridden waters at the mouth of the river, which now bears his ship's name. Gray, in addition to being the first navigator to enter the river's estuary successfully, laid claim to the entire region for the United States. Years later, during the 1844 presidential campaign, the dispute between the United States and Britain over ownership of the region (the United States claiming rights dating from Gray) and the position of the Canadian boundary would erupt, giving rise to the well-remembered slogan, "Fifty-four forty or fight."

Lewis and Clark also visited the region between 1805 and 1806 and returned to Philadelphia with scientific data and botanical specimens. Frederick Pursh made the botanical results of Lewis and Clark's collections known through his *Flora Americae Septentrionalis*—the first flora to account for plants collected west of the Mississippi River—and Pursh took specimens collected by Lewis and Clark to Sir Joseph Banks and Aylmer Bourke Lambert in London when he left the United States in 1811. Oddly enough, events in England involving both Banks and Lambert during the period Lewis and Clark were exploring the upper reaches of the Missouri River and Columbia River valley would result in David Douglas's dispatch to the same wilderness region some twenty years later. And two years before Douglas first saw the mighty, frequently salmon-choked Columbia, Philipp Franz von Siebold had sailed from Java to Nagasaki harbor and become enraptured with the rugged but picturesque Japanese landscape and the culture and refined manners of the inquisitive Japanese.

On Wednesday, March 7, 1804, some two months before Lewis and Clark left St. Louis on their two-year expedition, the inaugural meeting of the Horticultural Society of London took place in a Piccadilly bookshop. As early as 1801 John Wedgwood, son of the famed and wealthy potter Josiah Wedgwood and the uncle to be of Charles Darwin, had written to William Forsyth, King George III's gardener based in London, expressing his ambition to establish a horticultural society. As he took pains to explain, agricultural societies aimed at bettering agricultural practices and animal husbandry had been established throughout England; yet similar societies concerned with pomology, the improvement of garden vegetables, and ornamental horticulture were completely lacking. Wedgwood, then in Staffordshire, requested in a postscript that Forsyth seek Sir Joseph Banks's opinion of the plan, hoping Sir Joseph might give his blessings to the scheme.

Sir Joseph not only concurred with Wedgwood's sentiments but heartily agreed to support and back the idea and cooperate in the formation of a horticultural society. Other leading horticulturists were encouraged to join the original group, and soon after the initial meeting, the Horticultural Society of London began to grow and prosper. Almost immediately a small staff was employed, a charter setting out the aims of the Society was agreed upon, the nucleus of a library collection was assembled, a serial publication—the Society's *Transactions*—was launched, and a series of medals was struck to be given in recognition of service to the Society and the advancement of horticulture. Appropriately, Sir Joseph Banks was the first recipient of the Society's gold medal—awarded on June 5, 1811—for his services at Kew and to the Society.

As the Society prospered, the need for permanent rooms for meetings and exhibits—and especially the need for land suitable for a garden where trials could be made—was soon felt and discussed by the fellows of the Society. In 1821, the year after Sir Joseph Banks's death, thirty-three acres in Chiswick were leased from the Duke of Devonshire, and the Society entered a new stage

90. A view of Port Dick near Cook Inlet along the Alaskan coastline, a region first explored by Captain Cook in 1785 but revisited by Captain George Vancouver a few years later. It was on Vancouver's expedition that Archibald Menzies made the first collections of plants from California and the Pacific Northwest that were brought directly to English botanists.

of development, one that included a two-year course for practical and theoretical training of student gardeners. The Society had emerged as a clearing house for horticultural and botanical information for amateurs and professionals alike, as well as an educational center for young men intent on horticulture and gardening as a career. Through a system of examinations, standards could be set, and a cadre of well-trained, proficient young gardeners would receive diplomas and the virtual guarantee of employment on the estates and in the gardens of fellows of the Society.

Additionally, through a system of foreign or corresponding members, new plants were continuously received by the Society from distant lands. These were grown in the Society's garden and propagated for distribution to interested fellows. David Hosack in New York sent the American seckle pear, and consignments of chrysanthemums, camellias, tree peonies, azaleas, and other plants from China were received from John Reeves, an inspector of tea for the East India Company who was based in Canton. Reeves was a correspondent of Banks, who astutely urged that the Society offer Reeves corresponding membership. In that capacity Reeves sent additional plants, but most were sorrowfully lost during the long voyage to England. Reeves also sent to London a portfolio of paintings of Chinese plants prepared by native artists who worked under his direction. These illustrations, which depicted upwards of forty varieties of chrysanthemums and numerous camellias (Plate 12), aroused much comment and stimulated further interest among members of the Society in obtaining living material of the plants pictured.

The success rate of the Society in receiving plants from the Orient had been abysmally poor, and another corresponding member, Dr. John Livingstone, chief surgeon of the East India Company in China, calculated that the cost of each plant that arrived in England in living condition amounted to a staggering £300! Livingstone proceeded to suggest that a gardener should be sent to China to collect plants, properly prepare them for shipment, and accompany and care for his living cargo on the return voyage to Europe. The total cost estimated for such a venture, which would probably result in greater success than ever before achieved, would be in the vicinity of £200.

Accordingly, Joseph Sabine, industrious and ambitious secretary of the Society, raised the necessary funds for two promising collectors to book passage to China in the spring of 1823. John Potts, head gardener in the Society's Chiswick garden, duly set sail for China, but for

91. Blue and white jasper portrait medallion of Sir Joseph Banks, modelled in about 1776 by John Flaxman for the English firm of Wedgwood and Bentley. During his lifetime, Banks was without doubt the greatest champion of the study of science in all its branches and of natural history, botany, and horticulture in particular. The development of the gardens at Kew into collections of great botanical and horticultural significance and his active participation in the founding of the Horticultural Society of London (now the Royal Horticultural Society) were but two achievements of his illustrious career. The famous potter, Josiah Wedgwood, whose firm produced this medallion, numbered among Sir Joseph's personal associates.

David Douglas fate decreed otherwise. Political unrest in China at the time suggested that not all of the Society's efforts should be concentrated in that potentially volatile region. As an alternative, it was decided that Douglas, who had been recommended for the job by William Jackson Hooker, Regius Professor of Botany at the University of Glasgow, should undertake a sojourn in politically stable eastern North America.

With instructions to procure new varieties of fruit trees and vegetables as might be encountered there, Douglas set sail for America in June of 1823. In New York, Philadelphia, and other cities of the eastern seaboard the shy, painfully retiring Scotsman was welcomed with American hospitality and accorded every courtesy by fellow horticulturists. David Hosack and Thomas Hogg, who had recently emigrated from England to establish a florist's shop in the growing metropolis of New York, hosted Douglas in that city, while Henry Briscoe, a friend of Sabine's brother, welcomed the Scot in far distant Amherstburg in Canada.

In Philadelphia, shortly before his return to England, Douglas made the acquaintance of Thomas Nuttall. Recently appointed Professor of Botany at the Harvard Botanic Garden in Cambridge, Massachusetts, Nuttall was a well-known botanist, ornithologist, and explorer. Born in Yorkshire, he had come to North America in 1808 to study the natural history of the United States and had undertaken botanical exploration west of the Mississippi River shortly after the Lewis and Clark expedition returned home. Nuttall befriended the young Scot and personally escorted Douglas on a tour of Bartram's Garden.

92. "Exhibition Extraordinary in the Horticultural Room," a satirical cartoon engraved by George Cruikshank that poked fun at the Horticultural Society of London.

93. William Jackson Hooker (1785–1865), early mentor of David Douglas, who made many of Douglas's plant discoveries from the Pacific Northwest known to the scientific world in his *Flora Boreali-Americana*. Hooker became director of Kew Gardens and revitalized the institution's living collections as well as establishing its world-famous herbarium. On his death, Hooker was succeeded as director by his son, Joseph Dalton Hooker, a veteran plant explorer of the Himalayan region.

Before returning to England early in January of 1824, Douglas had been enormously successful in bringing together, largely through the generosity of American nurserymen and orchardists, a wide selection of new apple, pear, plum, peach, and grape varieties that had been developed in America. Plants of the Oregon grape (*Berberis aquifolium*) grown from seeds gathered by Lewis and Clark also crossed the Atlantic with Douglas on his return voyage to England, to be cultivated at Chiswick and eventually in gardens throughout the British Isles and all of Europe where the climate proved suitable.

The plants Douglas brought from America so pleased the members of the Horticultural Society that it was quickly determined that the Scot should be sent out again in the service of the Society. His second venture would also be to North America, but it would not be a "fruit trip." Instead, on this trip, which commenced on July 26, 1824, only six months after his return from New York, Douglas would sail via Cape Horn on the long voyage to the Pacific coast of North America, a region where the Hudson's Bay Company had recently established its western-most outposts and the area where the Oregon grape had been collected by Lewis and Clark. Plants described from that region in Pursh's *Flora*, coupled with Menzies's specimens, indicated novelties were to be had that would easily compensate for the expense of sending Douglas to the Pacific Northwest.

No other north-temperate region supports coniferous forests of such plenitude and magnificence as those of the Pacific Northwest. With mild winters and abundant rainfall approaching one hundred inches a year, these luxuriant woodlands suggest something akin to tropical rain forests transplanted to a northern clime. Stretching from northern California through Oregon and Washington and northward into British Columbia, this area was destined to become one of the world's most important lumber-producing regions. A variety of coniferous or cone-bearing trees—cedars, pines, hemlocks, spruces, and firs among them—combine to form the forested wilderness David Douglas explored as botanical agent of the Horticultural Society of London.

At the end of his first successful year in the region, Douglas decided to extend his stay for a second season of collecting, rather than return to England as planned. By the end of 1826, however, he had had his fill of the damp, wet, often soggy weather of the area. Moreover, problems with his eyesight, not to mention several narrow escapes from bodily injury, calamity, and prolonged privations, led Douglas to decide the time had come to return to England. Rather than sail directly from the Pacific coast, however, the collector waited until late March of 1827, when he joined a party of voyagers for a transcontinental trek to the Hudson's Bay Company's post at York Factory on Hudson Bay. On his return to London from Hudson Bay later in the year on board an English whaler, Douglas was accorded a hero's welcome by members of the Horticultural Society, who were more than pleased with the success of his self-prolonged journey.

After Douglas had spent a couple of years in London, during which time he attempted to edit his journal for

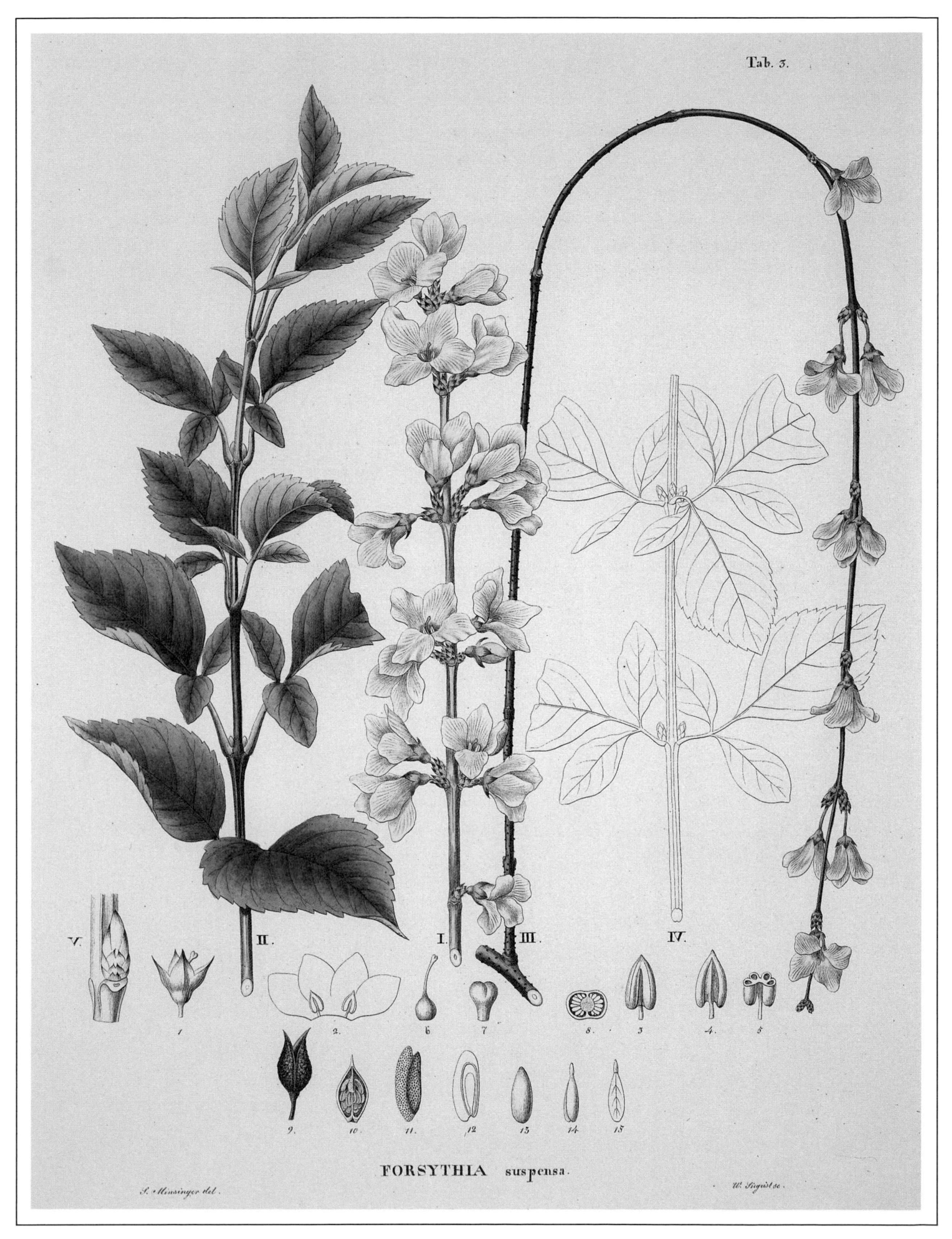

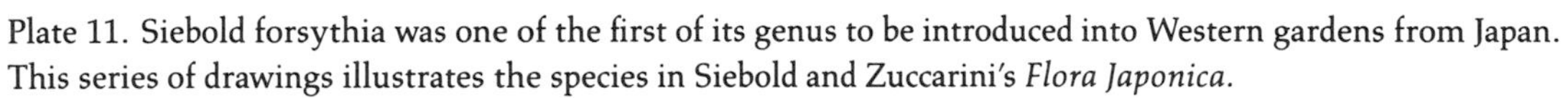

Plate 11. Siebold forsythia was one of the first of its genus to be introduced into Western gardens from Japan. This series of drawings illustrates the species in Siebold and Zuccarini's *Flora Japonica*.

Plate 12. Paintings of Chinese garden plants, such as these camellia flowers, enthralled members of the Horticultural Society of London and resulted in a great enthusiasm on the part of the membership for sending a collector to China to collect horticultural novelties. These paintings are from a collection—much like the one sent to the Horticultural Society in London by John Reeves—that was the property of Warren Delano, one of the early Boston merchants engaged in trade with China. The collection of 611 paintings was presented to the Arnold Arboretum in 1930 by Warren Delano's son, Frederic.

Plate 13. In England, magnificent cultivated specimens of the Douglas fir inspired awe and created an increased demand for exotic conifers in general.

Plate 14. The bleeding heart (*Dicentra spectabilis*) quickly became one of the favorite garden plants of the Victorian period and was widely planted in perennial borders. Robert Fortune was responsible for sending this species from China, and could he have capitalized on its sale, his fortune would have been assured.

94. When visiting Philadelphia, David Douglas met Thomas Nuttall (1786–1859), English botanist and ornithologist, who had succeeded William Peck as professor of natural history and director of the Harvard Botanic Garden. No stay-at-home, Nuttall preferred spending extended periods in the field and exploring first-hand for new plants and animals. An American species of smokebush (*Cotinus obovatus* Rafinesque) numbered among the ornamental shrubs or small trees Nuttall brought back from the Arkansas territory in 1819, and gave evidence of a floristic affinity of the region with the area of southern Europe and Asia where the Old World smokebrush is native. This photograph illustrates the multiple trunks of the American smokebrush in the Arnold Arboretum, where it grows in close association with the Eurasian species along Meadow Road.

95. David Douglas (1799–1834), intrepid horticultural and botanical traveler in the Pacific Northwest under the auspices of the Horticultural Society of London. Due to similar climatic conditions in the British Isles, many of the species Douglas introduced from the Pacific Northwest have become commonplace ornamentals in English and Scottish gardens. The sitka spruce, moreover, is widely planted in reforestation projects throughout northern Europe. Fewer of Douglas's introductions are hardy in eastern North America, although the inland race of Douglas fir and the sugar pine, two of his most important introductions, grow in the Arnold Arboretum and elsewhere along the Atlantic seaboard. The portrait of Douglas reproduced here was first published by J. C. Loudon in *Gardener's Magazine* for 1836 and accompanied an obituary of the plant collector, who had recently lost his life in Hawaii, then known as the Sandwich Islands.

publication, the Society decided that sponsorship of another trip to the Pacific coast of North America should be undertaken. Douglas, disliking city life and not enthusiastic about his editorial chores and his attempts in the herbarium to classify the dried specimens he had collected, was overjoyed at the prospect and agreed to represent the Horticultural Society for a third time as its collector in North America. Plans were quickly made, and Douglas left England on October 31, 1829, accompanied by his dog, Billy, his faithful companion.

On this trip Douglas again traveled extensively in the Pacific Northwest but went further afield—into Alaska, California, and the Sandwich Islands (now Hawaii) as well. As he had done on his previous trip to the area, he carefully packed dried specimens and packets of seed and periodically entrusted them to the care of any willing sea captain encountered who was sailing to England and the docks of London.

It was during Douglas's explorations in Alaska that the idea first occurred to him that his return to Europe and England should, as on his previous trip, include a transcontinental march. But this time it would be overland via Siberia, and plans were laid whereby Douglas and Billy would cross the Bering Strait, which separates Alaska from Siberia, and begin the trek westward toward Europe. But Douglas's plan was not to materialize. The collector's untimely and gruesome death on the slopes of Mauna Kea on Hawaii, where he fell into a pit-trap and was gored by a similarly ensnared wild bull, brought an abrupt end to the collector's activities and the steady stream of plants and seeds that had arrived at the headquarters of the Horticultural Society in London.

The results of Douglas's labors in the service of the Society, however, live on in gardens and parks around the world. And the botanical aspects of his work have set standards to the present day, particularly when one considers the hardships and difficulties under which he labored. One botanist wrote early in this century that "the extent and amount of this man's collections during the three seasons he spent in the Northwest almost surpass belief" (C. V. Piper, 1906, p. 13). These collections of dried specimens, moreover, provided the raw material on which his mentor in Glasgow, William Hooker, was to base an expanded flora of North America, one that would supplant and largely replace the earlier efforts of Michaux and Pursh. By current reckoning, Douglas successfully introduced into cultivation in Great Britain over two hundred species of plants.

The most abundant giant conifer of the region explored by the Scotsman is now commonly and appro-

96. Measuring one of the enormous conifers in the forests of Oregon. This engraving depicts a scene that occurred more than ten years after David Douglas had explored for plants in the Pacific Northwest when members of the United States Exploring Expedition under the command of Charles Wilkes came ashore in Oregon.

priately referred to with his name, and today it is widely planted as an ornamental tree across North America and Europe. Douglas fir (*Pseudotsuga menziesii* (Mirbel) Franco; Plate 13) was one of the first trees David Douglas encountered upon going ashore along the banks of the Columbia River. Quick to recognize the tree, which had already been given a scientific name by Aylmer Lambert based on a specimen collected by Archibald Menzies, Douglas was overawed by the sheer size and magnitude of these colossal giants. Individual trees attain heights today of 220 feet, yet during the last century enormous specimens were estimated by lumbermen to reach 350 feet above the forest floor. These estimates were never substantiated, however, and these individuals, if they did occur, have long since been toppled by lumbermen's axes, wedges, and long, two-man felling saws. One enormous individual measuring 295 feet, 7 inches, did find use as a flagpole at the Panama Pacific International Exposition of 1915. The largest individual David Douglas measured stretched 227 feet from its base to its apex and was 48 feet around. Today's reigning champion, which grows in Olympic National Park, measured 221 feet in height in 1945; at 4.5 feet from the base its circumference totaled slightly more than 45 feet.

Unlike most conifers, the trunk of the Douglas fir retains its massive girth for much of its length, tapering just below the summit of the crown. This characteristic, coupled with its resilient, strong timber, has made it the premier lumber tree in all of North America. The tree also differs in botanical details from the true firs as well as the spruces, hemlocks, and pines—the group to which Lambert mistakenly assigned it when naming Menzies' specimen *Pinus taxifolia*. The leaves or needles of Douglas fir are blunt-tipped and flat, and each is twisted at its base at the point of attachment to the stem. Occurring singly, not in groups or fascicles as in most pines, the needles occur around the stem, yet the twisted bases frequently orient the majority of needles in one plane, similar to those of many firs (species of the genus *Abies* L.) and spruces (species of *Picea* L.). The characteristic, pendulous cones are abundantly produced and allow for the identification of the tree far more easily than its vegetative characteristics. Usually between two and three inches in length, each cone is comprised of numerous, imbricated, reddish-brown scales, which bear the seeds. Protruding from between the scales are unique, elongated bracts, each of which is deeply three-toothed at its apex. Once seen, these cones with their conspicuous bracts immediately identify the tree.

In distribution, the Douglas fir is one of the most wide-ranging conifers of western North America, extending from Central British Columbia—at about fifty-four degrees of latitude—southward through the Pacific Northwest into the Sierra Nevada and the Coast Ranges of California. Dense, often pure stands of Douglas fir extend eastward to the Continental Divide in Alberta and carpet mountain slopes southward through the cordillera of the Rocky Mountains into Arizona and New Mexico. Small, isolated, outlying populations also occur in Texas and northern Mexico.

Trees of the inland populations rarely attain the height and dimensions of the trees of the Pacific coast, a fact of considerable economic significance to lumber companies and of botanical significance, too. As a consequence, the Rocky Mountain populations have been referred to a distinct variety (*Pseudotsuga menziesii* var. *glauca* (Mayr) Franco), the so-called blue or Colorado Douglas fir. While Colorado Douglas firs are of smaller stature and often produce smaller cones with reflexed bracts, many individuals intergrade with the typical form. In cultivation, the Colorado Douglas firs have proved hardier than stock from the Pacific Northwest, and the old Douglas firs growing at the base of Hemlock Hill near the Bussey Street gate in the Arnold Arboretum are of Rocky Mountain provenance.

Another conifer was actively sought and eventually found by David Douglas after he had seen its large, edible seeds in the pouch of an Indian encamped on the banks of the Columbia River. Seeing the enormous cone from which the winged seeds had fallen further increased the explorer's resolve to locate this new pine. His search finally ended far to the south of the Columbia near the headwaters of the Umpqua River in southwestern Oregon in October of 1826. Guided by an Indian of the Umpqua tribe, who recognized Douglas's quickly produced, rough sketch of the cone and seed, Douglas located several trees scattered in the forests of the region. Unlike the Douglas fir, this species of pine occurs as solitary individuals or in small groups, never forming pure, dense stands.

In his journal entry for Thursday, October 26, the Scotsman noted, "The large trees are destitute of branches, generally for two-thirds the length of the tree; branches pendulous, and the cones hanging from their points like sugar-loaves in a grocer's shop, it being only on the very largest trees that cones are seen, and the putting myself in possession of three cones (all I could) nearly brought my life to an end" (Davies, 1980, p. 103).

Not only do the large cones—up to a foot and a half in length—suggest the all but forgotten sugar loaves of yesteryear, but the sticky resin, which exudes and collects as irregular kernels between the platelets of the bark or wherever the tree has been wounded, has a sweet, sugary taste. The sweet resin became responsible for the tree's common name, sugar pine, although it should be avoided as a source of an edible sweetener; if more than a small amount of the resin is ingested, it acts as a powerful diuretic. The Latin, scientific name published for the tree by Douglas on his return to England in 1827—*Pinus lambertiana* Douglas—honored Aylmer Bourke Lambert, fellow of the Horticultural Society of London, patron of botany, and one of his era's foremost students of conifers.

A denizen of mountain slopes and the steep walls of ravines, the sugar pine occurs in the coniferous forests of Oregon southward in the Coast Ranges and in the Sierra Nevada of California and into northern Baja, California. With age, these monarchs of the forest develop picturesque, craggy crowns in which the wide-spreading limbs are held at almost right angles to the massive trunk. Some sixty-eight years after Douglas first described this species, John Muir wrote in *The Mountains of California* that "the Sugar Pine is as free from conventionalities of form and motion as any oak. No two are alike, even to the most inattentive observer; and, notwithstanding they are ever tossing out their immense arms in what might seem most extravagant gestures, there is a majesty and repose about them that precludes all possibility of the grotesque, or even picturesque, in their general expression" (J. Muir, 1961, p. 124).

In their habit, old sugar pines are reminiscent of ven-

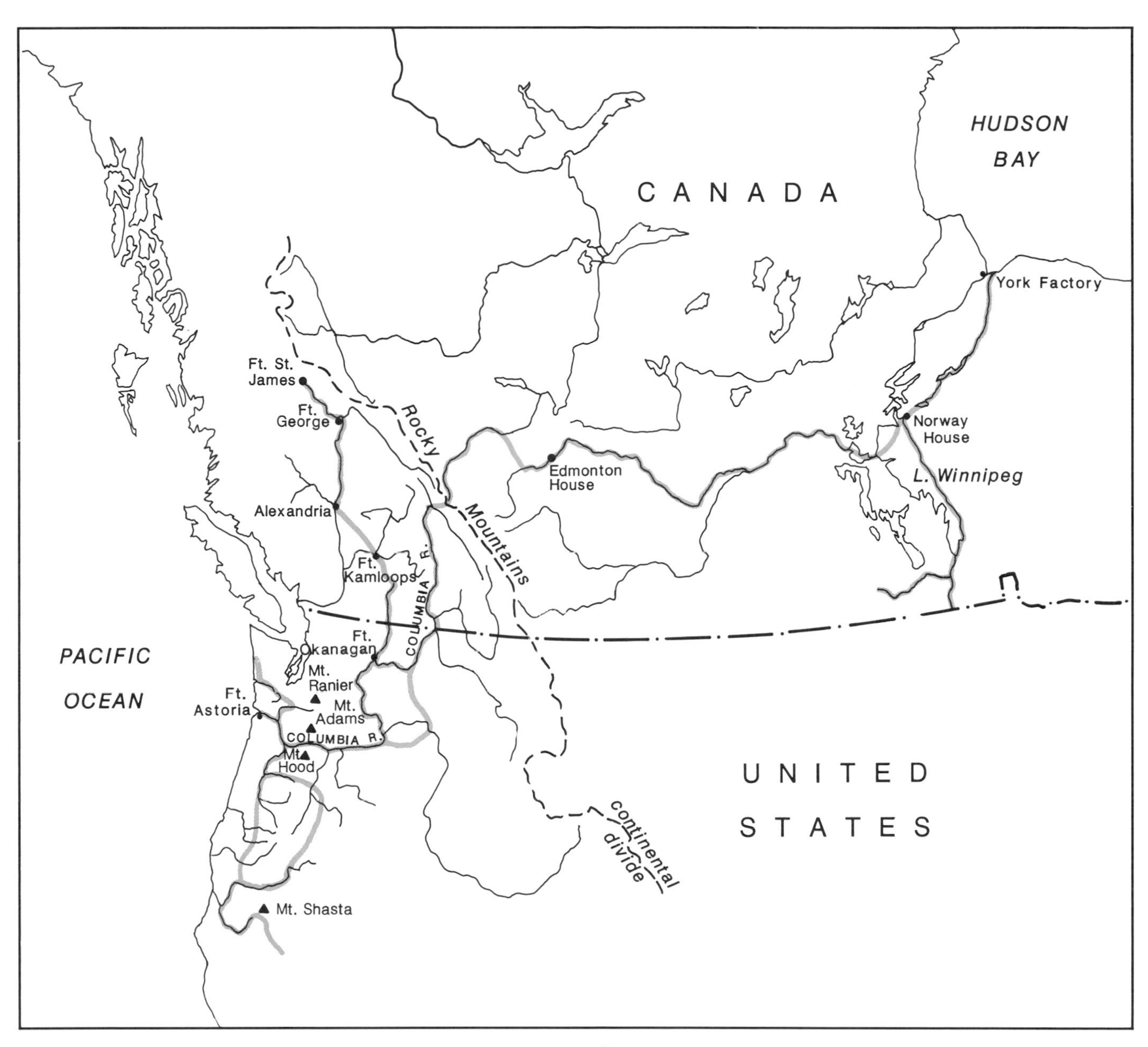

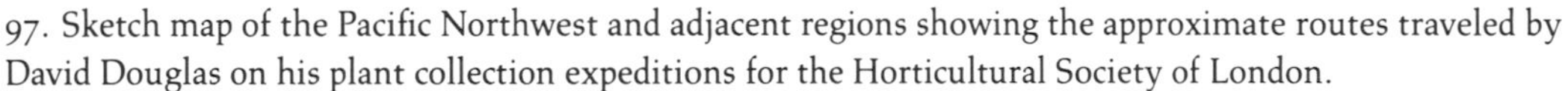
97. Sketch map of the Pacific Northwest and adjacent regions showing the approximate routes traveled by David Douglas on his plant collection expeditions for the Horticultural Society of London.

erable old specimens of eastern white pine (*Pinus strobus* L.), the familiar and commonplace pine of New England. Yet mature specimens of all but the tallest white pines could easily be accommodated beneath the outstretched branches of the crowns of sugar pines. In size, the sugar pine ranks as the largest and tallest of all the pines of the world. As David Douglas was the first to note, it is not uncommon for mature specimens of sugar pines to tower to heights of 150 to 200 feet, with the lowermost branches occurring 100 feet above the forest floor.

Sugar pines and eastern white pines are, in fact, closely related species. Both produce their needles in bundles of five; the morphology of their cones and the internal anatomy of their needles are similar; and both are susceptible to white pine blister rust, a fungal disease of European origin that has caused inestimable damage to populations of both species. Another western species, the so-called western white pine (*Pinus monticola* Douglas ex Lambert) also shares a close relationship with the sugar pine and the eastern white pine. Like the sugar pine, western white pine was first discovered, collected, and described by Douglas from the forests of the Pacific Northwest.

The list of coniferous trees associated with the name of David Douglas does not end here. The sitka spruce (*Picea sitchensis* (Bongard) Carrière)—a species now widely grown in northern climates as a lumber-producing species—the noble fir (*Abies procera* Rehder), silver fir (*Abies amabilis* Forbes), and giant fir (*Abies grandis* Lindley) all number among the conifers first discovered and introduced by David Douglas. So impressed and amazed were horticulturists at Douglas's introductions and the variety of species available from other regions that they promoted the development of collections limited to conifers to be grown in plantations especially devoted to them, called pineta.

Innumerable herbaceous annuals and perennials were sent back for trial in the Horticultural Society's London garden by the Scotsman as well. Many of these species—the California poppy foremost among them—have become subjects in herbaceous borders and are favorites with gardeners around the world. Fewer deciduous trees and shrubs are associated with Douglas's name, yet a maple introduced to England by him has proved hardy in the Boston region. Specimens of the vine maple (*Acer circinatum* Pursh) are included in the maple collection in the Arnold Arboretum, and plants also grow along Sargent Trail on Hemlock Hill. In association with the hemlocks and other conifers growing in that location, these small trees or shrubs appear almost as they do in their natural habitats in the moist, coniferous, coastal forests of British Columbia southward through Washington and Oregon into northern California. To a botanist or horticulturist unfamiliar with the species, vine maples—whether growing in their native haunts or in cultivation—are apt to be confused with one of the prized Japanese maples. The mix-up is understandable, inasmuch as vine maples are characterized by their beautiful, almost round but seven- to nine-lobed leaves, which are strikingly similar to those of Japanese maples. Indeed, the vine maple is the only species of maple native to an area outside of eastern Asia that is classified as belonging to the Japanese maple group.

While Philipp Franz von Siebold was actively seeking numerous varieties of Japanese maples that had been skillfully selected and propagated for generations by Japanese horticulturists, David Douglas was exploring a region inhabited by Indian nations with cultural traditions very different from those of the Japanese. Yet the two regions where these plantsmen concentrated their efforts supported in several instances closely related plant species that reflect a shared history and biogeographic continuity. This relationship is epitomized by the vine maple Douglas introduced into Europe and the Japanese maples introduced by von Siebold. An even stronger relationship between the Japanese and Chinese floras and the flora of eastern North America would become increasingly evident as continued botanical exploration of these regions bordering the North Pacific continued during the nineteenth century, fueled by both horticultural and scientific interests.

CHAPTER FOUR

East Meets West

Turning the Tables

The Treaty of Nanking brought the initial war between Great Britain and China to a close in 1842. Renewed disputes and hostilities erupted into open warfare between 1856 and 1858, and again in 1860 the two nations were engaged in armed combat. However, during 1843 and 1844, after the Nanking settlement brought a brief period of calm, commercial treaties with the British as well as the American and French governments were signed by official representatives of the emperor of the Celestial Empire. Another treaty, finally signed in Peking in 1860, followed in the wake of the continued hostilities.

In the end, when fighting had ceased, negotiations had been completed, reparations determined, concessions conceded, and the terms and conditions for future conduct had been forged, the tables had been turned. Commerce with China, which had previously been so completely controlled by the Chinese, was now managed and directed by the foreign treaty powers. The Imperial Maritime Customs, formed to oversee trade at the treaty ports and to collect import duties and tariffs for the Chinese government, was wholly staffed and operated by subjects of the treaty powers and under the jurisdiction of a British inspector-general. China was no longer mistress of her own household.

Under the terms of the Nanking treaty, Hong Kong was ceded to the British, and the ports of Shanghai, Ningpo (Ningbo), Foochow (Fuzhou), and Amoy (Xiamen)—designated treaty ports—were opened to foreign trade. Ten additional ports were opened in 1858, where Americans and Europeans could freely conduct their business with whomever they pleased and without Chinese intermediaries. The Hong merchants who had served as the requisite catalytic agents for foreign commerce at Canton were no longer necessary. Their traditional grip on the trade monopoly had been completely relaxed.

European and American supercargoes became free agents in China. In treaty ports they could buy or rent land, construct or lease buildings, and employ Chinese citizens as servants and translators or in other capacities as need arose. Their conduct while in China, moreover, was outside the jurisdiction of Chinese law; foreigners would henceforth be subject only to the statutes of the nations of which they were citizens. And in marked contrast with the confinement previous generations of traders had endured at Canton, foreigners could now freely roam the streets of the treaty ports, and sojourns into the surrounding countryside were possible. Under the terms of the treaty agreed to in 1858, citizens of the treaty powers could theoretically travel wherever they pleased in the interior of the empire, and the Yangtze River was opened to navigation by foreign vessels as far inland as Hankow (Hankou) in Hupeh (Hubei) Province. Navigation rights would later be extended to Ichang (Yichang), some one thousand miles inland, beyond which ocean-going vessels could not proceed because of the treacherous, churning current in the narrow gorge region upstream from Ichang. Numerous cities and towns along the banks of the Yangtze were simultaneously opened to trade, requiring the establishment of posts of the Imperial Maritime Customs inland in increasingly distant and remote localities from the coastal treaty ports.

Another radical change in the Chinese position, forced by the treaty of 1860, concerned religion. Despite the early success of the Jesuits in planting the first seeds of Christianity in the Celestial Empire, foreign missionaries and Christian converts had been banished from the country by imperial edict in 1724. From that time onward, missionaries had been denied access to the Chinese masses, and their activities had been confined to Macau and the foreign factories at Canton. Under these circumstances, Christian missionaries had become

98. This cartoon, entitled "A Little Tea Party," aptly depicts the awkward situation in which China found herself in relation to foreign powers during the middle of the nineteenth century. With pun intended, Britannia asks, "A little more gunpowder, Mr. China?" to which he responds, "O—no—tan—ke—Mum," as an attentive France demurely sips from her cup.

dependent upon Chinese converts to proselytize in the vast regions where foreigners were forbidden. But in 1860, foreign missionaries were again legally permitted to travel throughout China, and, surprisingly, they could establish missions wherever they chose.

Paradoxically, the altered positions toward foreigners that the Chinese were forced to accept were the ultimate result of justifiable attempts by the Chinese to curtail organized British and American smuggling operations. In this clandestine business, the product being sold to Chinese traders against the express wishes of the Chinese government was a highly valuable plant derivative. Chinese coffers were being drained in exchange for chests of a narcotic drug, opium, which is derived from a species of poppy. And the name given the first round of Sino-British hostilities reflected that fact. The Opium War of the early 1840s was effectively waged by the British to ensure the continuation of the extremely lucrative opium traffic by English and American traders. The continued bouts of warfare up until 1860 sought the same end. And by 1858 the weakened, militarily inferior, and humiliated Chinese government was forced to officially sanction the legal importation of the valuable plant alkaloid.

Not a native Chinese plant, the opium poppy (*Papaver somniferum* L.) had probably been introduced to China during the thirteenth century. Its use as a powerful narcotic, however, did not become widespread in all classes of society until the practice of mixing opium with tobacco and inhaling its fumes when smoked was introduced to the mainland from Formosa (now Taiwan) by the Portuguese in the seventeenth century. The Portuguese probably also first brought the tobacco plant—native to far-distant America—to the Orient. The Chinese gradually eliminated the tobacco altogether and smoked opium alone, and soon thereafter cultivation of the opium poppy became widespread in China.

Chinese authorities, alert to the increased use of opium, issued an edict in 1729 banning the smoking of opium. Nevertheless, the demand for the drug continued to grow and soon could not be satisfied through local production. After Robert Clive's military victories established British rule in India in 1757, the British East India Company seized a monopoly on poppy cultivation in the subcontinent; and by the nineteenth century enormous quantities of opium were being smuggled into Chinese waters on board British "East Indiamen." To help satisfy the Chinese appetite for the drug, smaller yet significant quantities came to China from the Middle East, largely aboard American ships.

Offshore, chests of opium were transferred from English and American ships to Chinese junks, and the

99. The poppy, *Papaver somniferum* L., is the source of opium, of which the alkaloids morphine and codeine are well-known derivatives. Despite the fact that the opium poppy was widely cultivated in eighteenth-century China, demand for the latex produced as an exudate on the incised, unripened capsules of the plants was so great that importation from India and the Near East became extremely profitable to merchants involved in the so-called China trade.

transactions were settled in silver. The junks easily navigated Chinese waters without suspicion, and their precious cargoes were offloaded and quickly transferred into the hands of dealers. Meanwhile, the holds of English and American ships, which had brought the drug to market, were being filled with Chinese teas and fine silks for lucrative export to Europe and America; payment could be made with the profits secured in the opium traffic.

By mid-century, when conflict between Britain and China over these smuggling operations broke out, Western interests in China had gone beyond mercantile gain and the conversion of her heathen masses to Christendom. Among other schemes, there were the plans of horticulturists and botanists in Europe, and even in far-distant America, to obtain new plants. The strong desire in Europe—where the descriptive aspects of both botany and zoology were flourishing—for further knowledge of Chinese natural history ensured that foreigners in that country would be on the lookout for new and unusual plants and animals. And soon after the treaty ports were opened in 1843 and new regions could be visited by outsiders, keen-eyed plantsmen were on the scene. This era gave birth to the professional plant hunter who would specialize in the Chinese flora, and one of the first of this breed to arrive in China was willing to risk life and limb to travel further than officially sanctioned.

Sing Wah

The small boat had come from Shanghai by way of the network of rivers and canals that form the highways and byways of travel on the broad, level plain of southeastern Kiangsu Province. After dark on the evening of the second day, the boat reached its destination, and by the light of the moon the boatmen moored the craft outside the walls of the ancient city of Soochow (now Suzhou). Here, they joined hundreds of other vessels of all possible descriptions that filled the broad, moatlike waterway surrounding the ancient metropolis, the Venice and garden city of China.

Early the next morning, after being carefully dressed by his servant, Sing Wah left the small cabin on the moored craft and stepped ashore. Accompanied by his man, he climbed onto the broad, hump-backed bridge that spanned the canal and connected the city with the surrounding countryside. Pausing on the bridge while other pedestrians entering and leaving the city pushed on, Sing Wah marvelled at the broad ribbon of water, which extends half the length of China, linking Hangchow (Hangzhou) in the south with Peking in the north. The Grand Canal—a waterway over seven hundred miles in length and an engineering feat that in complexity rivals the Great Wall itself—ranks as one of China's wonders. Sing Wah took a deep breath, glanced at his servant and other pedestrians, and continued on his way toward one of the city's fabled gardens. None of the Chinese had even paused to give him a second glance. Nor had any of the vagrant dogs approached to sniff at the hem of his long garment and attract attention to his person. His disguise was sufficient—none were aware that the first Scotsman to enter the walled city of Soochow, the cultural center of China, was in their midst on the morning of June 24, 1844.

On his return to Shanghai several days later, Sing Wah—with shaved head and braided, black pigtail—continued to adopt Chinese dress, not by choice but through necessity. For a midnight visitor had stolen Sing Wah's European-style clothing from the cabin of the boat on the first night of his trip as he lay sleeping. Now in the foreign settlement in Shanghai, the effectiveness of his disguise was tested and proven once again. A recent but close acquaintance, a Mr. Mackenzie, was astonished when he realized that Sing Wah was his newly made acquaintance, Robert Fortune, botanical and horticultural collector for the Horticultural Society of London.

When news of the Treaty of Nanking reached England in 1842, John Reeves had lost no time in suggesting that a botanical and horticultural collector should finally be sent to China. Reeves, one-time corresponding member of the Horticultural Society of London, had returned to England from China in 1831 and had become chairman of the Society's China committee. Reeves's enthusiasm for the undertaking was shared by the botanist John Lindley, Joseph Sabine's replacement as secretary of the Society, and little time was wasted in raising the required funds, locating a capable and promising although untried collector, and drafting the terms under which he would represent and work for the Society in China. With a salary of £100 per annum and a list of specific plants the Society hoped he could introduce, Robert Fortune embarked on board the *Emu* for China early in 1843.

On July 6, Fortune arrived in British-occupied Hong Kong to begin his career as a horticultural and botanical explorer. This Scotsman was destined to make four separate journeys to China between the years 1843 and 1860, with additional sojourns in India, the Philippines, and Japan. While his initial mission was for the Horticultural Society, his second and third trips were sponsored by the

100. An engraving entitled "Boats Used on the Rivers of China," which was published by Robert Fortune in his book, *A Residence Among the Chinese: Inland, on the Coast, and at Sea.* This work, which included Fortune's observations concerning the war with Great Britain and his comments on Chinese society and natural history, was published in London in 1857.

East India Company, which had developed botanical interests in China in addition to those surrounding the importation of opium. The demand for tea in Europe—particularly in England—was growing almost in proportion to the demand for opium in China. As a consequence, it was decided that tea plantations should be established in British India to help satisfy the demand. To ensure the success of the project, tea plants and a thorough knowledge of tea manufacture was essential. At the time, botanists were not even sure if black and green teas were produced from the same or different species. It was Fortune's job to travel to the finest tea-producing regions in China to obtain seeds and young plants of the best varieties and to observe and document the time-honored Chinese techniques in the manufacture of fine black teas. In this endeavor Fortune succeeded in sending 23,892 young tea plants and approximately 17,000 germinated seedlings to the hill country of northern India. Moreover, he traveled in India to ensure that the new plantations were properly cultivated and that the tea-manufacturing process followed Chinese traditions.

On his fourth trip to the Celestial Empire during 1858 and 1859, Fortune revisited the tea-producing regions he had explored for the East India Company, but on this mission he represented the government of the United States. While the development of tea plantations in the southern United States was destined to be a failure—in part due to the havoc caused by the Civil War—Robert Fortune played the instrumental role of sending tea plants from China to Washington, D.C., for the short-lived experiment. The unsuccessful plan was not a totally new idea, for André Michaux had first brought the tea plant to the Southeast from France. Tradition relates that sometime toward the close of the eighteenth century Michaux planted tea shrubs at Middleton Place near Charleston, South Carolina.

Fortune's success on all of his trips to the Orient was

101. These four postage stamps issued in the People's Republic of China in 1980 depict scenes in the *Liu Yuan*, one of the famous gardens of Soochow. In his book *Three Years' Wanderings in the Northern Provinces of China*, Robert Fortune wrote that "fine pictures, fine carved work, fine silks, and fine ladies all come from Soo-chow," the "Chinaman's earthly paradise."

enormous. New plants of signal horticultural significance, many representing new species previously unknown to science, were introduced into European gardens and conservatories as living plants and also prepared as dried specimens to satisfy botanical requirements. The Horticultural Society's garden at Chiswick served as the depository for the plants of his first mission, but commercial nursery establishments received consignments from his subsequent trips.

The botanical results of Fortune's booty were quickly made available to the scientific community by John Lindley and other botanists, who eagerly "worked up" his dried collections and described and often illustrated his novelties. The horticultural potential and the cultural requirements of the introduced species were also carefully monitored by growers, and the details recorded and made known in the horticultural press. Never before had so many Chinese plants been introduced into England, and fellows of the Horticultural Society and nurserymen in England, elsewhere in Europe, and in America anxiously sought propagating material for experimentation. Sometimes quickly, other times more slowly, Robert Fortune's discoveries were planted in the gardens and parks of Europe and North America.

Fortune's triumph in sending living plants from China and elsewhere in the Orient to England far exceeded any previous attempts. The key to his high success rate rested in his use of the recently invented Wardian case, a glass-walled box that prevented the evaporation of moisture sealed within. On his first trip out to China, Fortune experimented with the new cases, which were securely fastened to the deck of the *Emu*. Living plants had been established for transport in the cases before the ship weighed anchor in the Thames estuary. On the voyage to China the ship's course took the *Emu* through north temperate and tropical latitudes off the coasts of Europe and Africa and into the south temperate zone as it doubled the Cape of Good Hope. Gradually headed on a northeasterly course, the vessel passed back through tropical waters and finally ended its voyage in the warm temperate South China Sea. All the while, the English garden plants within the tightly closed cases withstood the vagaries of constantly changing conditions that occurred without. When the cases were opened, the plants were for the most part healthy and ready for establishment in gardens of the English stationed at Hong Kong.

In its simplest form the Wardian case constitutes a box with tightly glazed top and sides and a water-tight, boxlike bottom deep enough to hold a quantity of soil.

102. Among the British botanists who described the many new plants discovered in China by Robert Fortune, John Lindley (1799–1865) was perhaps the most productive and influential. Lindley began his career as an assistant in the library of Sir Joseph Banks but soon left Soho Square for the Chiswick garden of the Horticultural Society of London, where he began work as a clerk. Lindley rose through the ranks of the Society and served as secretary between 1858 and 1863. In 1832 Lindley earned a Ph.D. degree from the university in Munich and was one of the first botanists who trained at a German university and specialized in botany, not medicine. Serving as professor of botany at University College, London, Lindley authored many taxonomic studies as well as textbooks of botany that had great influence in popularizing the subject among the lay public.

103. This woodcut depicts Robert Fortune (1812–1880) at the age of fifty-four. Through his activities as a plant collector, and through his use of the Wardian case to transport living plants to Great Britain, many of the best-known Chinese garden plants were introduced to Europe and America.

Well-seasoned, rot-and-warp-resistant lumber is essential for its construction, yet its dimensions can vary, depending upon the size of the plants it is to carry. When the lid or door—usually the top or one side—is closed and sealed, the case becomes essentially air-tight and akin to a terrarium. It embodies a closed ecosystem to which nothing can be added and from which nothing can escape. Watering the plants during transport became unnecessary as a consequence, a particular boon since fresh water was usually a scarce commodity on long ocean voyages.

Notable among the many plants Fortune carefully established in Wardian cases for transport to England were numerous herbaceous perennials that in the hands of horticulturists and dirt gardeners have been propagated and shared with or sold to hundreds of thousands of others anxious to include these plants in their own perennial borders. The result today is their wide distribution in cultivation and their ranking as mainstays in flower gardens everywhere in north temperate regions. Balloon flower (*Platycodon grandiflorus* (Jacquin) A. de Candolle), Chinese anemone (*Anemone hupehensis* Hortorum ex Lemoine), and bleeding heart (*Dicentra spectabilis* (L.) Lemaire; Plate 14)—an all-time favorite that has become inextricably associated with Victorian gardens—are but three plants Fortune collected and first sent to Europe from China.

Of greater impact in the Arnold Arboretum and in regions enjoying climates that will support their growth are the woody Asian plants Fortune discovered and sent from eastern China. As a result of his keen eye and energetic collecting, many new trees and shrubs have been added to the varied list of plants available to today's landscape architects, designers, and amateurs alike who install plantings and garden in localities as far from the Chinese treaty ports as London and Boston. Regrettably, several of Fortune's significant discoveries have remained rarities not widely known to the gardening public. The reasons for their obscurity are linked, at least in part, with difficulties surrounding their propagation. For some, an added cause is a reluctance on the part of many commercial nurserymen to propagate and offer relatively unknown plants—for which there is little demand—in the horticultural market place.

The golden larch (*Pseudolarix amabilis* (Nelson) Rehder; occasionally incorrectly listed as *Pseudolarix kaempferi* Gordon) ranks as one of the unique conifers Robert Fortune discovered and introduced that has achieved only limited distribution in cultivation. It also has a narrow native range in China, and Fortune first

104. A Wardian case, as depicted by Nathaniel B. Ward (1791–1868) in his book on the subject.

met the tree growing in a small pot as the subject of peng-jin, the Chinese forerunner of Japanese bonsai. Years later he encountered the tree once again, this time growing in the open ground of a temple garden, and ample seed was harvested for shipment to England.

After viewing the large, old specimens of the golden larch that grow along the banks of Bussey Brook in the Arnold Arboretum (Plate 16), one is justified in questioning how this tree could adapt to growth in the limiting confines of a clay or ceramic pot. In the century the Arboretum trees have been growing, the largest have attained heights of upwards to fifty feet, and the largest limbs branch widely to form broadly pyramidal but open crowns. Compared with many other conifers, this rate of growth is slow, averaging only about six inches a year. Coupled with judicious pruning of the branches and roots, this minimal annual growth increment is an ideal attribute for a peng-jin subject. And the unique beauty of the foliage of the tree adds immeasurably to its charm as a potted plant.

But the true beauty and landscape value of the golden larch is best appreciated when mature specimens like those growing in the Arnold Arboretum are viewed over the course of the seasons. The growth habit is seen to best advantage during the winter months, when the twigs are leafless. Like the true larches (species of the genus *Larix* Miller), the golden larch is a deciduous conifer, which annually produces a new crop of needlelike leaves in spring and drops them at the close of the growing season in fall. Like both the ginkgo and the true larches, the branchlets of the golden larch are differentiated into long- and short-shoots. The majority of leaves or needles are produced at the ends of the curiously ringed short-shoots, where they occur in radiating, spokelike clusters. Because the numerous short-shoots are closely spaced along the long-shoots, the needles from adjacent short-shoots frequently overlap, giving the branches and the entire tree a soft, feathery appearance.

In spring the young needles are a delicate light green as they emerge from the bud. In fall they gradually turn from their rich summer green to a pale yellow highlighted by fading green. Gradually, they pass through increasingly dark hues of yellow until they attain a glorious golden or coppery yellow before falling. Such fall color is not normally associated with coniferous trees, and the appellation golden larch is well deserved.

Species of true larch are among the few other conifers that add shades and hues of yellow—in contrast with the summer greens of most conifers—to the autumn landscape before their needlelike leaves fall to the ground. Because of this shared characteristic associated with their deciduous habit, as well as similarities in leaf structure and branch differentiation into long- and short-shoots, the true larches and the golden larch evince a close botanical relationship. This affinity is implied by the botanical name of the golden larch genus, *Pseudolarix*—literally, false larch. The two groups differ significantly, however, in the structure and behavior of their seed-bearing cones.

Inasmuch as individuals representing species of the true larches grow on the slopes of the Bussey Brook valley adjacent to the golden larches, students visiting the Arnold Arboretum have easy access to trees of both genera for comparative purposes. The true larches annually produce a crop of seed-filled cones under two inches in length. The tightly imbricated, rounded, brown scales open slightly at maturity and thereby release the small winged seeds that have developed on their upper surfaces. After the seeds have been dispersed, the intact cones persist on the branches until they eventually fall to the ground beneath the trees.

By contrast, the seed-bearing cones of the golden larch measure up to three inches in length and are composed of numerous loosely spreading, greenish-yellow scales with pointed apices. In shape and overall appearance these cones have been likened to miniature artichokes. Every three or four years when they are produced in large numbers, almost every short-shoot is terminated

105. The conifer collection in the Arnold Arboretum was established on the south-facing slope above Bussey Brook as well as on the level ground at the base of Hemlock Hill. This view looks downstream toward the junction of Hemlock Hill and Valley roads.

by a cone, and they seemingly stud the branches of the trees, adding enormously to their decorative aspect. A full crop of these intriguing cones is rarely produced in succeeding years, and often, after a mast season (a season when an abundance of cones is produced), few if any cones are seen in the following year. Moreover, unlike the firmly attached cones scales of the true larches, the large, pointed cone scales of the golden larch detach easily from the cone axis once the long-winged seeds are mature. As a consequence, the interest these unique cones add to the beauty of the tree is transient. However, to those plantsmen who have experienced the subtle, changing beauty of the golden larch through the seasons and over the years, the tree's feathery delicacy and grace in the landscape is doubly rewarding when cones are produced in abundance; even without its mock artichokes it is a tree worthy of wider cultivation.

The Chinese fringe tree (*Chionanthus retusus* Lindley)—despite its common name, a widespread Asian species native to portions of Korea and Japan as well as a wide area of China—ranks as another beautiful ornamental shrub or small tree first sent to European gardens by Robert Fortune. Most individuals of this species in cultivation form large, rounded, multiple-stemmed shrubs. Oddly enough, the specimen growing along Chinese Path on the shoulder of Bussey Hill in the Arnold Arboretum has developed into an extremely shapely, vaselike tree suggestive of a small American elm. But here comparisons to the elm must end, for in addition to qualifying as a small, stately shade tree, the Chinese fringe tree ranks as a spring flowering ornamental that enjoys a second period of interest when its blue, olivelike fruits mature in late summer.

The flowers and fruits of the Chinese fringe tree could easily be mistaken for those of its New World counterpart, the American fringe tree (*Chionanthus virginicus* L.). But the impact this Asian tree, when in flower, makes on the landscape of the Arnold Arboretum is of a modified, arboreal version of the normally shrubby American species. Specimens of the American plant, which Mark Catesby sent to English gardens from Virginia over a century before Robert Fortune made known its Asiatic congener, grow adjacent to the Asian tree for easy comparison. The two species—one restricted to eastern North America, the other with an exclusively eastern Asian distribution—are members of a genus of only three or perhaps four species. Yet despite their sibling status, the two are easily distinguished. Over and above their difference in growth habit and technical details involving exacting measurements, their most obvious distinction relates to a behavioral quirk. The Asian species produces its inflorescences on growth of the current year, while its Occidental counterpart flowers from branchlets of the previous season's growth.

In 1844 a large shipment of seed of cryptomeria (*Cryptomeria japonica* (Linnaeus f.) D. Don) was sent by Fortune from Shanghai to England. The United States Patent Office received a similar shipment in 1858, and the next year young plants were growing in the five-acre Government Experimental and Propagating Garden between Four-and-a-Half and Sixth Street in central Washington. As a result, another unique Asian conifer became available to European and American horticulturists. Young plants of still another unusual conifer were transmitted to the English nursery firm of Glendinning of Turnham Green in 1846, and eighteen years later Fortune advertised seed of the same species for sale at

106. The seed-filled cones of the golden larch, one of which is depicted in this woodcut, have been likened to miniature artichokes.

107. The Chinese fringe tree is closely allied to the North American fringe tree, which is widespread from West Virginia and Ohio southward into Texas and Florida. A third species (*Chionanthus pygmaeus* Small), native to central Florida, has never been cultivated in the Arnold Arboretum, and it is not known if it would prove hardy in New England.

108. Another conifer introduced from China by Robert Fortune is the lacebark pine (*Pinus bungeana*). Fortune was greatly impressed by the beauty of the white trunks of mature specimens and published this illustration in his book entitled *Yedo and Peking*, which appeared in London in 1863.

twenty-five shillings an ounce. These precious seeds, which had been delivered by overland mail from Peking, had been collected from a large tree Fortune had located in 1861 in the mountains west of the Chinese capital. Thus, lacebark pine (*Pinus bungeana* Endlicher), exceptional for its white, exfoliating bark, was also added to the roster of interesting and diverse Chinese trees to be tested by cultivators outside the Orient.

A wide array of shrubs, including the double-file viburnum (*Viburnum plicatum* Thunberg), the old-fashioned weigela of Victorian gardens (*Weigela florida* (Bunge) A. de Candolle), greenstem forsythia (*Forsythia viridissima* Lindley), and the winter honeysuckle (*Lonicera fragrantissima* Lindley & Paxton), which produces its extremely fragrant flowers during the first days of spring, also became available to enthusiastic gardeners as a result of Fortune's success in China. The Chinese abelia (*Abelia chinensis* R. Brown), the curious hardy orange (*Poncirus trifoliata* (L.) Rafinesque), with its armature of stout thorns and its miniature orangelike fruits, and the beautiful common pearlbush (*Exochorda racemosa* (Lindley) Rehder; Plate 19, bottom), still too little known in gardens, were other shrubs that filled Fortune's Chinese cornucopia. Mahlon Moon, proprietor of the Moon Nursery in Morrisville, Pennsylvania, brought the pearlbush to America shortly after Fortune had introduced the shrub to England and was the first to cultivate the shrub in American soil.

New forms or cultivars of the moutan or tree peony (*Paeonia suffruticosa* Andrews; Plate 15, top) were included in Fortune's harvest, as were new camellias and an evergreen rhododendron (*Rhododendron fortunei* Lindley). Named for this intrepid man, Fortune rhododendron (Plate 15, bottom) was the first of the evergreen type, as opposed to species of the azalea group, to be sent from China. Little did Fortune realize—nor, for that matter, did the eager plantsmen who marveled at its beautiful foliage and large trusses of bluish-lavender flowers—that the rhododendron warehouse of the world would prove to be located within China's borders. They could not have predicted the multitude of hybrids that, utilizing this species' genetic mix, would result from the crosses of skilled rhododendron breeders on both sides of the Atlantic. One of these men, Charles Dexter of Sandwich, Massachusetts, was responsible for the many Dexter hybrids that grow at Heritage Plantation, in the Arnold Arboretum, and across North America today.

Last but not least in this greatly abbreviated catalogue of Robert Fortune's plant introductions to the West is the white-flowered form of wisteria (*Wisteria sinensis* (Sims) Sweet 'Alba'). This beautiful climber was the prize obtained on Fortune's sojourn in Soochow in 1844, a city then outside the boundary where foreigners could legally travel. Yet donning his Chinese disguise and passing under the alias Sing Wah, Robert Fortune gambled for the floral treasures that the garden city of China might provide, and his career hung in the balance. This venture, like all his other undertakings, ended in success and set the standards for future plant hunters who were to follow him to the flowery kingdom.

Sierran Gold

Gold fever had become an epidemic, as thousands of Americans headed overland for the gold fields of California during the year 1849. Others, determined to lay claims in the region where wealth could be had overnight, embarked from east-coast ports on the long sea voyage around Cape Horn. Additional argonauts crossed the Isthmus of Panama to shorten their voyage and hasten their arrival in California, the new El Dorado.

Mining camps sprang up everywhere along the rivers and streams in the northern Sierran foothills in the vicinity of Sutter's Mill on the American River, where James Wilson Marshall had discovered gold in 1847. Long-established San Francisco and recently settled Sacramento, entry points into California and the gold region, doubled and tripled their populations and overnight became tent cities. From all over America and the civilized world, would-be prospectors fanned out from these growing and thriving centers to test their luck at prospecting for gold.

One of the men to arrive in San Francisco from England during the eventful, chaotic summer of 1849 was Cornish-born William Lobb, who viewed the harbor with stupefaction. It was clogged with two hundred unmanned ships—the crews, smitten with gold fever, having abandoned them. Like the argonauts, Lobb (1809–1864) had traveled to California to seek elusive treasure. In the employ of James Veitch, nurseryman of Exeter, England, Lobb came to California specifically to collect the seeds of the many coniferous trees that had been made known to science and European horticulturists by David Douglas and his predecessor, Archibald Menzies. While Douglas had successfully introduced many of these majestic trees into cultivation in England, they were still rare and largely confined to the gardens of members of the Horticultural Society. Money could not buy them, yet an anxious clientele requested them, and the market was continuing to widen.

In response to the demand for exotic plants generally, the Veitch nursery firm had sent paid collectors to various regions of the globe to harvest and introduce novelties into England under their auspices. If the Horticultural Society could sponsor collectors, there was nothing to prevent a commercial establishment from doing likewise, so long as account balances were sufficient to fund the exploits. From his thriving business, James Veitch had money for these ventures, and he viewed them as investments in the future. Commercial rivalry was keen; prestige within the fraternity of nurserymen could be gained through the priority of introducing novelties, and large profits could also be garnered.

William Lobb was the obvious man for the California mission. During the years 1840 to 1844 he had successfully traveled in Brazil for the Veitch firm, and between 1844 and 1848 the costs of his extensive travels through Patagonia, Chile, Peru, Ecuador, Colombia, and Brazil had more than been repaid from sales of the novelties he had introduced.

Lobb's activities, once he arrived in California, are difficult to trace in detail, partly because the Veitch firm did not require that he keep a journal. As a result, there is no firsthand, written recounting of his travels. His collections must have been sufficient to please James Veitch, however, because Lobb was still in California in 1853, ever on the lookout for new and interesting conifers. Events outside of Lobb's knowledge, moreover, were soon to play into his hands.

A hunter, Mr. A. T. Dowd, employed by the Union Water Company of Murphy's, in Calaveras County, California, stumbled upon a grove of enormous trees in the Sierran foothills during the spring of 1852 while in hot pursuit of a large, old grizzly bear. When he returned to camp with his description of the fabulous trees, his campfellows scorned his story as a yarn. Soon after, however, Dowd lured several of his compatriots into following him back to the sheltered grove, where they, too, were astonished by the sheer size and girth of the giant trees (Plate 17). Word of the discovery soon spread from camp to camp, and by 1853 the news had traveled to San Francisco.

Among the other forty-niners who arrived in San Francisco in August of that climactic year was a thirty-six-year-old physician who had been born in New

109. James Veitch (1792–1863) succeeded his father and continued the family tradition established by his father as the proprietor of a nursery business at Mount Radford and Exeter. Under his sponsorship, the Cornishman William Lobb collected plants in South America, California, and the Pacific Northwest. Lobb's mission in the Pacific Northwest was to increase the availability of the conifers in cultivation that had been discovered earlier by David Douglas.

Hartford, Connecticut. Albert Kellogg had received his medical degree some years earlier in Charleston, South Carolina, and had undertaken further studies at Transylvania College in Kentucky. After spending some time in Sacramento upon his arrival in California, Kellogg returned to San Francisco, where he opened a pharmacy and accepted a few patients after hours. Intently interested in natural history, particularly botany, Kellogg collected plants in the region and soon made the acquaintance of other San Franciscans with similar interests. Familiar with scientific traditions on the east coast, Kellogg and six sympathetic colleagues met on April 4, 1853, to establish the California Academy of Sciences and thereby provide an institutional basis for scientific endeavor in the Pacific region.

Word of the association spread through the city, and Kellogg became known to many for his avid interest in plants and his botanical correspondents back east, which included John Torrey in New York and Asa Gray in Cambridge. Because of Kellogg's reputation, plants were frequently brought to him for naming, and his advice was sought concerning the potential medicinal use of Californian plants. Exactly when Kellogg was visited by Mr. A. T. Dowd—or perhaps one of Dowd's acquaintances who had been prospecting in the Sierran foothills southeast of Sacramento—is not known. But Kellogg did bring specimens and fabulous stories of a giant new conifer to a meeting of the fledgling Academy during the summer of 1853.

As chance would have it, one of the guests at the Academy meeting happened to be William Lobb, who was probably spending a brief respite enjoying the civility of city life before heading out again on another collecting foray for the firm of Veitch. Lobb must have paid rapt attention to the exciting news Kellogg brought before the small group gathered for the Academy meeting. Information concerning the whereabouts of the fantastic new tree must also have been shared and the specimens examined with great interest.

After the session at the Academy, Lobb wasted little time in leaving San Francisco for the Sierran foothills in the quest of "botanical gold." The whereabouts of the trees led the Cornish collector to an area now known as the Calaveras Grove, and there Veitch's agent went to work in earnest. Never before in all his travels had he seen such mammoth trees, many of them towering to heights of over three hundred feet. Equally amazing was the girth of the gigantic trunks near ground level; some measured thirty-five feet in diameter!

Wasting no time, William Lobb prepared botanical specimens that included vegetative shoots as well as mature cones, and eagerly gathered seeds, which he carefully packed in his saddle bags. As insurance, should the seeds fail to germinate, Lobb carefully dug two small seedling trees from the gravelly, granitic soil at the edge of the grove, somewhat removed from the outstretched limbs of the parent trees.

Once back in San Francisco, Lobb immediately booked passage to England, selecting the isthmus route across Central America as a time-saving measure. Regardless of how James Veitch might react to his premature return to England, Lobb was not about to entrust this conifer, unattended, to the vagaries of an ocean voyage. He would stake his continued employment and reputation in the gamble.

On Lobb's return to England, news of the remarkable conifer spread rapidly throughout botanical and horticultural circles. The two saplings had survived their long journey and were planted in the Exeter nursery of the Veitch firm. The seeds, too, were delivered to Veitch, and most appeared viable. They were sown immediately and germinated quickly. By the summer of 1854 the Veitch firm was offering seedlings for sale at the rate of two guineas each, six guineas for four, or twelve guineas a dozen; eighteen months after Lobb had returned home, saplings of these potentially giant trees were being planted across England.

The dried specimens and cones Lobb had collected in the Calaveras Grove—material needed to classify and describe the new tree properly—were entrusted to John Lindley at the Horticultural Society. With precise detail Lindley delineated the plant for the scientific community, publishing his description in the issue of the *Gardeners' Chronicle* that appeared on Christmas eve in December 1853. What better Christmas gift could a botanist offer to the British nation than to announce the discovery and introduction into cultivation of an enormous new tree that constituted a new genus named to commemorate the late, recently lamented Duke of Wellington? *Wellingtonia gigantea* was the name Lindley gave to the tree, and the specific epithet *gigantea* indicated both the service that Wellington had given his country and the size of the new tree.

The lead story of the issue included Lobb's firsthand impressions of the Calaveras Grove:

> From 80 to 90 trees exist, all within the circuit of a mile, and these varying from 250 feet to 320 feet in height and from 10 to 20 feet in diameter . . . A tree recently felled measured about 300 ft. in length with a

110. Within a few years of their discovery, the mammoth trees of the Calaveras Grove became a mecca for sightseers, and sections of the trunk of the first tree to be felled were exhibited widely in the eastern United States. Louis Agassiz, the famous Harvard zoologist, stated that "nobody who has any curiosity to see something of the wonders of nature ought to allow the opportunity of seeing a section of one of the big trees of California to pass." This woodcut illustrates one of the activities of an early group of tourists to the Calaveras Grove.

> diameter, including bark, 29 feet 2 inches at 5 feet from the ground; at 18 feet from the ground it was 14 feet; and at 200 feet from the ground, 5 feet 5 inches . . . The trunk of the tree in question was perfectly solid, from the sap-wood to the centre; and judging from the number of concentric rings, its age has been estimated at 3000 years . . . Of this vegetable monster, 21 feet of the bark, from the lower part of the trunk, have been put in the natural form in San Francisco for exhibition; it there forms a spacious carpeted room, and contains a piano, with seats for 40 persons. On one occasion 140 children were admitted without inconvenience. An exact representation of this tree, drawn on the spot, is now in the hands of the lithographers, and will be published in a few days. (Lindley, 1853, pp. 819, 820)

Word of these forest giants moved swiftly across California and all the way to the east coast as well, and the Calaveras Grove became an early California tourist mecca. A cabin was built there to accommodate guests soon after Dowd's discovery, but its few rooms had to be replaced by a full-scale hotel. One of the largest trees was felled, and its stump served as a dance floor, while its trunk was used as a bowling alley. The bark, stripped from the trunk, was sent to New York City, where in 1854 it was exhibited in the large Racket Court of the Union Club at No. 596 Broadway. The bark of another tree, christened "Mother of the Forest," also made the long voyage via Cape Horn for exhibition in New York. Finally, this exhibit was shipped to England and placed on view in London. Since no building in that city could accommodate its full height of 116 feet, the exhibit was dismantled for reassembly in the Crystal Palace at Sydenham, where it remained until December of 1866, when the palace was consumed by fire.

Living plants were also soon established on the east coast of North America. One argonaut, G. H. Woodruff, gathered the seeds from the cones of the giant conifer and filled a snuff-box with the tiny grains. Entrusting the snuff-box to an agent of the Pony Express and prepaying the shipping charges of twenty-five dollars, he forwarded the seeds to the nursery firm of Ellwanger & Barry in Rochester, New York. About four thousand seedlings resulted, and a few were sold to enthusiastic horticulturists on the eastern seaboard from Boston to New Jersey. But the vast majority were sold to nurserymen in England and throughout Europe, where the demand for the plants far exceeded the supply. Avenues of Wellingtonias were planted on estates across the face of Europe. While the prices Ellwanger & Barry could demand were high, they were at wholesale rates compared to those realized by English and continental nurserymen. Nonetheless, Ellwanger & Barry forwarded $1,030.60 to Mr. Woodruff in 1865 as his half-share in the profits from the seed he had supplied.

Now commonly referred to as big tree, giant sequoia, or, in England, Wellingtonia, this magnificent conifer, while neither the world's tallest nor oldest living species of tree, constitutes the most massive-trunked tree on the face of the earth. And today, because of loopholes in the rules of botanical nomenclature, its scientific name (*Sequoiadendron giganteum* (Lindley) Buchholz) continues to reflect its size, though it no longer commemorates the Duke of Wellington. Had Dr. Kellogg had his way, the generic name *Washingtonia* would have been used to commemorate George Washington, Revolutionary War hero and first president of the United States. Instead, after years of heated dispute, the generic name finally settled upon reflects the botanical relationship of the big tree with the redwood (*Sequoia sempervirens* (D. Don) Endlicher), the other forest giant intimately associated with the golden state. Redwoods are endemic to a 500-mile region of wet, fog-drenched forest along the Pacific coast in northern California and southwestern-most Oregon. By contrast, the big tree is restricted to the western slopes of the Sierra Nevada mountains at altitudes between five and eight thousand feet in a belt about 250 miles in length. In this region, winter snow is heavy and the summers are dry and warm. The moisture that does fall during the summer months results from often heavy, late-afternoon thunderstorms.

Only thirty-two scattered groves remain in this region. Soon after their discovery, trees growing on many acres were felled with explosives, wedges, and the woodman's axe. The destruction of groves was frequently complete and enormously wasteful. Ironically, because of the durability of their timber when in contact with soil, these majestic trees were turned into fence posts and vineyard stakes. They were also used for shingles and general building purposes. One individual big tree provided 600,000 board feet of lumber, an amount sufficient for the construction of 80 five-room houses.

The solitary big tree that dominates the slope of the conifer plantation above Bussey Brook in the Arnold Arboretum is a mere infant compared with the mature trees of the Calaveras and Mariposa groves in California. Yet the dimensions of this individual nonetheless provide students and casual visitors to the Arboretum with an idea of the stature and proportions of its congeners that grow three thousand miles distant in the Sierra Nevada. And like its distant brothers and sisters, the Arboretum specimen has a colorful history.

In its present location, the Arboretum specimen appears to be firmly rooted in the glacial soils of the Bussey Brook valley. Yet this tree attained most of its present height and girth growing in a sheltered and favorable site on an estate in neighboring Chestnut Hill. In 1972, the year in which the Arnold Arboretum celebrated its centennial, this conifer, monstrous by Boston standards, was carefully dug, transported to Jamaica Plain, and planted in its present site. Reminiscent of Gardner Greene's ginkgo on the Boston Common over a hundred years before, the gift of the big tree to the Arboretum filled a gap in its collection of conifers and symbolized the beginning of a new century of arboriculture at the Jamaica Plain institution. Never before had a big tree been successfully grown at the Arboretum, and the youthful specimen so carefully transplanted has fulfilled the hopes of donor and recipient alike in flourishing in its new location. If it reaches maturity at three thousand years, one hundred generations of Bostonians will have marked its continued growth and development.

William Lobb, the Cornish collector that made the big tree known to science and first introduced it into cultivation, returned to California in 1854, ostensibly to collect again for the Veitch nursery firm. But his trail became fainter as the years went by, and his shipments of specimens and seeds became sporadic and fewer in number. Lobb, probably recognizing that his active field days were nearing an end, formally resigned his employment with Veitch in the spring of 1857, and in 1858 the Cornish collector took up residence in San Francisco. He died there on May 3, 1864, and was buried at public expense in the Lone Mountain Cemetery. There his remains lay forgotten in the public plot until a group of Californian botanists, led by Alice Eastwood of the California Academy of Sciences, had his rediscovered gravesite moved to a more fitting location on South Ridge Lawn in 1927. In 1940, the Cornish collector's remains were again moved, and they now occupy a crypt at Cypress Lawn Cemetery.

William Lobb was not the only plant collector to experience firsthand the enthusiastic phenomenon of the forty-niners in their quest for wealth in the foothills of the Sierra Nevada. John Jeffrey (1826–1854?), born at Forneth, Clunie parish, in Perthshire, Scotland—a hamlet within a few miles of the birthplaces of Archibald Menzies and David Douglas—also traveled to the Pacific Northwest and California during the halcyon days of gold-rush fever. And like Lobb, Jeffrey was a paid collector sent to reintroduce the conifers made known years before by his compatriots Menzies and Douglas.

The idea of sending a collector to revisit the regions explored by Menzies and Douglas had occurred to George Patton (later Lord Glenalmond), who owned a large estate in Perthshire. On November 22, 1849, a meeting was called at the Royal Botanic Garden, Edin-

111. A view in the hemlock forest on Hemlock Hill in the Arnold Arboretum. While individual trees have been cut for timber over the years, this stand of the native eastern hemlock (*Tsuga canadensis*) is considered to be virgin, dating from pre-Columbian times.

burgh, to which gentlemen interested in the scheme were invited. By May 29, 1850, sufficient funds had been subscribed for the project, and on that day Jeffrey—then employed at the Botanic Garden and highly recommended for the post by Professor J. H. Balfour, the garden's director—signed a contract for a three-year stint as collector. The organization he would represent in the Pacific Northwest and California was known as the Oregon Association of Edinburgh.

Through prior arrangements made by the Association, Jeffrey was taken under the wing of Hudson's Bay Company officials; lodging, logistic support, and hospitality would be made available to the collector at the company's factories and outposts. Jeffrey sailed to York Factory on board the company's ship, *Prince of Wales*, which weighed anchor on June 6, 1850, and arrived in Canada on August 12. The fall, and the prospect of a long winter, faced Jeffrey, who took the opportunity of joining the winter express across the continent so as to arrive at the Pacific coast in early spring.

Jeffrey spent the season of 1851 in the Pacific Northwest retracing the routes of David Douglas and recollecting many of the plants first introduced by him for the Horticultural Society of London. By the fall of 1852 new country to the south beckoned, and new discoveries awaited the agent of the Oregon Association. In the vicinity of Mount Shasta a new pine was located, and later in the fall cones and seeds of the western white pine (*Pinus monticola* D. Don) were harvested in quantity. Jeffrey pine (*Pinus jeffreyi* Balfour), later named in the collector's honor by Professor Balfour, was discovered, and seeds of this distinctive new conifer, commonplace in the Sierra Nevada, were gathered for shipment to Edinburgh. Returning to Fort Vancouver for the winter, Jeffrey packed his booty for shipment to Scotland and entrusted his dried specimens and seed-filled chests to a captain of a Hudson's Bay Company ship that visited the fort on the mighty Columbia River. A few of Jeffrey's packages, sent via Cape Horn, were lost and never reached their destination.

The box containing Jeffrey's harvest gathered during the fall of 1851 did reach Edinburgh, and the seeds of the new conifers arrived in excellent condition. Included were seeds of the incense cedar (*Calocedrus decurrens* (Torrey) Florin), the most distinctive of the new conifers Jeffrey introduced into cultivation. Of narrowly pyramidal or almost fastigiate habit, well-grown specimens of incense cedar became elegant additions to the landscapes of Europe and North America in those regions where climatic conditions favor their growth. Specimens of this stately tree grow under the canopy of hemlocks and pines near the summit of Hemlock Hill in the Arnold Arboretum, and a solitary individual adds interest to a grouping of conifers above the right bank of Bussey Brook. This tree is of easy access, yet following Sargent Trail up Hemlock Hill to view the incense cedars growing there rewards visitors with the calm, vaulted, and cool magnificence of the forested summit. Walking this section of Sargent Trail, one can easily imagine that the incense cedars planted there are growing in their native habitat in the Sierra Nevada and not on a forested hilltop in New England's largest metropolis.

Additional conifers sent by Jeffrey included hemlocks from the Pacific Northwest, congeners of the Canadian hemlock that grows so luxuriantly on Hemlock Hill. Plants of both the western hemlock (*Tsuga heterophylla* (Rafinesque) Sargent) and mountain hemlock (*Tsuga mertensiana* (Bongard) Carrière) now grow in the Arnold Arboretum with the Canadian hemlock (*Tsuga canadensis* (L.) Carrière) and the Carolina hemlock (*Tsuga caroliniana* Engelmann)—the second eastern North American species. Jeffrey collected and sent seeds of both western species during the season of 1852, but by 1853—the third and final year of his contract with the Oregon Association—his field activities had diminished greatly.

Early in October of 1853 the collector arrived in San Francisco and undoubtedly heard the news of the giant trees that had been discovered southwest of Sacramento. By the time Jeffrey arrived in the bustling town, William Lobb had already left with his specimens, seeds, and the two seedling trees to carry the news to England. But the reports may have fallen on deaf ears, as the Scotsman was ill and did not bother to fetch his yearly mail, which at his request had been forwarded to the British Consulate. His last, scanty shipment of seeds and specimens was sent off from San Francisco in early January 1854, and shortly thereafter John Jeffrey, agent of the Oregon Association, disappeared and was never heard from again.

Some claim Jeffrey headed out on another collecting foray and died of thirst on the Colorado Desert. Others suggested that he was killed by hostile Indians, and some subscribed to the idea that a Spanish outcast murdered him for his mules and kit. It was also rumored that gold fever had claimed yet another victim and that the Scotsman was last seen panning for gold on the American River.

Errands in Japan

Temple bells rang, and signal fires fueled with the resin-rich timber of coastal pines sent columns of smoke skyward to alert the populace along the immediate coastline. One by one, pyres strategically placed farther along the rugged shore were torched in succession until the pungent smoke filled the air and stung the eyes of the inhabitants of Edo, twenty-seven miles inland at the head of the bay. Gongs and bells also reverberated throughout the metropolis as a further, audible warning of the approach of four black vessels flying the flag of a foreign nation. Two of the monstrous ships belched smoke even blacker than their hulls or the smoke of the fires signaling their arrival.

By midnight the atmosphere had cleared of smoke and haze. As the four ships rode at anchor at the mouth of the bay, the men on deck viewed the surrounding countryside with astonishment as it became illuminated by the bluish light of a giant meteor slowly arching across the heavens on a northeasterly course. Coinciding as it did with the arrival of the black ships, this astral phenomenon appeared to the Shogun in Edo as an omen. To scientifically minded Commodore Matthew C. Perry on board his flagship, *Susquehanna*, the event was of strictly astronomical interest; no thought of an omen crossed his educated mind. However, to many of the crew on board the four ships then constituting the

112 and 113. Anonymous Japanese artists were quick to capture caricature-like portraits of Commodore Matthew Perry (1794–1858) and Commander Henry A. Adams, captain of the fleet, once negotiations between the Americans and Japanese officials had been inaugurated. After the Kanagawa Treaty had been signed, Commander Adams sailed for America on the *Saratoga*, where a mail steamer was intercepted and the historic document was transmitted to Washington, D.C.

squadron of the United States Expedition to the China Seas and Japan, the meteor was optimistically viewed as a sign of ultimate success in their Japanese mission.

It was July 8, 1853, and the expedition was already eight months into its errand to the Pacific. Topmost on the agenda was a visit to Japanese waters and the transmittal of a letter to the emperor from Millard Fillmore, President of the United States, proposing diplomatic relationships between the two nations; it was hoped that a treaty would eventually open the isolated nation's ports to intercourse with the United States.

American ships already frequented the North Pacific in hot pursuit of oil-rich whales, and increasing numbers of American steamers were expected to cross the Pacific from Californian ports en route to Asian emporia in the years ahead. With the Mexican Cession of 1848, more than 500,000 square miles of territory had been added to the Union, and the United States—spanning the North American continent from the Atlantic to California on the Pacific—had achieved its "manifest destiny." Following the Treaty of Nanking in 1843, the frontier of commerce had been pushed across the Pacific to China, but remote, long-isolated Japan lay in the path of the most direct sea routes. American ships were bound to

114. This lithograph from a drawing by German-born Peter Bernhard Wilhelm Heine, Perry's master's mate, illustrates the banquet given on board the U. S. S. F. *Powhatan* to honor the Japanese officials engaged in the negotiations with Commodore Perry. This illustration was included in the three-volume report of the expedition, published by the United States government in 1856.

enter Japanese waters in search of fuel and provisions, and the victims of inevitable shipwrecks would need friendly assistance. It was time to ensure they would be received with respect and cordiality.

While Commodore Perry awaited a response from the frenzied Japanese officials, survey parties from each of the four American ships boldly took soundings and carefully charted the shore of the lower portions of Edo Bay. After four days of negotiations through intermediaries, a time was finally fixed when Perry might personally deliver President Fillmore's letter to a high-ranking imperial representative. Perry took that opportunity to address a letter of his own to the emperor. In it he indicated that, because of the gravity of the issues raised by the President's proposals, he understood that a reply would require time. As a consequence, he would leave Japanese waters for the time being, but would return the following spring to receive the emperor's response. He promised to return, moreover, with a larger squadron, and before departing—in an early example of gun-boat diplomacy—the four black ships moved farther up, rather than down Edo Bay, as the survey parties continued their strenuous but all-important work.

Perry was true to his word, but returned to Japan earlier than expected. Rumors circulated that a Russian squadron was expected to visit Japan with identical objectives. To preclude the possibility of a successful Russian mission, Perry's fleet left its winter anchorage in Chinese waters at the end of January. Toward the middle of February 1854, his squadron—three steam frigates and six vessels under sail—entered Edo Bay. Ten days of haggling with Japanese officials ensued concerning where the squadron should anchor and where negotiations could take place. In these dealings—all conducted

115. One of the high points of the festivities surrounding the agreement between Perry and the Japanese is illustrated in this lithograph entitled "Delivering of the American presents at Yokuhama," after a drawing by P. B. W. Heine. Many agricultural implements were included among the gifts, and demonstrations of their various uses were staged. Also included was a fully operational miniature steam engine, and a railroad was erected to the surprise and amazement of the Japanese dignitaries. Note, also, the telegraph line.

through intermediaries—Perry held firm in his resolve to be received by his Japanese counterpart as near to Edo as possible. For their part, the Japanese tried in vain to urge Perry to anchor his squadron as far distant as possible from the capital city.

In the end, Commodore Perry proved to be the master strategist in this diplomatic game of chess. No opportunity was lost by the American seamen in continuing the soundings of Edo Bay that had been initiated the previous summer, and in a bold move Perry ordered the squadron to anchor nearer to Edo. Aware of the commodore's wishes, the Japanese quickly agreed that negotiations could be held on shore at Yokohama, directly opposite the new anchorage. Finally, on March 8, with an escort of about 500 officers, marines, and seamen, Perry went ashore to initiate the long-sought negotiations. On March 31 the Treaty of Kamagawa was signed, and Japan took the first steps toward reopening its doors to the West after more than two centuries of isolation.

During and upon completion of the successful negotiations, a period of diplomacy by revelry and then celebration ensued. As outward tokens of friendship and respect, gifts were exchanged, and the commodore hosted an extravagant banquet for the Japanese aboard the *Powhatan*. Toasts were drunk with champagne, whiskey, and sake, and the entertainment included an authentic minstrel show. On shore near the hastily constructed treaty pavilion, agricultural implements brought from America were demonstrated, and a telegraph line linking Yokohama and Kamagawa demonstrated American technology. As further evidence of American industry, a fully operational scale model of a steam engine with tender and passenger car, complete with

116. A native of Fairfield, Connecticut, Charles Wright (1811–1885) was a veteran botanical collector when he joined the United States Surveying Expedition to the North Pacific Ocean in 1853, which sailed under the command of Cadwallader Ringgold. John Rogers, second in command, was given full authority for the expedition in August of 1854, when naval physicians determined that Ringgold was not fit to command.

track, provided the Japanese diplomats with their first train rides and simultaneously added to the carnival-like atmosphere of the proceedings.

In return, the Japanese fêted members of the squadron and tried not to be outdone in their expressions of friendship and hospitality. Opportunities were even provided for firsthand observations of Japanese life and customs. Dr. James Morrow (1820–1865), in charge of the American agricultural exhibit, and S. Wells Williams (1812–1884), Protestant missionary in China who joined the expedition in Canton to serve as Chinese translator, took every opportunity for shore leave to collect plants in the vicinity of Yokohama. Their modest but interesting collection of dried specimens eventually came into the hands of Williams's boyhood friend, Asa Gray (1810–1888), then Fisher Professor of Natural History at Harvard College and Director of the Botanic Garden in Cambridge. There the Harvard botanist examined and named the plants. A brief account of the botany based on the collections was prepared for inclusion in the official three-volume report of the Japan expedition.

Williams and Morrow's specimens were the first Gray had received directly from Japan, yet in 1856, the year Perry's report of the Japan expedition was published, Gray received a far larger collection of Japanese plants. This second harvest had been gathered by the veteran field botanist Charles Wright, who had a long-standing working relationship with Gray. Wright had previously supplied the noted botanist with rich, beautifully prepared collections from the Texas–Mexican boundary region. His new haul from Japan resulted from his participation, at Gray's urging, as botanist on the United States Surveying Expedition to the North Pacific Ocean during the years 1853 to 1856. For no sooner had Perry been successful in negotiating a treaty with the Japanese than Congress saw the necessity of sending another American squadron to the area to test the effectiveness of the infant treaty and to continue the all-important navigational surveys of Chinese and Japanese waters.

By the time Wright returned home to Fairfield, Connecticut, and traveled to Cambridge with his Japanese collections, Asa Gray was at the peak of his long and distinguished career, having achieved the status of dean of American botanists. If a scientific record based on Wright's collections was to be prepared for an expedition report, Gray was the only American qualified for the job. Overworked and besieged with requests for assistance from governmental agencies, foreign and American colleagues, and amateur naturalists alike, Gray nevertheless took time from his research on the distribution of

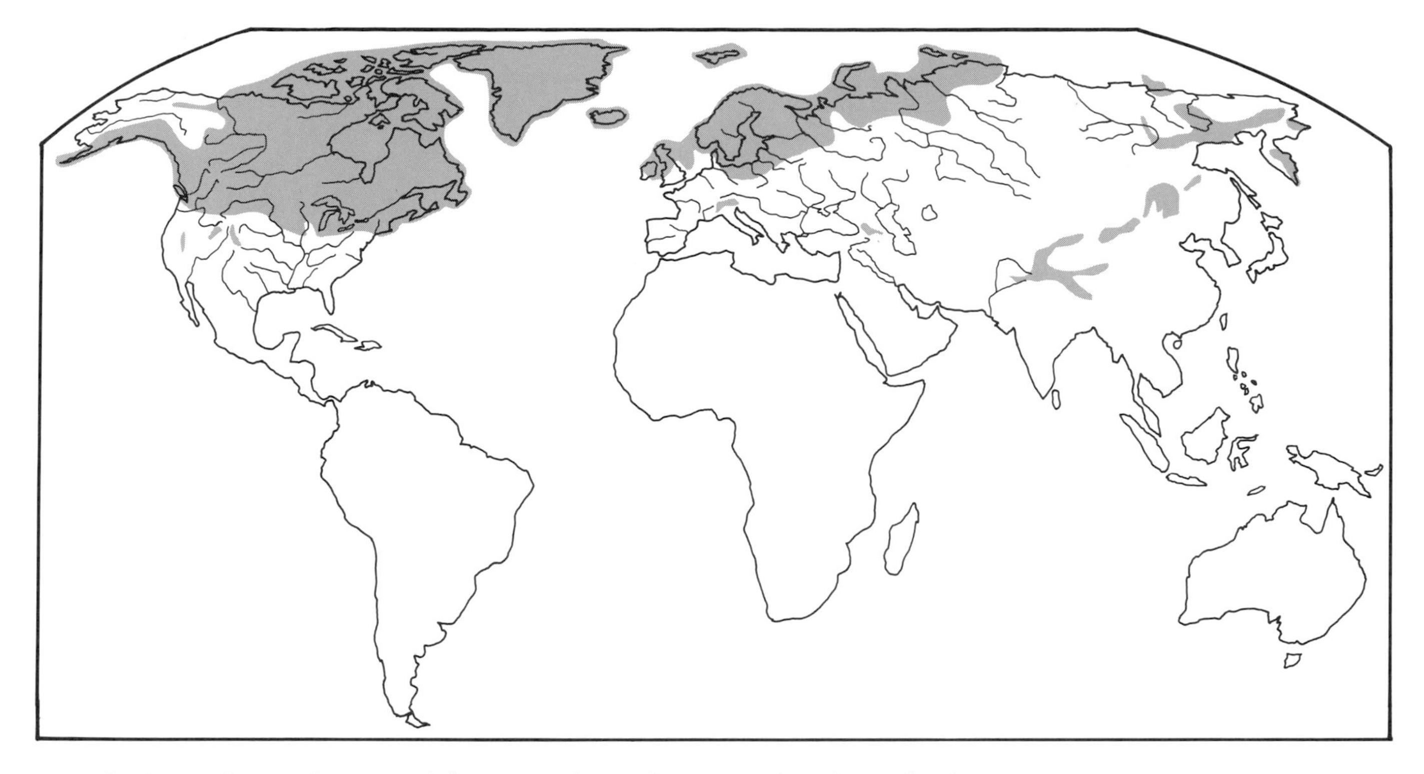

117. Sketch map showing the extent of glaciation in the Northern Hemisphere during the Pleistocene.

North American plants to sort through Wright's Japanese plants. To his astonishment, he recognized many of the exotic plants collected in far-off Japan as identical to those growing in the forests and fields of eastern North America. Here was proof of a floristic relationship Gray had suspected ever since he had reviewed Siebold and Zuccarini's elegant *Flora Japonica* years before. Even Williams and Morrow's collection had included specimens that caused a sense of déjà vu in the mind of the experienced botanist.

Temporarily setting aside his other work, Gray delved more deeply into Wright's Japanese plants. By 1859 a manuscript was ready for publication, and the results of his scholarship could be presented personally to members of Boston's American Academy of Arts and Sciences. Based on Wright's collections, Gray had evidence that the flora of Japan was extraordinarily similar to that of eastern North America. Many of the same genera, and even identical species, were represented in these two widely separated regions. Yet these same plants were not known to occur in the intervening regions of North America or Eurasia. How might this phenomenon of widely disjunct distributions be explained?

In seeking an explanation that might account for floristic similarities between Japan and eastern North America, Gray drew upon geological evidence of events that had occurred during the ice ages of the Pleistocene. He also astutely incorporated the concept of common origin for the two floras and thereby borrowed a key tenet from the theory of evolution through natural selection that his frequent correspondent, Charles Darwin, had proposed and discussed in his letters to Gray. Darwin was working on his "big book" at the same time Gray was working on the Japan problem, and Darwin's *Origin of Species,* a work that would forever change humankind's perception of nature and itself, was also published in 1859.

In brief outline, Gray hypothesized in his 1859 publication that before the glacial epoch the flora of the North Temperate Zone had been relatively homogeneous, extending in a more or less undisrupted belt across North America and Eurasia. During the Pleistocene, when the glaciers began their advance southward from the polar region, great areas that had supported this flora eventually became covered by glacial ice. As the glaciers advanced, the temperate flora migrated southward—where that was possible—keeping pace with the south-

118 and 119. The vegetative morphology of these two species of *Pachysandra* illustrates the close similarities of two taxa that occupy widely separated ranges. The Allegheny spurge (*Pachysandra procumbens* Michaux, shown above) is native to eastern North America from Kentucky southward into Florida and Louisiana, while the Japanese spurge (*Pachysandra terminalis* Siebold & Zuccarini, below) is widespread throughout Japan and also occurs in central China. The American species was first described by André Michaux and was brought into cultivation by John Fraser in about 1800. The Asian species was introduced from Japan in 1882 and has become widely cultivated throughout eastern North America as an evergreen groundcover. Approximately 120 genera of plants exhibit this pattern of distribution, sometimes with only one species in eastern North America and one in eastern Asia, other times with several species in one area and one or a few in the other. In still other instances numerous species occur in both areas.

Plate 15 (top). The moutan or tree peony (*Paeonia suffruticosa*) has been called the king of flowers, and its hundreds of varieties originated primarily in Chinese gardens. Robert Fortune was responsible for introducing many forms into cultivation in Europe, although the first tree peony had been received at Kew as early as 1787, as the result of a request for the plant made by Sir Joseph Banks to a Mr. Duncan, a medical officer of the East India Company stationed in China.

Plate 15 (bottom). *Rhododendron fortunei* was one of Robert Fortune's introductions that was first described by the botanist John Lindley in the pages of the *Gardeners' Chronicle*, a highly successful and popular weekly newspaper published in London. Lindley based his description on plants raised from seed in the nursery of a Mr. Glendining of Turnham Green. At the time of the rhododendron's discovery, Fortune was able to report that only one other species (aside from those of the azalea group) was known from the Chinese flora. Today, the genus is known to be represented in China by upwards of three hundred species, many of which are now in cultivation.

Plate 16. Among the deciduous conifers, the golden larch ranks as the most beautiful. Four mature specimens in the Arnold Arboretum grow near species of the true larch.

ward displacement of the temperate climate. In areas where barriers prevented southward migration, the flora was severely decimated or was completely destroyed as the climatic conditions deteriorated.

The displaced pockets of temperate flora that persisted during the full glacial periods eventually recolonized the ice-scarred areas as the glaciers melted and retreated northward. But in much of western North America and Eurasia the climate was sufficiently altered by a series of other geologic phenomena—mountain building in particular—that the old temperate flora could not reclaim its former territory without drastic adaption through evolution. In these regions new species evolved to replace the preglacial, or Arctotertiary, flora. As a consequence, the Arctotertiary species became discontinuous in the postglacial world. In eastern North America and eastern Asia, the old Arctotertiary flora had either escaped the vicissitudes of the Pleistocene or was able to reclaim its former territory without drastic adaptation. Thus, the similar floras in the two regions today constitute relicts of the preglacial flora that once encircled the globe.

Asa Gray's hypothesis and Darwin's theory of evolution through natural selection were mutually compatible and supportive, and Gray became Darwin's leading American proponent in the debates that raged concerning Darwin's theory. Through his paper based on Wright's Japanese collections Gray also firmly established botany as a scholarly discipline on American soil.

Asa Gray's recognition of similar floras in eastern Asia and eastern North America was also destined to have an enormous practical impact on horticulture in the latter region. Evidence would soon accumulate in American gardens that many plants native to Japan could also grow in eastern North America, where their congeners were native. Shortly after Japan had been opened to limited trade under the terms of a new treaty hammered out in 1858 by Townsend Harris, first United States consul at Shimoda, a diverse array of Japanese plants began to enter the nursery trade in the West. In this traffic Americans joined European plantsmen in the race to exploit the rich Japanese flora and to send home a bountiful harvest of Japanese plants. American horticulture, like botany, was coming of age.

Untold Windfalls

Kousa dogwoods, hiba arborvitaes (*Thujopsis dolabrata* (Linnaeus f.) Siebold & Zuccarini), Fuji-yama rhododendrons (*Rhododendron brachycarpum* G. Don), seedlings of the sawara cypress (*Chamaecyparis pisifera* (Siebold & Zuccarini) Endlicher), crabapples with semidouble flowers (*Malus halliana* Koehne var. *parkmannii* Rehder), and umbrella pines were among the first Japanese plants that arrived in Boston directly from Japan. F. Gordon Dexter, returning to New England from the Orient in 1861, agreed to take responsibility for this ligneous cargo on the seventy-day passage from Yokohama to Boston and to personally deliver the plants to Francis L. Lee of Chestnut Hill. This unique collection had been carefully assembled and established in Wardian cases for transport to Boston by Dr. George Rogers Hall, then a resident of Yokohama.

During Dexter's absence from the states, Confederate artillery had bombarded Fort Sumter in the harbor of Charleston, South Carolina, and the Civil War had erupted. Francis Lee, about to respond to President Lincoln's call for troops by enlisting in the Union Army, was forced to entrust the nurture of the totally new plants to someone other than himself. He chose Boston's most celebrated horticulturist, Francis Parkman, his friend, former Harvard classmate, and Jamaica Plain neighbor.

When Parkman returned to Boston in the fall of 1846 after a summer of arduous adventure exploring the Rocky Mountain region, he was ill and physically exhausted. He had temporarily lost his sight—a recurring impairment that alternated with periods of poor vision—and suffered from headaches and an injured knee that severely restricted his mobility for the rest of his life. With assistance from his sister, the historian nonetheless began to dictate *The Oregon Trail* and plan for the numerous other historic accounts he would eventually author.

Turning to horticulture as an avocation, Parkman directed, from his wheelchair, a small grounds staff at his summer home on the shore of Jamaica Pond. Their labors and Parkman's plans transformed the three-acre site into a horticultural wonderland. The collection of roses alone consisted of over one thousand plants, and other horticultural novelties vied for the admiration of visitors. Lee knew that his new Japanese plants would be pampered under Parkman's supervision, and their growth and horticultural attributes would be duly noted and communicated at meetings of the Massachusetts Horticultural Society.

One of the evergreen Japanese conifers, in particular, caught the eye of Parkman, who probably gave it the protection of a greenhouse before deciding to test its hardiness out-of-doors during a New England winter.

120. Francis Parkman (1823–1893) contributed greatly to the development of American horticulture. Best known as a rosarian, Parkman published his *The Book of Roses* in 1866, but the diversity of plants he grew in his garden testified to broad horticultural interests. It was in Parkman's garden on the shore of Jamaica Pond near the present-day Arnold Arboretum that many species indigenous to Japan were first successfully cultivated in North American soil.

Parkman may also have coined its common name, umbrella pine (*Sciadopitys verticillata* (Thunberg) Siebold & Zuccarini), to denote the spokelike arrangement of its glossy green needles. Not a pine at all, this unique tree has been placed by botanists in its own family—the Sciadopitaceae—and, like the ginkgo, it has no close living relatives. Miraculously, it too has survived from the remote geological past.

Fossils provide evidence that these trees once grew over a wide area of Eurasia and formed an important component of European forests. Brown coal deposits in Germany from the mid-Tertiary are frequently characterized by the remains of the leaves of umbrella pine, attesting to its former abundance. It also once grew in Greenland and Canada, but today the single, extant species is restricted in nature to forests occurring between three and six thousand feet in elevation in the mountains on the Japanese islands of Honshu, Shikoku, and Kyushu.

The Japanese umbrella pine has proven hardy in the environs of Boston, and a grove of fifty-year-old trees has firmly established the species in the collections of the Arnold Arboretum. These individuals produce cones on a nearly annual basis, and as they mature the lower limbs die, exposing the trunks of the trees to view. Young trees rarely produce cones and usually retain their lower limbs; consequently, the cinnamon-brown bark of the trunk is obscured by the dense whorls of the dark green, almost plasticlike leaves. As young trees, umbrella pines grow slowly and symmetrically, forming shapely, evergreen cones that are highly prized as specimen trees in the gardens of those fortunate enough to grow them. When plants can be located in the nursery trade, the prices they command reflect the esteem in which they are held. At the Arboretum, a younger generation of these trees accent the plantings in front of the Hunnewell building and illustrate their landscape use.

Another Japanese tree that Parkman was the first to grow in North America along the shore of Jamaica Pond would have aroused the interest of Asa Gray at the Harvard Botanic Garden in Cambridge. It is likely, moreover, that Parkman was aware of Gray's interest in the flora of Japan. Word of the debates at the meetings of the Cambridge Scientific Club and American Academy of Arts and Sciences that had raged between Gray and Professor Louis Agassiz in 1859 had surely reached the historian's ears. These spirited discussions had been spawned by Gray's hypothesis concerning the close relationships of the floras of eastern North America and Japan and Darwin's theories of evolution. Gray had argued for the descent of species in the two regions from common ancestors, while Agassiz had attempted to defend the multiple origins of related forms. Gray's reasoning, based on his empirical assessment of factual evidence,

121. A view of Francis Parkman's house and garden on the shore of Jamaica Pond. The house is no longer standing, and although the land is now part of Boston's Emerald Necklace, there is no trace of the garden that once flourished on the site.

122. The umbrella pine constitutes one of the finest conifers available for landscape use in eastern North America. Young plants are slow growing and achieve a symmetrical, conical habit.

won the day and paved the way for the debate over Darwinism, which would occupy scientific center stage in the decades ahead.

When the Japanese kousa dogwood (*Cornus kousa* Hance) first came into flower on the shores of Jamaica Pond in the middle of June, Parkman was confronted with the same sense of déjà vu Gray had experienced when he sorted Charles Wright's brittle, dried specimens of Japanese plants in the herbarium. The morphological similarity of the Japanese species with the flowering dogwood (*Cornus florida* L.) of the eastern United States, which had flowered earlier in May, became immediately evident. Parkman was growing in his Jamaica Plain garden two closely related species from opposite sides of the world (Plate 18). Here was living proof of the distributional phenomenon Asa Gray had recognized and had gone far to explain. The similarities between the two species, moreover, could be comprehended based on the concept of descent from a common ancestor in the remote past.

Like its eastern North American congener, the ornamental attributes of the kousa dogwood depend largely on the four white, leaflike bracts that subtend the small, tight clusters of true flowers. Held erect on long pedicels, the abundantly produced clusters and their associated bracts stud the branches of the shapely trees in June and appear like thousands of miniature, creamy-white pinwheels hovering above the trees' outstretched branches. The bracts contrast abruptly with the bright green of the leaves, yet some individuals produce such an abundance of flower clusters that the foliage is almost completely obscured from view by the associated bracts. These disease and pest-resistant trees are one of the most valued ornamental subjects available for planting wherever a small tree is required.

While the Asian and North American dogwoods are undeniably related, they differ from one another in several ways, and each is classified as a distinct species. Among other differences, the fruits of kousa dogwoods differ from the individually borne seeds of the flowering dogwood, each of which sports a bright red seed coat. By contrast, the seeds developed from each flower cluster of the kousa dogwood are embedded in the flesh of a red, strawberry-like compound fruit. The weight of the fruit eventually pulls the initially erect pedicel downward, and the mature fruits hang suspended along the leafy branchlets.

One very plausible explanation for the difference in fruit types between the Oriental and Occidental species relates to their means of dispersal in nature. In the for-

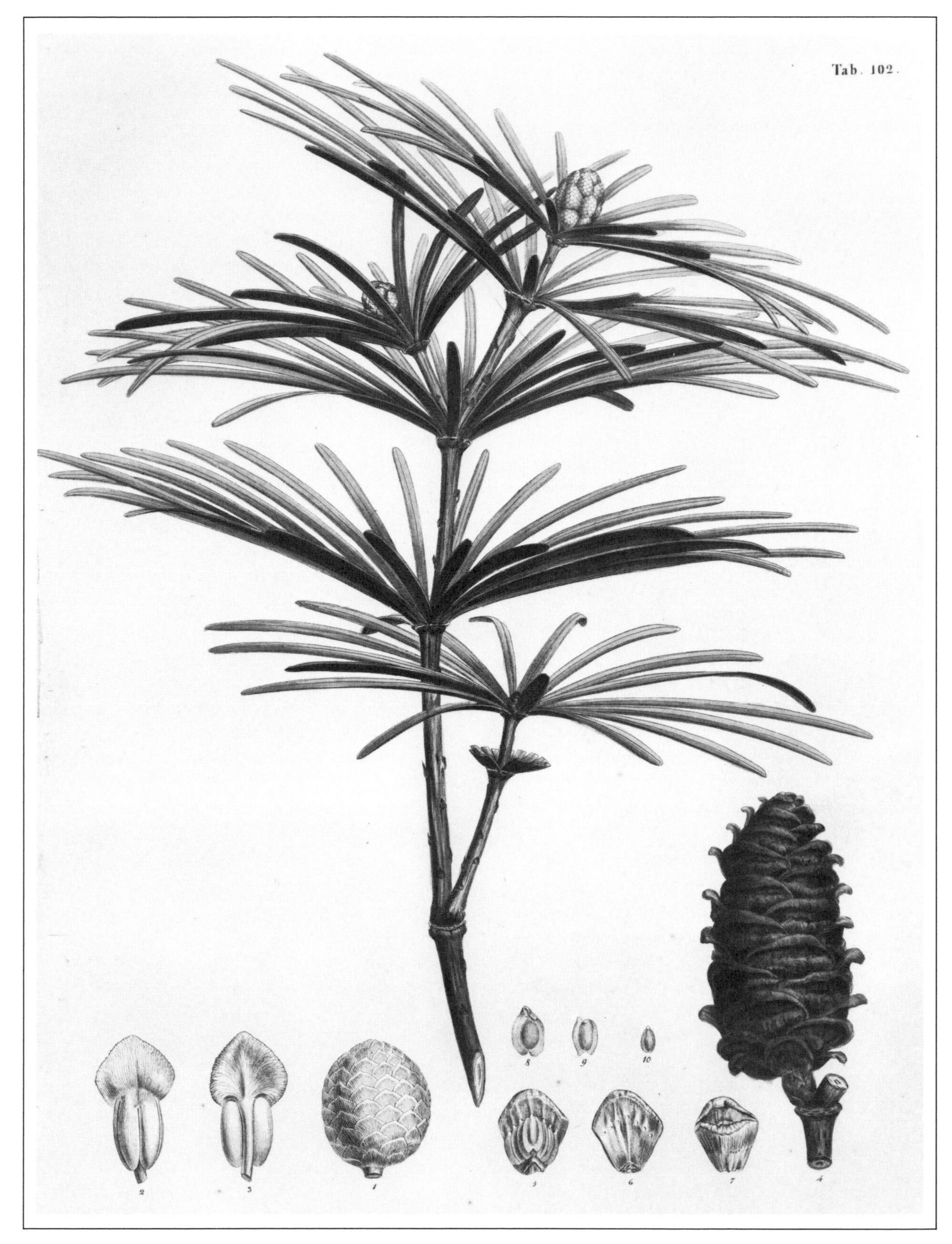

123. This botanical drawing of the umbrella pine is from Siebold and Zuccarini's *Flora Japonica* and clearly illustrates the whorled arrangement of the so-called leaves.

ests of eastern North America the small fruits of the flowering dogwood are the right size for birds, which eat them and then disperse the seeds after they have passed unharmed through their digestive systems. In Japan and China, where the kousa dogwood is now known to occupy a wide range, monkeys, particularly macaques, are denizens of the same regions, and the larger, bright red, strawberry-like fruits appeal to these arboreal acrobats. These seeds also pass unharmed through the animal's digestive system and are dispersed prepackaged with primate fertilizer. Had New World monkeys occurred in the same regions as the flowering dogwood and not been blind to the color red, our native species might have evolved fleshy compound fruits similar to those of the kousa dogwood. Conversely, had monkeys not occurred in Asia, kousa dogwood fruits would probably be simple and their seeds dispersed by birds.

In April of 1862, a year after Francis Parkman received his horticultural windfall, a letter was published in *The Horticulturist or Journal of Rural Art & Rural Taste*, one of the leading horticultural periodicals of the day, which had been founded by Andrew Jackson Downing. Under the title of "Japanese Trees," the notice was signed by Parsons & Co., Flushing, March 20, 1862, and the column began:

> A few days since, while sitting in our office, there walked in a gentleman, with an intelligent face, and frank, pleasant manner, introducing himself as Dr. Hall of Japan, whom we had for some time known by reputation . . . He informed us that for the past two years he had resided in Yokohama, and being greatly interested in trees and plants, had, for his own amusement, collected in his garden all of any interest which Japan contained . . . Expecting to return home this year, he had also collected a large quantity of seeds of trees and plants, many of them unknown either in Europe or this country. These plants and seeds he had brought with him, except some six Wardian cases yet to arrive, and proposed to place them all in our hands for propagation and culture. (Parsons and Co., 1862, p. 186)

Relating the arrival of the Wardian cases, the article continued,

> If you have ever seen the eagerness with which a connoisseur in pictures superintends the unpacking of some gems of art, among which he thinks he may possibly find an original of Raphael or Murillo, you will have some idea of the interest with which all, both employers and propagators, surrounded those cases while they were being opened. (Parsons and Co., 1862, p. 186)

Among the "originals" transported to Long Island in the glazed cases were the first plants of additional Japanese trees and shrubs that are now mainstays in landscapes in New England and across North America and Europe. Included were plants of the familiar kobus magnolia (*Magnolia kobus* De Candolle) and the now more ubiquitous star magnolia (*Magnolia stellata* (Siebold & Zuccarini) Maximowicz, in America first known and offered for sale as *Magnolia halleana* Parsons), both prized for their abundantly produced white or pink flowers (Plate 19, top left) that herald the arrival of spring. In the Arnold Arboretum both of these precocious flowering species grow near the Hunnewell Visitors' Center and in April annually provide one of the earliest floral displays of spring.

Ten garden forms of the sawara cypress (*Chamaecyparis pisifera* (Siebold & Zuccarini) Endlicher)—each selected and maintained by Japanese horticulturists—and plants of the beautiful hinoki cypress (*Chamaecyparis obtusa* (Siebold & Zuccarini) Endlicher) were exposed to the fresh North American air from within the humid confines of the Wardian cases. Saplings and seeds of a new elmlike tree, the Japanese zelkova (*Zelkova serrata* (Thunberg) Makino)—destined a century later to be widely planted in American cities and towns as a replacement for native American elms ravaged by Dutch elm disease—provided living evidence of new trees to be found growing in Japan. Seeds of Japanese umbrella pines filled a small sack, and several horticultural forms of Japanese maples (*Acer palmatum* Thunberg), Japanese wisterias (*Wisteria floribunda* (Willdenow) De Candolle), and many others, including the raisin tree (*Hovenia dulcis* Thunberg), rounded out the shipment. While most had been described in the floristic accounts of Thunberg in the eighteenth century or later by Siebold in the nineteenth century, none had ever before been available to North American horticulturists. To the zealous plant propagators of the Parsons firm and American horticulturists of succeeding generations, several have become the botanical equivalents of canvasses by Raphael and Murillo.

Ironically, among the horticultural treasures Hall brought back to the United States, one plant in the shipment was to become more comparable to the Norway rat brought to America by the first European explorers than to any work of art by an Old World master. This partic-

124. The flowers of Hall's honeysuckle (*Lonicera japonica*) produce a cloyingly sweet fragrance, which perfumes large regions of the South during the late spring and early summer. Despite its sweet fragrance, the plant has become a pernicious weed that has invaded thousands of acres of woodland along the eastern seaboard of the United States from New Jersey southward.

125. The American consulate and port of Shanghai as they appeared in the 1850s when George Rogers Hall was a resident of the American concession.

ular plant proved so well adapted to the climate and growing conditions of a portion of the eastern United States that it has become a pernicious weed that plagues foresters and naturalists throughout much of the Southeast. Initially referred to by horticulturists as Hall's honeysuckle (*Lonicera japonica* Thunberg), this vigorous, twining climber is now more frequently known as Japanese honeysuckle, or simply honeysuckle. This last name is most usual, particularly in the regions where the plant has invaded thousands of acres of woodlands on the Piedmont and Coastal Plain and literally overwhelmed the native vegetation. For many generations of Southerners its flowers have perfumed the air and provided drops of sweet nectar to be sucked from the bases of its tubular corollas. It has also provided untold hours of sweat and frustration on the part of those who have attempted, most often in vain, its eradication. So widespread and pervasive has it become that only its name suggests it Japanese origin.

If he were alive today, Dr. Hall might be satisfied that his name has generally become disassociated with this plant. He would undoubtedly have preferred to leave the Japanese honeysuckle in his Yokohama garden. Little did he or the staff of Parsons Nursery realize that the woodlands of much of the eastern United States from Pennsylvania southward would be forever changed by offspring of a plant carefully transplanted from a Wardian case to a nursery row on Long Island in March of 1862.

At North Farm, Dr. Hall's Rhode Island estate on the shores of Narragansett Bay—now a condominium development—a venerable, multistemmed Japanese yew (*Taxus cuspidata* Siebold & Zuccarini) planted by Dr. Hall on his return from Japan dominates one corner of the old garden. This tree is now over 30 feet tall and over 130 feet in circumference. A bronze plaque at its base indicates that it ranks as the first Japanese yew to be planted in North American soil. It was certainly not the last, for in northern regions of the United States this species has become the signature shrub of the modern-day urban and suburban landscape. While the Japanese honeysuckle has invaded southern woodlands, the Japanese yew has achieved the status of the quintessential landscape shrub in northern cities and towns.

Japanese yews constitute one of the mainstays of the American nursery industry. Countless thousands of balled and burlapped individuals annually fill the sales areas of bona fide nurseries as well as hardware stores, supermarkets, and other retailers who attempt to capture a part of the spring market for bedding annuals, perennials, and landscape trees and shrubs. Plants of this species that have been used in foundation plantings alone probably number in the millions. All too frequently, yews are yearly clipped and shaped with pruning shears and hedge clippers into rounded balls, boxlike cubes, and cones. All across New England—like chessmen standing sentinel at entryways or guarding gravesites in suburban cemeteries—the Japanese yew is omnipresent and contributes to the monotonous repetition of suburbia.

When not pruned to within an inch of its life but allowed to grow and develop naturally, the Japanese yew assumes a pleasing, widely branching habit. Its growth rate is slow, but a good size will eventually be achieved unless pruning shears are resorted to. Its lustrous, dark green needles contrast with the abundantly produced seeds, each embedded in a bright red, fleshy aril-like covering, adding to the ornamental aspect of the plant (Plate 19, top right). Another attribute that recommends its judicious landscape use is its tolerance of light shade. When a dark evergreen is needed in such a location, the Japanese yew should rank high on the list of candidates.

But who was Dr. Hall, recently of Japan, who brought the Japanese yew to North America, who sent cases of exotic plants to his friend Francis Lee, and who generously offered the Parsons Nursery horticultural treasures from his Yokohama garden? A native Rhode Islander, George Rogers Hall was born near Bristol in March of 1820 and graduated with the class of 1842 from Trinity College in Hartford, Connecticut. After graduation, Hall matriculated with the Harvard Medical School class of 1846. Once his medical education was completed, he sailed for China and the new opportunities that awaited enterprising Yankees in the wake of the Opium War. Settling in the foreign compound in Shanghai, Hall formed a partnership with another physician, John Ivor Murray, and in 1852 the two medics opened the Seamen's Hospital, with beds for twelve patients.

As the number of foreign vessels calling at Shanghai increased, Hall's medical practice flourished and the hospital staff was enlarged to include another physician and an apothecary. But despite the influx of American and European seamen requiring medical attention, the venture realized only small profits. Compared with the fortunes being made in commercial ventures, the hospital business hardly repaid the efforts involved.

Leaving the hospital and his medical practice behind, Hall joined with two friends in a business enterprise. His new partners were Edward Cunningham and David Oakes Clark, both from Milton, Massachusetts. These young New Englanders had been encouraged to enter the

126. A painting by a Chinese artist of George Rogers Hall's Shanghai residence in the coastal city where Hall founded the Seaman's Hospital.

China trade by an old hand in the business, Robert Bennett Forbes, long-time Milton resident and partner in Russell & Company.

Hall's decision to give up his medical career was undoubtedly a difficult one, yet pressing financial need forced his hand. In 1850 he had returned to the United States to marry Helen Beal, daughter of a Kingston, Massachusetts, lawyer. Together they returned to Shanghai, and in the space of four years three sons were born to the young physician and his wife. With the Taiping Rebellion looming on the horizon—an internal revolt that nearly saw the overthrow of the Manchu dynasty, one which was fueled by government corruption and a socioeconomic decline that had worsened in the wake of the Opium War—Mrs. Hall left Shanghai with their three sons and returned to America in 1854. Sadly, the youngest son, George Rogers, Jr., died on board ship. With a young family to support, Hall decided to remain in China long enough to make his fortune before returning to the States to rejoin his wife and family.

It was at this time that Dr. Hall turned to business interests, and with his friend Cunningham first visited Japan on Cunningham's schooner yacht, the *Halcyon*. Dealing in fine Chinese and Japanese curios—porcelain, lacquer work, bronzes, jade, and ivory—brought considerable profits, but even more money could be made through speculation in gold and silver. A considerable fortune was accumulated in this way, and toward the end of his Oriental sojourn Hall decided to establish himself in Yokohama, where direct access to the recently opened Japanese market was possible.

In Japan, George Rogers Hall's latent interest in plants emerged, and he diligently set about assembling a collection of Japanese species in his Yokohama garden. Many plants were obtained from Philipp Franz von Siebold, who had returned to Nagasaki and his beloved Japan in 1859, the same year Hall moved to Yokohama. Robert Fortune visited Dr. Hall when he traveled to Yokohama in 1860, and arrangements were made whereby Fortune's collections could be held in the physician's garden until they could be planted in Wardian cases for shipment to England. Without a doubt, the Scots collector shared some of his collections with the Yankee physician turned trader and plantsman.

And so it was that a first New England-bound shipment of Japanese plants arrived in Boston in 1861, and a second larger consignment arrived in 1862 when Hall returned home to Rhode Island to be reunited with his family. George Rogers Hall was the first American to send a wide assortment of Asian plants to eastern North America, where most were destined to join their New World relatives in landscapes across America.

127. George Rogers Hall (1820–1899) of Bristol, Rhode Island, the physician turned trader who first sent living plants from Japan directly to New England.

CHAPTER FIVE

Coming Full Circle

The View from the Top

At ten o'clock on the beautiful, clear morning of Tuesday, September 11, 1860, a party of eight Englishmen reached the summit of the towering volcanic cone and peered into the dry, lava-strewn crater. To celebrate their ascent, they drank the health of Her Most Gracious Majesty the Queen in champagne, hoisted the Union Jack on a makeshift staff, and after a twenty-one-gun salute fired with revolvers, shouted "God Save the Queen" from the mountain top.

The view commanded from the summit extended to the sea in the east, and five gemlike lakes, the metropolis of Edo, and numerous small towns and villages could be identified. Rice paddies and avenues of towering cryptomerias on the plane below appeared in miniature, and to the south, west, and north, range upon range of forested mountains stretched to the horizon.

News of the first ascent of Mount Fuji by a party of foreigners reached the horticultural world in England three months later. The *Gardners' Chronicle* for December 22, 1860, published extracts of a letter written by John Gould Veitch, young scion of the Veitch family of nurserymen, relating details of the exploit, and in an accompanying article their correspondent sketched the outlines of the vegetation and its zonation on the majestic mountain. Veitch also contributed for publication the diary he kept on the twelve-day venture, and it quickly appeared in a January issue in 1861. In it the young Englishman recounted in vivid detail each day's progress and the route taken, as well as the vicissitudes suffered by the adventurers in flea-infested Japanese way-stations by night.

The English party that had joined Japanese pilgrims in climbing the sacred mountain was headed by Rutherford Alcock, British consul in Japan. Six members of his staff at the legation in Edo accompanied him, and by a stroke of good fortune John Gould Veitch, who had arrived at Yokohama only two days before the expedition departed, had been asked if he cared to join the group. Without giving the matter a second thought, Veitch elatedly accepted the invitation, whereupon Alcock appointed him "Botanist to her Britannic Majesty's Legation at Jeddo." In the letter relating these events Veitch exclaimed, "As you may imagine, I at once grew six inches taller on my appointment" (J. G. Veitch, 1860, p. 1126).

Veitch had arrived on the scene in the knick of time. He had come up the Inland Sea from Nagasaki, where he had arrived on July 20 and where he had already gathered together a large assortment of plants for eventual introduction to his father's Chelsea nursery. Desiring to make Alcock's acquaintance with the hopes that the diplomat might persuade Japanese officials to lift travel restrictions in the vicinity of Edo—constraints that had confined him to a ten-mile radius of Nagasaki—the young collector had a chance meeting with Alcock soon after his arrival in Yokohama and the opportunity to join the consul's party.

While Japanese officials accompanying Alcock's expedition required that the main roads be followed and that little time be wasted, Veitch did manage to collect along the way. Coupled with collections he would make later in the fall in the vicinity of Hakodate in northern Honshu, and combined with the plants already assembled and awaiting shipment in Nagasaki, his booty was considerable. His haul, moreover, was increased through visits to nurseries in the vicinity of Edo and further supplemented by seeds gathered at his request by Japanese collectors.

By the end of November, Veitch had established his plants in Wardian cases and was prepared to depart for England. His had been a whirlwind trip of only four months' duration, yet on his return to England he carried with him no fewer than seventeen new species of conifers, as well as seeds and plants of many of the same

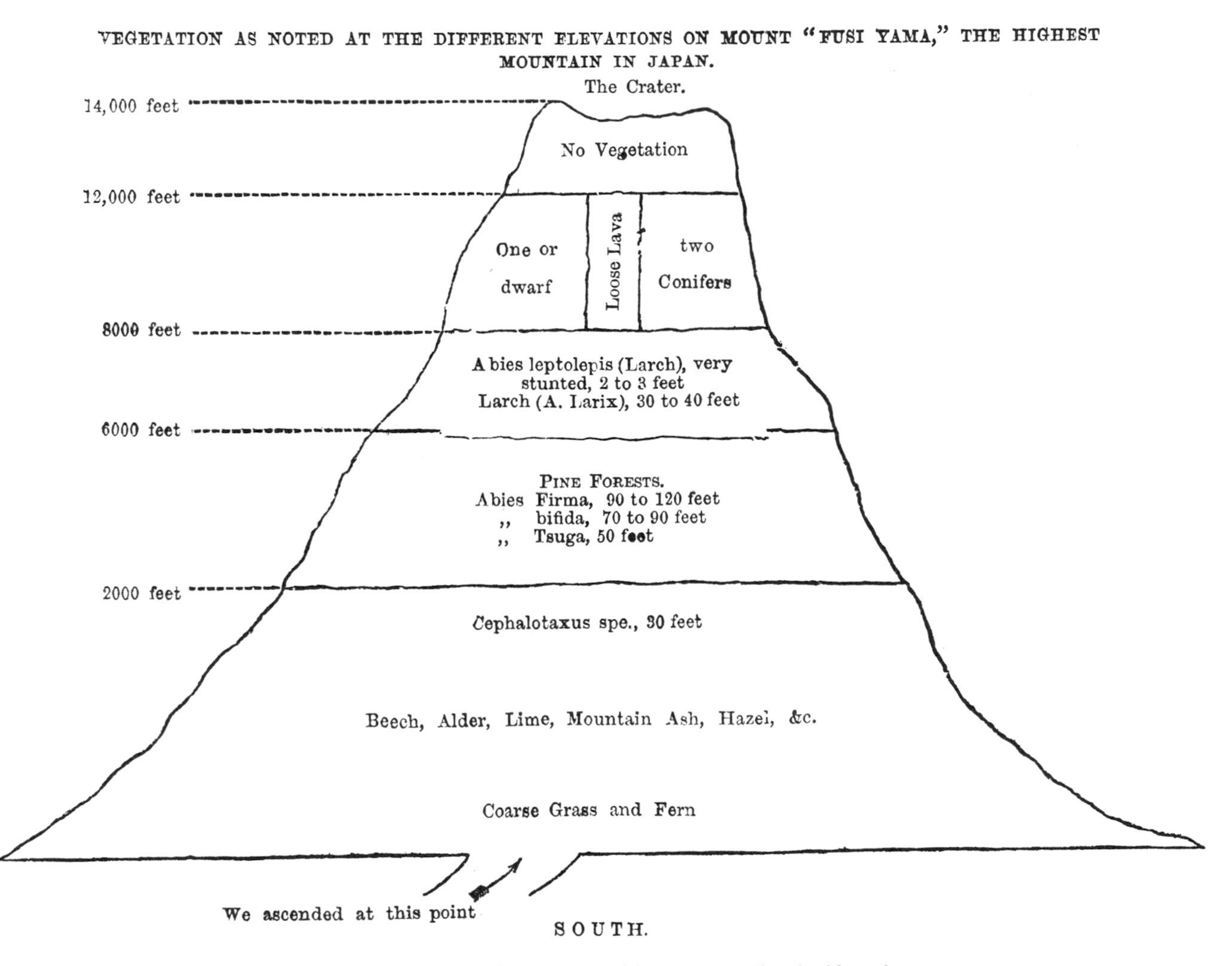

128. Schematic vegetation map of Mount Fuji, showing the zonation of forest types, sketched by John Gould Veitch and based on his observations made on the ascent of the sacred mountain in 1860. This sketch—probably the first to show the ecological associations occuring on Mount Fuji—was included in an article entitled "Notes on the Vegetation of Japan," which was published in the *Gardeners' Chronicle and Agricultural Gazette* that appeared on December 22, 1860.

129. John Gould Veitch (1839–1870), scion of the famous family of English nurserymen, was the first family member to personally undertake plant exploration. In addition to his travels in Japan, Veitch traveled in the South Seas in search of orchids and other tender exotics for culture in English conservatories and glass houses.

horticulturally significant plants that George Rogers Hall was bringing to the United States at about the same time.

Unique to Veitch's collection was the first introduction into the West of the Japanese hemlock (*Tsuga diversifolia* (Maximowicz) Masters). Individuals of this species now share the favorable Hemlock Hill habitat with the American species of the genus in the Arnold Arboretum. Veitch fir (*Abies veitchii* Lindley), named to commemorate the young traveler who first brought herbarium specimens of this species to the attention of the botanical world, also grows in the Arboretum's conifer collection, where it joins Fraser fir (*Abies fraseri* (Pursh) Poiret) from the southern Appalachians. A short-lived tree, Veitch fir nevertheless grows well in the eastern United States and is characterized by needles that exhibit two silvery-white bands on their lower surfaces.

The plant destined to become Veitch's most popular introduction, however, was neither a majestic Japanese conifer nor a choice flowering shrub but a high-climbing deciduous vine that now clads and conceals the faces of countless brick and masonry buildings across Europe and North America. Boston ivy (*Parthenocissus tricuspidata* (Siebold & Zuccarini) Planchon; Plate 20, bottom), less frequently known as Japanese creeper, ranks as the most valuable vine ever introduced into Western horticulture. Its common name indicates the prevalence of this plant in Boston, yet in part it is a misnomer. For Boston ivy is no ivy at all but belongs to a genus classified as a member of the Vitaceae or grape family. The genuine ivy or English ivy (*Hedera helix* L.) is an evergreen climber that belongs to the ginseng family, the Araliaceae. The Ivy League—whose members are Harvard, Yale, Princeton, Pennsylvania, Columbia, Brown, Cornell, and Dartmouth—got its name from the "hallowed halls" covered with Boston ivy that dominate the campuses of these academic institutions. And while some schools have recently investigated removing the vines from their buildings because they erode the mortar between the bricks, students and alumni generally agree that their alma maters should retain the green-clad walls of the Ivy League.

Individual stems of Boston ivy can attain lengths approaching thirty feet, yet they are entirely self-supporting by means of small, adhesive discs at the tips of the tendrils opposite most leaf stalks. Upon coming in contact with a surface, these glandular tips secrete a gum that soon hardens, firmly bonding the seven or so small discs of each tendril to the substrate. The tendril itself soon shortens its length by forming a tight spiral,

130. *Abies veitchii* was named by John Lindley to commemorate John Gould Veitch and his services to horticulture. This Japanese fir was one of the magnificent conifers Veitch introduced from Japan.

thereby pulling the stem of the vine even closer to the wall on which it is suspended. Vine and wall become cemented together, as anyone who has attempted to keep pace with these fast-growing vines by checking their unwanted advance across window screens and door frames can attest. The strength of the vegetable glue manufactured by the Boston ivy intrigued Charles Darwin, whose experiments showed that an individual disc could support up to two pounds' weight. Most tendrils produce between five and ten discs, so that each tendril can support upwards of twenty pounds of vine.

Another hallmark of the Boston ivy that increased its horticultural repute is the wonderfully bright shades of crimson its deeply three-lobed leaves acquire each fall. When the leaves finally fall to the ground, small clusters of bluish-black, grapelike fruits, which have developed over the summer months, are exposed to view. The leaves sever their connections with the vine at the base

131. Without a doubt, John Gould Veitch's most popular introduction from Japan is now known in North America as Boston ivy. This high-climbing vine—not a true ivy at all—thrives in the growing conditions present in New England, and this 1916 photograph of the administration building at the Arnold Arboretum provides evidence of its luxuriant growth. Many buildings in Boston are similarly festooned, and despite its Asiatic homeland, the association of the plant with the New England metropolis gave rise to the climber's common name.

of the leaf blades, and the long, outward-reaching leaf stalks or petioles remain attached to the stems for a week or two before they, too, fall to the ground after a hard frost. In Boston and other east-coast cities, these persistent stalks serve as perches for house or English sparrows and members of the large flocks of starlings that annually congregate before migrating southward. Like Boston ivy, both of these highly successful species of birds are exotics that have become thoroughly naturalized in North America. The house sparrow was introduced into North America from Europe via Brooklyn, New York, during the winter of 1850–1851, while the starling joined the North American avifauna when one hundred individuals were released in Central Park in New York City in 1890. Without the petioles to serve as convenient perches, the Boston ivy and its grape-like fruits would be inaccessible to these feeding aliens, and the seeds of this Japanese climber might not be dispersed. Oddly enough, seedling plants of the Boston ivy are rarely found, despite the wide dispersal of its seeds across North America.

Doubtless encouraged by the results of his brief Japanese mission, John Gould Veitch undertook a second voyage to introduce new plants into cultivation in Europe. This journey took him to Australia and the South Sea Islands between 1864 and 1866. Returning to England with a bountiful harvest of foliage plants suitable for cultivation in conservatories, the traveler married and took on additional responsibilities in the Chelsea, Coombe Wood, Feltham, and Langley branches of his father's growing nursery empire. Sadly, shortly after the birth of his son, James Herbert Veitch, who was destined to follow in his father's footsteps to Japan, John Gould Veitch fell victim to tuberculosis, from which he died at the age of thirty-one in 1870. If he had lived to undertake further horticultural exploration, our indebtedness to the industry of the man who introduced the vine that now characterizes American academic landscapes would undoubtedly be even greater.

Japanese Customs

On the eve of the Civil War in the United States, Japan was fast becoming a mecca for European botanical and horticultural explorers. In addition to John Gould Veitch, Carl Maximowicz arrived in the island nation in 1860 and botanized for the next three and a half years. Chests and crates of seeds and dried, pressed specimens prepared by Maximowicz and his valued Japanese assistant, Tchonoski, were destined for St. Petersburg in the explorer's native Russia. Although Maximowicz attained his reputation primarily as a botanist specializing in the Japanese flora, the beautiful red-vein enkianthus (*Enkianthus campanulatus* (Miquel) Nicholson) numbers among his few introductions into Western horticulture from Japan. From Germany came Max Ernst Wichura for a brief four-month sojourn, and in 1861 Richard Oldham was sent out from the Royal Botanic Gardens at Kew. Sir William Jackson Hooker, who had publicized so many of David Douglas's introductions years before, had become director of Kew twenty years earlier and had transformed the gardens into a first-class scientific institution. In addition to the extensive living collections, Sir William had established Kew's herbarium and library and its laboratories of economic botany. He, too, was anxious to have specimens from Japan, both living and dead, for the continually expanding collections under his care.

Little known among the horticultural travelers stopping in Japan was an American, English by birth but of Scottish ancestry, who was sent on a diplomatic mission by President Abraham Lincoln. Thomas Hogg arrived in Japan on August 22, 1862, in the official capacity of United States Marshall. Hogg held that position for eight years, and toward the close of his residence in Japan he became involved with the Japanese customs service. In this official capacity he promoted reforms in the service and particularly in the assessment of taxes that would be more equitable to Japanese businessmen. As a result of this work, Hogg returned to Japan again in 1873 at the request of the Japanese government for a two-year stint, again working with the customs service. This undertaking afforded Hogg an added opportunity to travel widely throughout the island nation and to become acquainted with numerous high-ranking officials, many of whom, like so many Japanese, shared Hogg's interest in plants and horticulture.

That Hogg had an interest in plants and eagerly set about arranging for Wardian cases filled with novelties to be shipped from Japan to New York came as no surprise to his close acquaintances. For Hogg's father, Thomas Hogg, Sr., was a Scottish-born gardener and horticulturist who in 1822 immigrated to New York, where he established a nursery business. With him came his wife and two young sons, James and Thomas, Jr., who joined their father in the nursery enterprise, which was located at 23rd Street and Broadway and was known as the New York Botanical Garden.

When Thomas, Sr., died in the fall of 1854, James and

132. Thomas Hogg (1820–1892), United States consul in Japan and adviser to the Japanese customs service. While the majority of plants Hogg introduced from Japan have remained horticultural favorites, one, the kudzu (*Pueraria lobata* (Willdenow) Ohwi), a high-climbing, semi-woody vine, has escaped from cultivation in certain areas of the southeastern United States and rivals or perhaps surpasses the Japanese honeysuckle as an invasive weed. In the Arnold Arboretum, where the vine is killed to ground level each winter, kudzu can be seen growing on the chain-link fence at the edge of the employees' parking lot at the side of the Hunnewell building.

Thomas, Jr., took over the family business. Despite the fact that Thomas was destined to spend ten years in remote Japan, he retained an interest in the family firm, which James operated in his brother's absence. Thomas, hoping to fulfill a useful purpose for the concern, sent Japanese plants to his brother in New York and by so doing was responsible for the introduction of several outstanding species into American gardens, where in cultivation they frequently joined native American congeners. Hogg also sent consignments to Samuel Parsons's firm at Flushing, on Long Island. His generosity to this rival nurseryman was undoubtedly a gesture made in a sincere attempt to ensure the successful introduction of as wide an array of Japanese plants as possible.

One of the most sought-after small trees for year-round landscape appeal is the Japanese stewartia (*Stewartia pseudocamellia* Maximowicz), a plant Thomas Hogg sent from Japan to his brother James in New York City. As indicated by the Latin name of this plant, the flowers of the Japanese stewartia are like those of a single camellia, as are those of the two American representatives of the genus (*Stewartia malacodendron* L. and *Stewartia ovata* (Cavanilles) Weatherby). The beauty of the large flowers of all three species is attributable to their five creamy-white, scalloped-shaped petals and the central boss of numerous yellow or purplish stamens. Yet as a year-round ornamental, the Japanese species has an added attribute not shared by its occidental cousins.

Pedestrians in the Arnold Arboretum frequently request directions to the shoulder of Bussey Hill, where several fine, old specimens of the Japanese stewartia grow side by side with American and other Asian representatives of the genus. Individuals also grow below the Center Street wall near the Center Street gate, and a solitary tree thrives in an open, sunny location at the edge of the meadow near the Hunnewell building. During late June and early July their flowers are produced along the branches, but it is in late October and through the winter months that their added attribute is seen to best advantage. In fall, after the leaves have turned a deep burgundy-red—or, on some plants, a bright yellow or reddish-orange—the leaves fall and the trunks and branches of the trees are exposed to view. Unlike both American species, which produce a tight, finely fissured bark, Japanese stewartias produce a smooth, mottled, exfoliating bark that gives the sinuous trunks and limbs a piebald appearance. Coupled with the numerous dark brown, five-valved capsules that persist on the curiously zigzagged branchlets, the Japanese stewartia provides

133. The Japanese stewartia (*Stewartia pseudocamellia*), a small tree whose ornamental value is doubled by its exfoliating, mottled bark as well as its camellia-like flowers, was sent to New York from Japan by Thomas Hogg.

ornamental interest to the winter landscape as well as that of summer and fall.

Native to a wide area of eastern Asia from southern Hokkaido in northern Japan and southward through the rest of that country as well as in Korea, China, Taiwan, and the Philippines to the south, the Japanese snowbell (*Styrax japonica* Siebold & Zuccarini) was included by Hogg in a Wardian case destined for North America in 1862. Almost simultaneously, Richard Oldham sent plants of the same species to Kew in England. In Japan and Korea snowbell is a common shrub or small tree in the temperate forests, where it usually grows in moist areas in sunny openings and along stream banks.

In cultivation, Japanese snowbells usually develop into large shrubs or many-branched small trees about thirty feet in height and often nearly as wide. A wonderful specimen almost a hundred years old grows adjacent to the Center Street beds in the Arnold Arboretum and was one of the first plants of Japanese origin to be planted in the Arboretum. In late May and early June this venerable tree produces an overabundance of pendulous white flowers in which the five petals form a bell-like corolla. In the center of each flower, ten stamens with yellow anthers surround the long style and together appear like a miniature clapper. Close examination reveals that the petals are densely covered with small, starlike hairs, which collectively suggest a frosty coating.

The Japanese snowbell lends a delicate grace to any landscape; but, like so many other first-class plants, it is infrequently seen in North America outside of botanical gardens and arboreta. Its rarity may relate to difficulties nurserymen have encountered in attempts to germinate the abundantly produced seeds; yet softwood cuttings dipped in a rooting hormone—a procedure familiar to all plant propagators—root readily in a rooting medium when placed under intermittent mist.

The same treatment has also proven successful in attempts to propagate a second species of snowbell sent to the United States from Japan in 1862 by Thomas Hogg. Plants of the fragrant snowbell (*Styrax obassia* Siebold & Zuccarini), however, are even less frequently encountered in general cultivation than the Japanese snowbell (*Styrax japonica*). Yet as a small landscape tree the fragrant snowbell provides an elegant boldness, in decided contrast with the delicacy of the Japanese snowbell. For the leaves of the fragrant snowbell are generally twice the size of those of the Japanese snowbell and roughly orbicular in outline; in length they can measure up to six and a half inches, and their width can be almost as large. Moreover, the many-flowered inflorescences of the fragrant snowbell comprise drooping, terminal racemes that equal the largest leaf blades in length; those of the Japanese snowbell are smaller and rarely consist of more than five flowers. The close relationship of the two species, however, is confirmed by the similar structures of their beautiful white flowers and their dry, globose, one- or two-seeded fruits, each subtended by a persistent, cuplike calyx.

Two additional Japanese plants—first described, respectively, by Thunberg and by Siebold and Zuccarini—awaited introduction into the Western hemisphere by Thomas Hogg. The first, a large, deciduous shrub that attains a height of twenty feet at maturity, is commonly known as sapphire berry (*Symplocos paniculata* (Thunberg) Miquel) from the intense blue color of its ripened fruits. Yet those familiar with this species are aware that the beautiful ornamental quality of these fruiting shrubs is of very brief duration. Birds quickly devour the ripened fruits—technically drupes—and the unique display can disappear during the course of a single day. A more prolonged period of ornament, however, occurs in spring, when the shrubs produce thousands of fragrant white flowers in small, airy panicles that literally cover the plants and frequently obscure the small oval leaves.

The sapphire berry is now known to occupy a wide range in eastern Asia, from the Himalayan region through China to Korea and Japan. As one of approximately 350 species placed in the genus *Symplocos*, it is one of the few species that ranges northward into temperate latitudes. A congener in North America, the horse sugar or sweetleaf of the southeastern United States (*Symplocos tinctoria* (L.) L'Héritier de Brutelle), is also a denizen of the Temperate Zone; the remaining species of the genus are largely confined to tropical and subtropical regions of both the Old and New Worlds.

By contrast, Thomas Hogg's remaining introduction is classified as a member of a genus consisting of only two species, both of temperate Asia, and the genus comprises a distinct plant family, the Cercidiphyllaceae. The katsura tree (*Cercidiphyllum japonicum* Siebold & Zuccarini) is widely distributed throughout Japan but is most frequently encountered in the forests of northern Honshu. The species also occurs in the forests of central China, and at one time it may have had a more continuous distribution in the intervening region of eastern China. That the genus enjoyed a wide distribution around the northern hemisphere during the Upper Cretaceous and Tertiary periods is evidenced by fossils representing extinct species of the genus from sites in Europe and North America. Like the ginkgo, the genus *Cercidiphyllum* was extirpated from the floras of these regions during the Pleistocene but survived in Asia to inhabit the forests of China and Japan.

In the floras of both of these Asian countries the katsura tree attains the greatest height and girth of any broadleaved deciduous tree, and in Japan it is one of the country's most valuable sources of timber. Its fine-grained, soft, light wood is especially sought for cabinet work and paneling. By contrast, the old specimens of the katsura trees growing along Meadow Road in the Arnold Arboretum have failed to achieve great height and girth in the hundred or more years they have been growing there, and in both Europe and North America the tree is grown not for its timber but as an unusual shade tree.

In cultivation katsura trees usually develop a multi-stemmed habit, and a combination of intangible characteristics lends a decidedly exotic aspect to older individuals. These trees have an oriental appearance—if a plant can be said to look oriental—unlike any deciduous or evergreen tree associated with the forests of North America or Europe. Close inspection will reveal that the branches are differentiated, like those of the ginkgo, into long-shoots and stubby short-shoots on which most of the broadly heart-shaped leaves are borne. Careful inspection of the leaves, each finely crenate at the margins, will show that they, too, are of two types. Short-shoot leaves have five to seven main veins originating from the base where the blade meets the long petiole. Long-shoot leaves, however, have a solitary main vein originating at the leaf base, with secondary veins branching from along its length.

Not only are the shoots and leaves of two types, but the trees themselves are sexually distinct, producing either staminate or carpellate flowers on short-shoots before the leaves appear in early spring. The carpellate flowers consist of two to six carpels, each subtended by a bract, while the staminate flowers comprise between fifteen and thirty-five dull purplish-red stamens and two to four subtending bracts. Lacking a showy perianth, these rudimentary flowers nonetheless give color to the leafless branches of the trees in late March and early April.

As leaves emerge from the buds, they appear a muted purplish-red; as they expand and attain their full dimensions, the leaves assume a mat green on their upper surfaces, while the lower surfaces provide the cool contrast of a grayish-green. However, the petioles, which hold the leaves at a slightly upright angle from the shoots, retain the purplish-red coloration of the embryonic leaves. In fall the leaves change color once again, turning clear yellow to a warm apricot-orange. Additional interest is provided in the fall and winter months by the clusters of small, seed-filled follicles (Plate 20, top), which have been likened to bunches of small bananas, that develop from the flowers at the ends of short-shoots on carpellate trees.

From these curious seed vessels Thomas Hogg gathered the seeds he sent to America, thereby establishing the genus *Cercidiphyllum* in the cultivated flora of the United States. Little did he realise in the 1860s that representatives of the same genus had flourished as native species in the forests of North America before a succession of Pleistocene glaciers transformed and shaped so much of the American landscape and simultaneously caused the decimation of the diversity of its preglacial flora.

New England Philanthropists and a Boston Brahmin

In December of 1868, as Thomas Hogg was about to begin the final year of his initial eight-year residence in Japan, James Arnold, a successful and wealthy merchant of New Bedford, Massachusetts, died, leaving a portion of his estate in trust. By 1873, when Hogg returned to Japan at its government's request to serve in the customs service, the President and Fellows of Harvard College had agreed to serve as permanent trustees for Arnold's bequest and to devote the income of the fund to the establishment and support of an arboretum. The arboretum was to be called the Arnold Arboretum, and an agreement had been reached with Arnold's three designated trustees which provided that the arboretum would be located in the Jamaica Plain section of Boston.

The site selected for the new arboretum comprised 137 acres of the glacially scoured but topographically diverse land known as Woodland Hill. This property had been the former farm of Benjamin Bussey, a philanthropic businessman and gentleman farmer of an earlier generation who had left to Harvard his estate, along with a fund, for the establishment of an agricultural school. Bussey had died in 1842, the year Asa Gray took charge of the Harvard Botanic Garden in Cambridge. Little had been done with the land in the intervening years, although seven of the 394 acres had been set aside for the use of the Bussey Institution in 1869, and an impressive stone building had been erected to house classrooms, offices, and laboratories. A staff for the school had also been assembled, and Francis Parkman—still engaged in his masterful historical studies as well as his horticultural pursuits—was named first professor of horticulture in 1871.

134. Asa Gray brought American botany onto an equal footing with its European tradition and made contributions to the study of plants that had world-wide implications. Gray was also responsible for reinforcing and expanding the botanical establishment at Harvard University.

The concept of adding an arboretum to the botanical establishment at Harvard pleased Asa Gray, who understood only too well the space limitations of the seven-acre Botanic Garden and the space requirements for growing plantations of trees. Two of Arnold's trustees also eagerly backed the idea. One—George Barrell Emerson—was a student of Massachusetts trees, and the other—John James Dixwell—was developing his Jamaica Plain estate into a horticultural showplace. What Gray disliked was the proposed location of the new arboretum. He had urged that it be adjacent to the Botanic Garden in Cambridge, not six miles distant on the other side of the Charles River in Boston. But the price of land in Cambridge, the proximity of the Bussey Institution, and the immediate availability of the Bussey acreage—already owned by the college—were factors that ultimately determined the site of the new institution.

135. Portrait of Charles Sprague Sargent (1841–1927), first director of the Arnold Arboretum, drawn in charcoal by his more famous cousin, John Singer Sargent, the well-known American artist.

When James Arnold's trustees and Harvard Corporation officials completed their negotiations, the indenture establishing the Arnold Arboretum was executed on March 29, 1872. The primary mission of the infant arboretum was embodied in its indenture, which stated that "as far as is practicable, all the trees, shrubs, and herbaceous plants, either indigenous or exotic, which can be raised in the open air" would be grown, and this all-encompassing collection should "be raised or collected as fast as is practicable." With these obligations, the new establishment would soon be playing a major role in the introduction of new trees and shrubs into North American and European gardens from remote, poorly known, or totally unexplored corners of the Northern Hemisphere.

Once committed to the new arboretum, Harvard officials had to put the plan into action. The Bussey land was secure, but the income from the Arnold bequest—roughly three thousand dollars a year—was scant, and a staff had to be assembled for this novel undertaking, a university-affiliated living museum of woody plants for education and research. Surely the most important appointment was a person to head up the new institution, one who would also claim the chair of Arnold Professor, a post created by the legal arrangements. The Arnold Professor would simultaneously direct the arboretum and "teach the knowledge of trees" in Harvard College.

Once again, Asa Gray's advice was sought, and an unlikely candidate, Charles Sprague Sargent, was nominated and duly appointed. Sargent had graduated near the bottom of his Harvard class in 1862 and had enlisted in the Union Army shortly after finishing his college education. Upon his honorable discharge from the army in 1865, Sargent postponed a career decision, preferring to spend the next three years traveling in Europe. When he finally returned to Boston in 1868 at the age of twenty-seven, the young man agreed to manage his father's Brookline estate, Holm Lea, on a full-time basis.

The fact that Sargent was a scion of an old, established Boston family of considerable wealth had allowed for the luxury of his leisurely continental tour after the war, but he was probably expected to follow in traditional footsteps and enter one of the family's business or banking concerns upon his return from Europe. Yet his move to Holm Lea, hardly an overt career decision, which probably caused eyebrows to be raised throughout his large family, was a turning point in the young Brahmin's life. With a large estate to manage, Sargent's interests necessarily turned to arboriculture, horticulture, landscape design, and the science of botany.

136. A view of the grounds at Holm Lea, the Sargent estate in Brookline, Massachusetts, depicting plantings of rhododendrons around a pond.

In making his choice to devote his energies to maintaining and developing his father's estate, Sargent did receive encouragement, support, and advice. His own family circle included Henry Winthrop Sargent, whose Hudson River estate, Wodenethe—the old Saxon word for "woody promontory"—was already a horticultural wonderland, and Horatio Hollis Hunnewell, who was aggressively developing his estate in nearby Wellesley, Massachusetts. There, Hunnewell was the first American to follow the European precedent of devoting acreage to coniferous plants and developing a comprehensive collection in a pinetum. Henry Sargent, moreover, had been strongly influenced and guided by his close friend and near neighbor, Andrew Jackson Downing, author of the influential book *The Theory and Practice of Landscape Gardening* and founding editor of *The Horticulturist or Journal of Rural Art & Rural Taste.* Closer to home, Francis Parkman was a source of inspiration and information, as were many other local members of the flourishing Massachusetts Horticultural Society, which had been founded in 1829 as an offshoot of the long-established Massachusetts Society for Promoting Agriculture.

Across the Charles River in Cambridge, Asa Gray provided the inspiration of a professional botanist and was a constant source of scientific information. After thirty years of running the Botanic Garden, Gray had practical advice to offer and provided an entrée to numerous botanical and horticultural colleagues in Europe and America. Gray was also on the lookout for a promising candidate to take over his administrative obligations so that he might concentrate his energies in the herbarium. Surprisingly, in 1872 Sargent was appointed director of the Harvard Botanic Garden and simultaneously replaced Francis Parkman as professor of horticulture in the Bussey Institution. On November 24, 1873, Sargent added to his list of Harvard appointments the directorship of the fledgling Arboretum. It was to this challenge that Sargent would devote himself for the remainder of his life. Realizing the unique opportunity provided by the Arboretum directorship, he resigned his other Harvard posts in 1879.

Collections had to be built from scratch: living plants had to be obtained and dried specimens and books had to be gathered to document and facilitate the study and proper classification of the plants soon to be growing in the living collections. Again, Asa Gray provided an opportunity that propelled Sargent in the space of a few years into the position of America's leading dendrologist. Asked to recommend someone capable of surveying American forest resources for a report to form part of the tenth United States census, Gray suggested Sargent. With an official appointment, Sargent was free to visit the forested regions of the United States to gather data and make firsthand observations of the condition of America's arboreal resources. A network of correspondents across the length and breadth of the land was also established, and requests for information as well as for seeds and specimens—the former for trial in the living collections, the latter for the growing Arboretum herbarium—were soon being dispatched in great numbers from the Brookline desktop of C. S. Sargent.

As a result of his wide-ranging field work for the census report and his wide circle of correspondents, Sargent almost single-handedly established a living collection of American trees at the Arboretum's Jamaica Plain site. Furthermore, through correspondence with the leading botanists and nurserymen in Europe, European and other exotic species began to be planted in the collections that had been systematically superimposed on the Arboretum grounds by Sargent in partnership with the landscape architect Frederick Law Olmsted. Olmsted had won the competition for the design of New York's Central Park and at the time was at work developing Boston's park system. The Arboretum was seen by Olmsted, and later by Sargent, as an integral "jewel" in the "Emerald Necklace," the chain of parks that Olmsted envisaged linking Boston's Public Garden with Franklin Park.

By 1878 William S. Clark, one-time professor at Amherst College and third president of the Massachusetts Agricultural College—now the University of Massachusetts—was sending seeds to Sargent from Hokkaido, the northernmost island of Japan. On temporary leave from the Massachusetts institution, Clark (1826–1886) had traveled to Japan at the request of Japanese officials to help establish the Imperial College of Agriculture at Sopporo, where he served as founding president. Quantities of seeds of Japanese trees, the first received directly from Asia at the Arboretum, including those of the katsura tree and the Japanese climbing hydrangea (*Schizophragma hydrangeoides* Siebold & Zuccarini), arrived in Boston as a result of Clark's generosity. And by 1882, a correspondent in China was sending seeds he had gathered from the Western Hills in the vicinity of Peking.

Sargent's Chinese correspondent was a physician attached to the Russian legation in Peking, where he resided between 1866 and 1884, with an absence of two years spent in Europe. During his residence in China, Emil Bretschneider became a serious sinologist and ama-

137. *Schizophragma hydrangeoides* is a high-climbing vine with a very similar appearance to *Hydrangea anomala* subsp. *petiolaris*. The former, however, produces sterile or neuter flowers with one bract, as opposed to the four bracts of neuter flowers of the climbing hydrangea. Both species can be seen growing on the brick walls of the Hunnewell Visitors' Center at the Arnold Arboretum.

138. Emil Bretschneider (1833–1901) became fascinated with the flora of the Peking region during his thirteen-year residence there and established a correspondence with botanists at the Jardin des Plantes in Paris and with C. S. Sargent at the Arnold Arboretum. Many of the plants raised from seeds he provided prospered in the collections of the Arnold Arboretum, and some still exist today. Bretschneider was made an honorary member of the staff of the Imperial Botanical Gardens in St. Petersburg (now the Komarov Botanical Institute in Leningrad) in recognition of his contributions to botany and horticulture.

VOL. I. NO. 1.

GARDEN AND FOREST

A JOURNAL OF HORTICULTURE LANDSCAPE ART AND FORESTRY

FEBRUARY 29, 1888.

PRICE TEN CENTS.] Copyright, 1888, by THE GARDEN AND FOREST PUBLISHING COMPANY, LIMITED. [$4.00 A YEAR, IN ADVANCE.

139. The masthead of the first issue of *Garden and Forest*, a weekly journal established by Charles Sargent in 1888. Over the course of the journal's ten-year history, articles in its pages documented the growth of the collections and the overall development of the Arnold Arboretum. Other articles and notices kept its readers abreast of conservation and forestry issues in the United States as well as botanical and horticultural developments from around the world.

teur botanist and among other contributions added significantly to scientific knowledge of the northern Chinese flora. After his retirement he returned to St. Petersburg, and shortly before his death in 1901 he published his monumental *History of European Botanical Discoveries in China.* Many of the seeds he collected in the Western Hills were forwarded to Sargent for trial in the climate of Boston, and other consignments were addressed to Kew, the Jardin des Plantes in Paris, and the botanical gardens in Berlin and St. Petersburg.

One lot received by Sargent contained the tiny, easily germinated seeds of a widespread northern Asian rhododendron, now sometimes known as the Korean rhododendron (*Rhododendron mucronulatum* Turczaninow) because of its frequency in the flora of the Korean peninsula. The first of the azalea-type rhododendrons to flower in spring, the Korean rhododendron, with its large, rosey-purple flowers, has become an extremely popular small garden shrub. It is perhaps best known in the American nursery industry by a selection known as 'Cornell Pink'. Another small cloth bag held the capsules and seeds of the hairy lilac (*Syringa pubescens* Turczaninow), while seeds of a second lilac, a late-flowering species (*Syringa villosa* Vahl; Plate 21), added to Sargent's excitement on receiving Bretschneider's package from Peking. These would be grown and added to the collection of lilacs—destined to become one of the Arboretum's most comprehensive and noteworthy collections—that Sargent was in the process of establishing on the slope above Bussey Road.

Mindful of the research and educational roles the youthful Arboretum should play in the realm of botany and horticulture, as well as in conservation issues that had become so apparent during his forest census project, Sargent established *Garden and Forest* as a vehicle for bringing recent developments and noteworthy news to public notice. In the pages of this weekly publication, subtitled *An Illustrated Weekly Journal of Horticulture, Landscape Art and Forestry,* Sargent documented the development of the Arboretum under his charge and monitored the growth and development of plants growing there, including Bretschneider's Chinese lilacs. Of these the issue of June 26, 1889, reported:

> *Syringa villosa* has, now that the plants are thoroughly established, and of large size, flowered here more abundantly than it ever has before. It is certainly an ornamental plant of the first-class and one of the most important introductions of late years among flowering shrubs. It has, moreover, the merit of flowering late, long after the flowers of other Lilacs with long-tubed corollas . . . have faded. (J. G. Jack, 1889, p. 309)

Of *Syringa pubescens* Sargent recorded in 1891:

> [It] is one of those plants which improve with age. The large plants in the neighborhood of Boston, raised in the Arnold Arboretum from seed sent by Dr. Bretschneider from Pekin, have flowered better this year than ever before, and have proved that this species is one of the most beautiful Lilacs in cultivation. The individual flowers are not large, and the clusters are smaller than those of other species; they are produced, however, in the greatest profusion, and quite cover the branches. The flowers are at first a delicate rose-color, but, before fading, become almost white; they are deliciously fragrant. (C. S. Sargent, 1891, p. 262)

Other species received at the Arnold Arboretum from the Russian sinologist included beautiful flowering pears (*Pyrus phaeocarpa* Rehder f. *globosa* Rehder and *Pyrus bretschneideri* Rehder, the latter named to honor Bretschneider), the wonderful shrubby, downy cherry (*Prunus tomentosa* Thunberg) that produces bright red, edible fruits, and a mountain ash (*Sorbus discolor* (Maximowicz) Maximowicz) with white rather than the usual orange or red fruits. A new linden, the Mongolian linden (*Tilia mongolica* Maximowicz), and a close shrubby relative of the arboreal lindens, the unusual grewia (*Grewia biloba* G. Don), added to the botanical interest of Bretschneider's collections. And to Sargent's astonishment, the large majority of Bretschneider's consignments thrived under the climatic conditions of the Boston Basin. Bretschneider, who followed with interest and pride the successes and failures of the plants raised from the seeds he had sent to Europe and America, recorded that "the best results of cultivation were obtained in the Jardin des Plantes and in Arnold Arboretum" (Bretschneider, 1898, p. 1049).

By 1892 it was clear from the Arboretum's expanding living collections that a large proportion of the woody plants that had thus far been received from northern China and Japan could survive and even flourish when cultivated in eastern North America. Fully aware of Asa Gray's theory concerning the relationships between eastern North America and eastern Asia, Sargent wanted to capitalize on the theory's horticultural significance and the practical implications of the similar climates in the two widely separated regions. Gray, moreover, had expanded upon his theory before his death in 1888 and had extended his comparisons of floristic data to include much of China as well as Japan.

By the summer of 1892 work at the Arboretum was proceeding nicely. Trees forming the backbone of the collections were established on the Arboretum hills and in its valleys, and staff occupied the newly completed administration building inside the Arborway gate. Constructed with funds provided by Horatio Hollis Hunnewell, the new building offered ample room for the expanding library and herbarium, as well as space for museum exhibits. With his census report behind him and with work on his new project, *Silva of North America*, well under way, Sargent decided the time had come for the Arboretum to take an active role in the investigation of the Asian flora. The Arboretum would send a collector to Japan, and eventually to China.

As a veteran of field work across North America, Sargent decided that he would travel to Japan to personally inaugurate the Arboretum's first Asian mission. Sargent enjoyed the prospect of exploring the forests of Japan for "horticultural delicacies," and seeing firsthand the Japanese counterparts of eastern North American trees. In Japan, a country that had so readily adopted many Western ways, travel from north to south would be facilitated by the modern rail system; consequently, a good portion of the island nation might be visited in a relatively short period of time. China, on the other hand—a vast territory by comparison with Japan, and still governed by feudal war lords—would have to wait for another year, or another man.

Crossing the continent on the Canadian Pacific Railroad, Sargent embarked for Yokohama from Vancouver on August 7, 1892. Philip Codman, his nephew, went along as Sargent's traveling companion, and once in Japan they left Tokyo (formerly Edo) behind and immediately headed north. By the middle of September they were on the northern island of Hokkaido, where a courtesy call was made at the Imperial Agricultural College at Sopporo, which had been established by Sargent's Massachusetts associate, William Smith Clark. There, Sargent and Codman made the acquaintance of Professor Kingo Miyabe, who provided assistance in the form of a Japanese guide. A long-standing friendship was established between the two professors, and Miyabe

140. James Herbert Veitch (1868–1907), son of John Gould Veitch. Veitch's explorations in Japan were but one stop on a two-and-a-half-year, world-wide tour, and his published travelogue, *A Traveler's Notes*, records his experiences in India, Malaysia, Australia, and New Zealand, as well as Japan.

(1860–1951) soon numbered among Sargent's foreign correspondents. Over the years to come, the Japanese botanist would respond to Sargent's repeated requests for seeds from Hokkaido plants.

With a knowledgeable young man named Tokubuchi recommended by Miyabe leading the way, the two Yankees began collecting in the rich forests of Hokkaido. Ironically, they were not alone in the pursuit of a rich harvest of seeds and specimens, and Sargent and Codman soon learned that another collector was in the vicinity.

James Herbert Veitch—son of John Gould Veitch, who had come to Japan in search of plants thirty-two years earlier—joined the two men from Boston in Sopporo, and the threesome, with Tokubuchi as their guide, traveled together southward to Honshu. There they made an ascent of Mount Hakkoda, spending several cold, miserable nights in straw huts near the summit. But the collecting was good, and the talk centered on plants, botany, and the possibilities of exploration in China. Young Veitch was able to relate the impressions of Charles Maries concerning the Middle Kingdom and of his unfortunate experiences there as a plant collector.

Maries had been a foreman in the Chelsea nursery of the London branch of the Veitch firm, and in 1875 Sir Harry Veitch asked if he would like to cast his lot among the plant hunters in Japan. Maries jumped at the chance and found himself in Japan for most of the next three years. He broke his travels in the island nation, however, with two sojourns in China. On his first Chinese venture Maries traveled up the Yangtze River to the Lu Shan mountains, where he fell victim to sunstroke so severe that it completely incapacitated him for two months. On his second attempt, Maries managed to penetrate one thousand miles inland to Ichang on the Yangtze, but at that outpost he was robbed by bandits and suffered a severe beating. With relatively little to show for his efforts, Maries quickly returned to Japan, where travel was easy and the Japanese people were, by and large, friendly and hospitable.

Charles Maries's travels in Japan were far more successful, and the Stratford-upon-Avon native was the first plant explorer to tramp the forests of the northern island of Hokkaido. Due to their desire to see Maries fir (*Abies mariesii* Masters), discovered by Veitch's collector in 1878, Sargent and James Herbert Veitch made the ascent of Mount Hakkoda in 1892. This stately fir was successfully introduced to England by Maries, as was a curious actinidia (*Actinidia kolomikta* (Maximowicz & Ruprecht) Maximowicz), a climber that frequently has its leaves partially variegated pink and white. The handsome Nikko maple (*Acer maximowiczianum* Miquel, frequently listed as *Acer nikoense* Maximowicz), a species with trifoliolate leaves that turn brilliant shades of yellow, red, and purple in the fall, was also successfully introduced into England from Japan by Maries. Of his scantier Chinese collections, the most noteworthy plant was the beautiful Chinese witch hazel (*Hamamelis mollis* Oliver), a congener of the eastern North American witch hazel (*Hamamelis virginiana* L.). In fact, Maries returned to England so disillusioned by his Chinese experiences that he is reported to have claimed that all Chinese plants of horticultural merit had long since been introduced into cultivation in the West.

James Herbert Veitch was nonetheless anxious to pursue horticultural exploration in China and take his chances with the difficulties of travel and the reputed capriciousness of the Chinese. For his part, Charles Sargent hoped to have a share of any seeds or plants the young nurseryman might gather in China for trial in New England. When the Englishman parted company with the men from Boston to return to Hokkaido, it was agreed that Veitch would keep Sargent informed of his intentions to visit China. Sargent and Codman continued southward toward Tokyo, stopping en route at the famous shrine city of Nikko. After ten weeks in Japan, the twosome left Yokohama for America in early November hoping to arrive in Boston before the Christmas holidays. Included in their luggage were the seeds of roughly 200 species of plants and 1,225 collections of dried botanical specimens to be added to the Arboretum herbarium and to be shared on an exchange basis with other botanical institutions.

The seed from which grew the magnificent specimen of Japanese snowbell that can be seen today in the Arnold Arboretum was brought back to Boston by Sargent and Codman during the late fall of 1892. This species was already known in American horticultural circles through its earlier introduction from Japan by Thomas Hogg. The Nikko maple, on the other hand, was new to North America, although Charles Maries had found this tree for the Veitch firm in England in about 1878. A wonderful specimen of the Nikko maple grows in the maple collection near the Arborway wall in the Arnold Arboretum and dates from Sargent and Codman's 1892 gathering.

But other seeds carried in Sargent's luggage had been collected from species never before attempted in cultivation in either America or Europe. Two of these—Sargent crabapple (*Malus sargentii* Rehder) and Sargent

141. Between 1877 and 1880 Charles Maries (ca. 1851–1902) explored the forests of Japan as a horticultural collector for the English nursery firm of Veitch. Maries also made two brief journeys to China and penetrated as far inland as Ichang on the Yangtze River in Hupeh Province. Suffering difficulties on both of his Chinese sojourns, Maries returned to England and reported that few Chinese plants remained to be introduced into Western horticulture. After leaving the employment of the Veitch nursery, Maries settled in India, where he lived out his life and held posts in two botanical gardens.

142. One of the regions of Japan visited by Sargent and Codman on their tour during the fall of 1892 was the mountainous Nikko area northwest of Tokyo. Famous for its shrines, mineral hot springs, and natural beauty, the region supports a rich forest flora. Shown in this illustration are hemlock forests on the slopes above Lake Yumoto.

cherry (*Prunus sargentii* Rehder)—were totally new to botanical science. A third species, the torch azalea (*Rhododendron kaempferi* Planchon), was long-established in botanical literature and commemorates Englbert Kaempfer, the German physician-botanist who resided at Deshima over two hundred years before Sargent visited Japan. Strangely, this wonderful shrub—the mountain or hill azalea of Japan—had not previously been introduced into cultivation in Europe or North America, despite its widespread distribution in nature from Kyushu in the south to central Hokkaido in the north. As its common name in the West implies, the low, twiggy shrubs of the flame azalea light up the spring landscape when the plants are in flower. In the Arnold Arboretum, mass plantings of this species growing under the oaks along Meadow Road and under the hemlocks at the base of Hemlock Hill are particularly attractive in late May and early June, when the hot pink to red flowers cover the low, rounded contours of these Japanese shrubs.

Earlier in the year, during April or sometimes in late March, the silvery, hairy buds of the precocious-flowering magnolias swell rapidly during the first warm

143. Charles Sargent returned to the Arnold Arboretum from Japan with seeds of a crabapple that came to bear his name. Sargent crab (*Malus sargentii*) is valued by horticulturists as the smallest of the crabapples, since the trees grow only to about eight feet in height. Shown here are its small red fruits.

144. Despite the fact that the torch azalea (*Rhododendron kaempferi*) is very widespread and a common shrub on sunny slopes in the mountains of Japan, it was not introduced into Western gardens until 1892, when Charles Sargent collected its seeds in a temple garden and from a naturally occurring population on the shores of Lake Chuzenji in the Nikko region. As is evident in this photograph, the shrubs are extremely floriferous.

145. Among the evergreen species of holly hardy in New England, the long-stalk holly (*Ilex pedunculosa*) is perhaps the most beautiful and best adapted to landscape use. As in all hollies, individual plants produce either staminate or carpellate flowers, and at least one staminate plant should be planted among carpellate plants to ensure pollinization and the production of fruits. Most hollies offered by reputable nursery firms have been propagated asexually so that the "sex" of individual plants is known.

days of spring and open to allow the preformed flowers held within to expand and cover the otherwise naked branches of these small trees. Of the several species included in this group, one—the star magnolia (*Magnolia stellata* (Siebold & Zuccarini) Maximowicz)—had been brought to American shores by George Rogers Hall in the 1860s. Seeds of a second species, the kobus magnolia (*Magnolia kobus* De Candolle), had been sent to the Arnold Arboretum from Hokkaido in 1876 by William Smith Clark. But the third, the anise-leaved or willow-leaved magnolia (*Magnolia salicifolia* (Siebold & Zuccarini) Maximowicz), a tree of narrowly pyramidal habit, was not brought into cultivation until seeds of that species collected on Mount Hakkoda by Charles Sargent in 1892 were germinated at the Arnold Arboretum.

Fine specimens of anise-leaved magnolia grow along Goldsmith Brook near the Hunnewell building, and other individuals punctuate the spring landscape between Willow Path and the Arborway wall. A spectacular spring display is also created annually by an isolated specimen growing in the low sink area across Meadow Road from the maple collection. While the flowers are often damaged and browned by late frost, during those springs when frosts fail to occur the floral display can continue for up to two weeks before the white, spatula-shaped tepals fall to the ground. In late summer and fall, the fruits—technically cone-like aggregates of follicles—become dry as they reach maturity. In opening, the walls of each follicle pinch the two bright red seeds held within and thereby force them outward. Still attached to the follicles on thin, threadlike strands, the seeds hang suspended from the ripened cones, presenting themselves to foraging birds.

The bright red fruits of the long-stalk holly (*Ilex pedunculosa* Miquel) also stand out conspicuously against the rich green, glossy leaves of this fine evergreen. As the common name indicates, the clusters of berries—technically drupes—are held above the leaves on long stalks or pedicels up to two inches in length. Unlike the leaves of the familiar English and American hollies (*Ilex aquifolium* L. and *Ilex opaca* Aiton, respec-

Plate 17. This painting of the California big tree was executed using early photographs taken in the Sierra Nevada foothills.

Plate 18. As this painting illustrates, the kousa dogwood (with pointed bracts) is morphologically similar to the flowering dogwood of eastern North America. In the American species the seeds are produced individually (upper left), while in the Asian plant the seeds are embedded in a fleshy, strawberrylike fruit (bottom right).

tively), the leaves of the long-stalk holly lack prickly spines and are smooth along their margins, and the plants have the added attribute of extreme hardiness.

Sargent, on finding this heavily fruiting holly in the mountains of central Japan, was surprised to learn that the Japanese did not cultivate the plant. He returned to Boston with a quantity of seed, and long-stalk holly has been included in the collections of the Arnold Arboretum since that time. It now grows in gardens across North America and Europe. At Christmas time the large specimens flanking one corner of the Hunnewell building provide a festive, seasonal greeting of red and green. And the Christmas season of 1892 must surely have been a festive one for the Sargents at Holm Lea. A successful expedition to Japan had been completed, and the fortuitous meeting with James Herbert Veitch in Sapporo had given the prospect of a share in the plant materials this representative of the famous family of nurserymen hoped to collect in China. It was with great optimism and determination that Charles Sprague Sargent returned to the Arnold Arboretum in the first days of the New Year.

Fragile Evidence from China

The first volume of *Plantae Davidianae*, an elegantly produced work in large format that included forty-five illustrations—hand-colored in some copies—came off the presses in Paris in 1884. A second volume, devoted to the plants of "eastern Tibet," followed in 1888. The careful work of Adrien Franchet of the Muséum d'Histoire Naturelle and the Jardin des Plantes, this botanical treatise summarized and described upwards of 1,500 species of plants new to science. The descriptions outlining the botanical features of these and other included species had been obtained from painstaking observations and comparisons of the thousands of dried, pressed specimens that had been received at the museum from China over a thirteen-year period that commenced in 1862.

Not from the maritime provinces of China, the floras of which were relatively well known by the 1880s, these unique, beautifully prepared, and carefully selected specimens were the first to arrive in Paris from remote regions of Mongolia and from isolated areas extending from central China along the Yangtze River to far western China on the Tibetan Marches. Included in this remarkable collection were new cotoneasters, astilbes, rhododendrons, roses, and scores of other plants, some comprising totally new genera! But how had these specimens come to the Muséum, which had been erected along one edge of the Jardin des Plantes, and who had been responsible for collecting them in the first place? The answer to the second question was evident in the title of Franchet's two-volume opus, which incorporates the name David in the possessive Latin form, *Davidianae*. An answer to the first query requires a brief biographical sketch of the man who gave Western botanists their first glimpse of the phenomenal diversity and beauty of the western Chinese flora.

Jean Pierre Armand David, a Basque who had been born on September 7, 1826, in the small French village of Espelette on the western edge of the Pyrenees, was the man responsible for the beautiful dried plant specimens. As a youth, David developed two loves that would be the focal points to which he would devote his life and career: natural history and the Church.

After completing his schooling, David entered the Lazarist brotherhood in November of 1848 and in 1851 was ordained for the priesthood as a Lazarist or, more properly, as a member of the Congregation of Priests of the Mission. After a tenure of almost ten years at a Lazarist school in Italy, where David was able to further his interests in natural history by teaching science, the energetic priest was finally given the opportunity to realize a desire he had frequently expressed and for which he had long prayed. In 1862, two years after the Chinese had been forced to agree to allow missionaries to travel throughout their land, Abbé David received an assignment at the Mission of Lazarists at Peking, where again he was to teach science and natural history, this time to Chinese boys enrolled in the mission school.

Almost at once after his arrival in China, the young priest began his geological, zoological, ethnographic, mineralogical, and hydrographic observations in the immediate vicinity of Peking. Throughout the years, birds occupied his attention perhaps more than any other group of living organisms, although none of the other classes of animals escaped his scrutiny. While zoology claimed his greatest interest, plants were certainly not neglected. He assembled materials for a small natural history museum at the Lazarist school—a collection that later became the property of the Chinese emperor—to aid him in his teaching. Second, he amassed collections to be sent to the Muséum d'Histoire Naturelle in Paris.

Abbé David's shipments were more than welcome at the Muséum and in fact were sent at the request of museum scholars who had been anxious to meet David and enlist his help prior to his departure for the Middle

146. The Abbé Armand David (1826–1900), a Lazarist missionary in China and expert in natural history, was discharged from his normal clerical responsibilities to allow for three extensive journeys of biological exploration to remote regions of China. The dove tree, a species that excited European botanists and horticulturists alike, was discovered by Père David, and its generic name, *Davidia*, was given to honor his industrious collecting activities.

Kingdom. Scientists from all of the Muséum's departments had given the enthusiastic cleric commissions to collect material, and the specimens and artifacts that began to arrive in Paris as David went farther and farther afield astonished their recipients with their uniqueness, careful preservation and scientific significance. If a young missionary could gather such a wonderful array of new materials during his free-time ramblings, the scientists argued, consider the material that might be added to the museum shelves if he were permitted to collect full time and travel even farther from the mission at Peking.

The director of the Muséum, Henri Milne-Edwards, totally agreed with his colleagues' views and decided to determine if something might be done to free Abbé David's time for additional scientific investigations. Along with representatives of the Muséum's various departments, Milne-Edwards requested the French Minister of Public Instruction, Victor Duruy, to intercede on their behalf with the Superior General of the Lazarists. Specifically, Milne-Edwards asked that Abbé David be freed of his teaching and other mission obligations and be allowed to travel to Mongolia on a scientific mission. To everyone's surprise and delight, the Lazarist Superior General agreed, and a "mission scientifique" to the northern edge of the Chinese empire was undertaken by the Basque churchman, the first of three great journeys David would undertake to inaccessible regions of China.

Of his three trips—the first to the Mongolian frontier, the second to the edge of Tibet, and the third to Shensi (Shaanxi) and Kiangsi (Jiangxi) provinces—the second, which occupied the years between 1868 and 1870, proved to be the most important botanically. On this strenuous sojourn David traveled from Peking to Shanghai by sea. Next he went by boat up the Yangtze River to Ichang, and, continuing westward, he finally arrived in Chengtu (Chengdu), the capital of Szechwan (Sichuan) Province, after nine months of travel. Finally, after the Chinese New Year's celebrations had come to an end in late February of 1869, David left Chengtu on the final leg of his outward trip. He traveled overland to his destination, arriving at the remote outpost of Mupin (now Baoxing) in a tiny principality on the Tibetan borderland as the first spring flowers were coming into bloom. There, surrounded by mountain peaks ranging to 15,000 feet, David made his headquarters at a small mission school operated by the Société des Missions Etrangères, the oldest of the missionary societies in the Roman Catholic Church, which had been established in Paris in 1658. Abbé Dugrité, the priest in charge, and a small enclave

147. *Pinus armandii,* a pine native to western and central China and southward into Burma, was discovered in Shensi Province in 1873 by Abbé David, who sent dried specimens to the Muséum d'Historie Naturelle in Paris. The species was not introduced into cultivation, however, until 1895, when Père Farges, another French missionary in China, forwarded seeds to the Vilmorin Arboretum at Les Barres in France. Augustine Henry sent seeds from Yunnan Province to the Royal Botanic Gardens at Kew in 1897. The cones produced by *Pinus armandii* are among the most beautiful of all pines. Before the scales have opened widely, when the wingless seeds are released, the cones are a distinctive emerald green color and appear to be carved of jade. A member of the five-needled group of pines, which includes the white pine of New England, this Asian species is most closely allied to *Pinus flexilis* James, the Rocky Mountain white pine of western North America.

of French missionaries warmly welcomed the traveling naturalist into their midst and asked for news of the outside world.

David spent six months exploring in the mountains surrounding Mupin, constantly gathering scientific specimens and noting his observations in his diary. During the course of his trip he was the first Westerner to record observations of the giant panda, and, miraculously, he arranged for a living specimen to be transported down the Yangtze River from Szechwan and to be shipped to the Jardin d'Acclimatation in Paris. On its arrival in the French capital, the panda created a sensation as a public exhibit. Sadly, like many of its kind, it did not survive long in captivity.

In the eyes of Adrien Franchet, David's dried specimens of wonderful new species of rhododendrons and roses were as exciting as the panda. And a new genus named in the naturalist's honor, the genus *Davidia* (Plate 22), also came to the Muséum as a result of David's strenuous labors. Despite his prolonged bouts with illness and the potential danger of hostile Chinese, many of whom had never seen a Westerner before, David had collected specimens that attested to a far richer flora in western China than anyone could have predicted.

In 1885, the year after Franchet published *Plantae Davidianae,* Sir Joseph Dalton Hooker, the director of the Royal Botanic Gardens at Kew, responded to a letter received at the herbarium with a Chinese postmark. The correspondence thus inaugurated shortly before Hooker (1817–1911) retired as director was to result in another flood of botanical specimens from China. These would surpass even Abbé David's contributions, and in this instance the Kew herbarium would be the depository. Unaware of David's activities in the Tibetan Marches and elsewhere in the Chinese empire, Augustine Henry was also largely unschooled in botany and natural-history subjects when he addressed his first few inquiries to Kew. He had turned to botanizing almost out of desperation as an antidote to outright boredom, and he sought the advice of experts in his new avocation.

Scottish by birth but Irish by heritage and upbringing, Henry had left Ireland because of the depressed Irish economy and political unrest surrounding the question of Home Rule. In this move he followed in his father's footsteps; the elder Henry had left famine-ravaged Ireland during his youth to join the forty-niners in California. Failing to find his fortune there, he had tried his luck a second time in the gold fields of Australia before returning to Ireland via Scotland, where the young man married and Augustine was born.

148. The treaty port of Shasi on the banks of the Yangtze River at Ichang in Hupeh Province. Ocean-going steamers could navigate the Yangtze River as far inland as Ichang, and consequently it became an important commercial center. While the city was one thousand miles inland from the coast, an office of the Imperial Maritime Customs was required, because of the brisk trade that was conducted there, and it was here that Augustine Henry was posted beginning in 1882.

With a masters degree from Queen's College, Belfast, a smattering of Chinese, and enough medical training to qualify, the younger Henry sought adventure and success by joining the Imperial Chinese Maritime Customs Service in 1881 as Assistant Medical Officer. His first post in China was in Shanghai, where the ordered life in the British compound included polo matches and the recently popularized activity of lawn tennis. But in March of 1882 Henry left behind the niceties of Shanghai life when he was reassigned to a customs office in the interior. There he joined a few colleagues stationed at Ichang on the banks of the Yangtze River, one thousand miles inland from Shanghai. Above Ichang the mighty Yangtze courses through five treacherous gorges for a distance of almost two hundred miles, and west-bound junks and sampans had to be hauled upstream against the churning current by teams of fifty or more men pulling tow ropes along narrow footpaths etched into the precipitous, canyonlike walls of the gorges. Because of the numerous rapids and swift current, even downstream travel was extremely hazardous and required skillful navigation. The five defiles constituted a bottleneck for the considerable commerce between the river's upper and lower reaches. Yet the river was the artery linking the fertile farmlands of the Red Basin of Szechwan and the salt mines and mineral deposits of western China with the large cities and commercial ports on the broad Coastal Plain of eastern China. For transporting goods,

149. A view of the Yangtze River in the gorge region upstream from Ichang. This photograph was taken near the boundary of Hupeh and Szechwan provinces and gives a good indication of the topography of the region.

the difficulties of river navigation were preferable to overland routes in the primeval, mountainous borderland between Hupeh and Szechwan, where travel was difficult in the best of seasons and almost impossible during the rainy season, when the narrow, steep tracks became slick with mud.

The routine in the customs office at Ichang left the men stationed there with plenty of leisure time; office hours were ten to four, with an hour for the midday meal. Not content to pass his idle hours playing cards or tennis with fellow customs officers or hunting for game, Henry began to fill his spare time with investigations of local plants used in Chinese medicine. A brisk trade in medicinal herbs was conducted at Ichang, and the customs officials were responsible for levying tariffs on identifiable commodities. With encouragement from his botanical correspondents at Kew, to whom he turned for identifications and advice, Henry soon threw out a wider net and began to collect any plant he did not recognize. Advice and support was also forthcoming from Henry Hance, an official in the British Consular Service in Hong Kong, who had made the flora of China his hobby ever since he arrived there in 1844, shortly after the Opium War. The result was that during his first season—with the help of a Chinese man employed for the purpose—Augustine Henry obtained a thousand collections, which were carefully numbered and dispatched to Kew in November of 1885. Stunned by both the quality of the specimens and the novelties included among them, Daniel Oliver, then Keeper of the Herbarium at Kew, professed that "this collection is one of the most important which we have ever received from the interior of China" (quoted by S. Pim, 1966, p. 37; 1984, p. 29).

William Thiselton-Dyer, Sir Joseph Hooker's son-in-law and his successor in 1885 as director at Kew, was also enthusiastic to the extent that he contacted Sir Robert Hart, Inspector General of Chinese Customs, to request that Henry be given a leave from the service in order to collect more widely in the mountains west of Ichang. This appeal was eventually granted, and with modest financial support from Kew Henry undertook two extensive forays during 1888—the first south and the second north of the Yangtze.

On these prolonged trips Henry penetrated the mountainous no-man's land on the Hupeh-Szechwan border. There, in the lush, undisturbed forests where golden monkeys moved effortlessly through the forest canopy, Henry and his Chinese assistants risked life and limb to procure collections to fill their makeshift presses. Repetitiously they replaced the damp papers in the presses with dry ones until the plants themselves were dry. The specimens that resulted proved to represent an amazing botanical diversity in which novelties outnumbered known species. New maples, viburnums, cotoneasters, hollies, rhododendrons, lilacs, lilies, and honeysuckles, to mention but a few, were included in Henry's sampling

150. Augustine Henry (1857–1930) spent nineteen years in remote towns and villages in China as a medical officer in the Chinese Imperial Maritime Customs Service. During these years his correspondence with botanists in Europe and America provided a needed link with the outside world. Henry wrote from Mengtse in Yunnan, "This place is isolated to the extreme, and it takes such a time to receive letters." In another letter he claimed, "Letters are the only stimulant of a healthy kind an exile has to cheer him up in moments of depression, and to remind him that there is a fair world on the other side of the globe, where men and women live . . . As one grows older the less one seems to like isolation." He found the long solitude difficult and wrote, "In such places as this letters are esteemed more than gold, more than tobacco—they are the only little joys we have. It is very hard to live in such places as this, and we only go through it by the aid of letters and a certain doggedness which is acquired by practice. Russian officers in similar situations in Siberia commit suicide at the rate of 5 per cent. annually, I have been told" (A. Henry, 1889–1898).

of the Chinese flora. In all, upwards of five hundred new species, twenty-five new genera, and an entirely new family of plants—the Trapellaceae, comprised of two species of aquatic plants from China and Japan—were based on Henry's collections from Hupeh and Szechwan.

The botanists at Kew were overwhelmed, and articles in the pages of English botanical journals were soon announcing Henry's discoveries to the scientific community. Daniel Oliver became so intrigued with one of Henry's specimens—one that had been collected near the hamlet of Ma-huang-pou in May of 1888—that he spent hours attempting to determine its identity. His search in the Kew herbarium was unfruitful; and, in order to determine if the plant had already received a name, the botanist was forced to resort to paging through an accumulation of books and journal articles concerning Chinese plants. Consulting Franchet's *Plantae Davidianae*, Oliver found the answer. Henry had chanced upon the same species Abbé David had collected in Mupin in 1869, on which the genus *Davidia*, honoring David, had been based by the French botanist Henri Ernest Baillon. Henry's specimens came from a locality roughly a thousand miles east of Mupin, and fruiting branchlets were also found among his dried collections. Henry, too, had been so impressed by the beauty of the flowering tree that he had sent two of his coolies to gather fruiting specimens in the fall of 1888. Oliver devoted a special article to a description of the previously undescribed fruits and commented that the "*Davidia* is a tree almost deserving a special mission to Western China with a view to its introduction to European gardens" (D. Oliver, 1891, *pl. 1961*).

151. Sir William Turner Thiselton-Dyer (1843–1928) succeeded his father-in-law, Sir Joseph Dalton Hooker, as director of the Royal Botanic Gardens, Kew. When Sir Harry Veitch asked for a recommendation of a man who could undertake field work in China, Thiselton-Dyer suggested Ernest Henry Wilson, who had recently won a Kew diploma at the Royal gardens.

In England during 1889 on home leave, Henry was received at Kew as a celebrity. He returned to China knowing that his collections were both significant and appreciated by the botanists at Kew. But he returned to China with misgivings as he faced a new assignment on Hainan. Henry continued to send specimens from that off-shore island, yet he was not as enthusiastic about the tropical flora that the climate there supported. After a stint on Formosa (Taiwan), Henry was again transferred, this time to the isolated outpost of Mengtse (now Mengzi) in Yunnan Province. There, his collection activities increased; and while he was stationed in Yunnan, Thiselton-Dyer began to urge Henry to send seeds of plants suitable for cultivation in the British Isles. To one of these requests Henry responded:

> With respect to seed, I will do what I can, especially later on when I shall have less plant collecting to do in our immediate neighborhood. But it really is a difficult matter collecting seeds—one arrives on the ground too early or too late . . . Money is not what is wanted, but time, oceans of time. Nothing astonishes people at home so much as the fact, a real fact, that in countries like China, you cannot do everything with money. Patience is more valuable. (Quoted by E. C. Nelson, 1983, p. 31)

In another letter Henry recommended:

> In regard to seed collecting it is not a question of money, but of finding some one with the time on hand and the requisite intelligence and energy, and this is very difficult to find indeed.
>
> I would suggest, so great is the variety and beauty of the Chinese flora and so fit are the plants for European

climate, that an effort ought to be made to send out a small expedition . . . and what I would recommend is that a man be selected, who has just finished his botanical studies at Cambridge University . . . The locality I would suggest is the mountain range separating Szechwan from Shensi or thereabouts—the expedition starting from Ichang in April and covering two seasons.

> In conclusion, I can see now that there were hundreds of interesting plants which I might have noticed earlier in my plant collecting, if I had the experience or the genius or the teaching. If you ever again come across a budding collector like what I was when we began correspondence some years ago, please insist on him being more than a mere collector: and perhaps you will help to develop a naturalist. (Quoted by E. C. Nelson, 1983, p. 32)

And commenting specifically on the davidia, Henry echoed Daniel Oliver's sentiments:

> Davidia is worth any amount of money. I only saw one tree of it, but doubtless there are others in the district . . . I assure you that if I could do anything by writing myself to the Ichang Consul I would do it; I know the ways of people in outposts. You will draw them, if you make the offer exciting. Davidia is wonderful. (Quoted by E. C. Nelson, 1983, p. 31, 32)

Welcome letters reached Henry in China not only from England but from the United States as well. Charles Sargent learned of the Irishman's botanical activities and was soon badgering the customs official for duplicate herbarium specimens and also requesting seed collections for trial in the Arnold Arboretum. When Henry answered Sargent's appeals for seeds in much the same vein he had responded to Thiselton-Dyer, Sargent asked a more direct question. Would Henry consider leading an Arboretum-sponsored collecting expedition in China?

Time had passed, and in 1898 Sargent was still searching for a man to represent the Arnold Arboretum in China. The possibility of a joint venture with the Veitch nursery firm, one in which James Herbert Veitch would serve as the collector, had been tabled when Sir Harry Veitch, uncle of Sargent's erstwhile traveling companion in Japan, had quashed his nephew's traveling plans early in 1893 before the young Veitch had even returned to England from his world tour. The elder Veitch, head of the family firm, expressed his sentiments concerning the matter in a letter to Sargent dated January 10, 1893:

> It will not in my opinion be well for him to be too long absent from England or he will lose too much of his touch with this business. Should he not be here for another 12 to 20 months after his completing his present journey which will have lasted 2 years and ¼ at least, it would be taking him out of the business for about *four years* and this would be a very serious matter indeed . . . Again, too much traveling is apt to unfit a man for settling down again to the routine of work. My nephew even before leaving home found routine work irksome & I am desirous that this feeling should not overpower him. It was partly in the hope that he would overcome this feeling that induced me to let him go abroad. Again if he is to become one of us here he must come and take his place. (AAA, Sargent correspondence)

With this discouraging development, Sargent persisted in attempts to lure Augustine Henry into his scheme in the years that followed. The Irishman was already a seasoned collector who was familiar with conditions in China, Sargent reasoned, but again the Harvard professor was to face disappointment. Henry was primarily interested in collecting botanical specimens rather than living plants, and he had obligations to fulfill as a customs official. Moreover, after nearly eighteen years in China, he was looking forward to returning to England on home leave in 1900. In his frequent letters from China, Henry nonetheless outlined his thinking concerning a botanical expedition in western China and gave Sargent advice concerning costs, routes, and the kind of man needed to undertake the task. Henry also suggested that the successful candidate visit him once in China so that he could personally give information and advice to the new collector. With regret that Sargent's offer had not come earlier in his career when he would have accepted the challenge gladly, Henry responded to Sargent's final proposal in the negative, but softened the blow with encouraging words. A portion of his letter to Professor Sargent dated November 14, 1899, read:

> I regret that I do not see my way to thinking of your proposals in the light of accepting them. But I do not feel by declining that I am really depriving you of the great field of investigating the Chinese flora. You will be sure to find a man. I trust that you will do so, and that you will have many years to live to devote yourself to the flora of China. It is by far the most interesting one on the globe—& to an American must offer even more than to a European—as China in so many ways,

its great rivers, mountains, climate, etc., seems to be a counterpart of the United States.

In conclusion, I must again express to you my best thanks for the kindness and liberality of your offer; so much esteemed by me, as coming from you, one of the most distinguished of living botanists. (AAA, Sargent correspondence)

Augustine Henry did take home leave in 1900 and never returned to China. Instead, the veteran of the Chinese Maritime Customs Service pursued schooling and a career in forestry and dendrology and distinguished himself in both fields. In collaboration with H. J. Elwes, Henry authored *The Trees of Great Britain and Ireland*—a monument of research and scholarship that stands as the British counterpart of Sargent's *Silva*—and together with Elwes established the forestry school at Cambridge University. In later life Henry was the first to occupy the Chair of Forestry at the College of Science in the University of Dublin. In that position Henry spearheaded the campaign to reforest Ireland's denuded landscape, a need that became evident during his years in China. For everywhere that Henry had traveled in the Middle Kingdom he had been dismayed by the poverty and low standard of living that followed in the wake of forest destruction in that heavily populated land. He reasoned that Ireland's economic plight might be improved through the establishment of plantations of trees that would provide needed forest resources.

The Making of a Naturalist in Western China

In October 1899, shortly before leaving China for the last time, Augustine Henry welcomed a visitor in the remote village of Szemao (now Simao) in Yunnan Province. An Englishman who had never before left his native shores, Henry's visitor had departed from Liverpool on April 11 on board the *Pavonia* for the transatlantic crossing to Boston. Arriving in New England on April 23, he spent the next five days exploring the Arnold Arboretum, meeting Professor Sargent, and discussing seed collection and transport with Jackson Dawson (1841–1916), the plant propagator at the Arboretum. Recommencing his global journey, the young man traveled by train across the North American continent to San Francisco, where he sailed for China on May 6. At the conclusion of an uneventful crossing, his ship dropped anchor at Hong Kong on June 3, where letters from friends and supporters in England awaited his arrival.

Almost two months into his journey, he finally reached the Asian mainland. The young Englishman's employer had strongly urged that his first priority be to visit Augustine Henry in Yunnan Province, some one thousand miles inland from Hong Kong. Four long, tiresome months would be required for Ernest Henry Wilson (1876–1930) to cover that relatively short distance and arrive at the outpost near the Burmese border in southwestern China where Henry was stationed. During that four-month period Wilson had ample opportunity to test his mettle as a traveler in the face of long, frustrating delays, outward hostility on the part of the native peoples, and the vicissitudes of travel and lodging in remote regions. These difficulties were all compounded and made more frustrating by Wilson's total ignorance of both the written and spoken Chinese language.

Wilson's route to Szemao, necessarily circuitous, required that he travel southward from Hong Kong to Tonking in French Indochina. From that city, travelers destined for Yunnan were forced to ascend the Red River in small steamers to Laokai (Lao Cai), a jungle town on the Chinese frontier. Small, native boats carried one from there to Manhao, where the overland portion of the journey by mules and sedan chairs commenced; wheeled vehicles had yet to make their debut in this remote area. Moreover, hostilities against foreigners had erupted throughout the border region, and news of murders and attacks on missions, custom houses, and French surveying parties engaged in field reconnaissance for a railroad route from Hanoi to Kunming conspired to make travel impossible. After an almost two-month delay at Laokai, where Wilson was strongly suspected of being an English spy, he undertook the last leg of his long journey to Szemao.

In the company of a Chinese interpreter and a native cook, Wilson reached Mengtse and pushed onward for Szemao, seventeen days distant. Traveling with a caravan of mules loaded with supplies for the customs officials at Szemao, Wilson commented, "Being unable to speak any Chinese, [I] traveled very much as a parcel and enjoyed the trip" (Wilson, 1917, p. 284). When he finally arrived at his destination, Augustine Henry was not among the few Europeans who welcomed the weary traveler. Henry was botanizing in the neighboring hills and did not return until the following day. But once the two men had met, Wilson and Henry set about the business Wilson had traveled for six months to accomplish.

152. Ernest Henry Wilson (left) and Charles Sprague Sargent, whom Wilson referred to as the "autocrat of autocrats, but the kindliest of autocrats," pose in front of flowering specimens of *Prunus subhirtella* Miquel, the higan cherry from Japan, which was first introduced into North America by the Arnold Arboretum. En route to China on his quest for the dove tree, Wilson traveled by way of Boston to confer with Professor Sargent and Jackson Dawson, plant propagator at the Arnold Arboretum.

Wilson surely shared his employer's instructions with Henry, who could have predicted with accuracy their content. "The object of the journey is to collect a quantity of seeds of a plant the name of which is known to us. This is the *object*—do not dissipate time, energy, or money on anything else" (Wilson, 1917, p. 275). These were Wilson's instructions, written on folded foolscap under the printed heading of The Royal Exotic Nursery, James Veitch & Sons, Chelsea.

Sir Harry Veitch, after seeing Augustine Henry's dried specimens of the davidia or dove tree in the herbarium at Kew, had had a change of heart concerning sending a man to China and agreed that the introduction of the dove tree into cultivation was worthy of a special expedition. Sir Harry was willing to risk his firm's funds and prestige in sending a man in Charles Maries's footsteps, despite the fact that some knowledgeable plantsmen maintained that almost every worthwhile plant in China had already been introduced into Europe. To this end he had asked Thiselton-Dyer to recommend a man for the

153. Sir Harry James Veitch (1840–1924), head of the Chelsea branch of the family's nursery empire, hired Ernest Henry Wilson to travel to China in search of the dove tree.

job, and the director of Kew, with Henry's advice uppermost in his mind, had suggested Wilson, a twenty-two-year-old graduate of the diploma course at Kew, who was intent on pursuing a career as a science teacher. Unbeknown to Thiselton-Dyer or Veitch, Wilson had the makings of a naturalist and fit the description of the man Henry had suggested in his earlier letters to Thiselton-Dyer and Sargent.

Henry must have been overjoyed that someone had finally taken his suggestions seriously and sent a man on the mission he himself felt unable to undertake. Before sending Wilson off on his quest for the dove tree, the customs official delighted in giving advice concerning collecting techniques, travel in China, currency values, Chinese customs, and how best to handle men hired for the expedition. He urged that Wilson begin his explorations from Ichang early the following spring, and to aid in locating the one specimen he personally had discovered years before, Henry gave Wilson a sketch map of the western Hupeh region, pinpointing the approximate

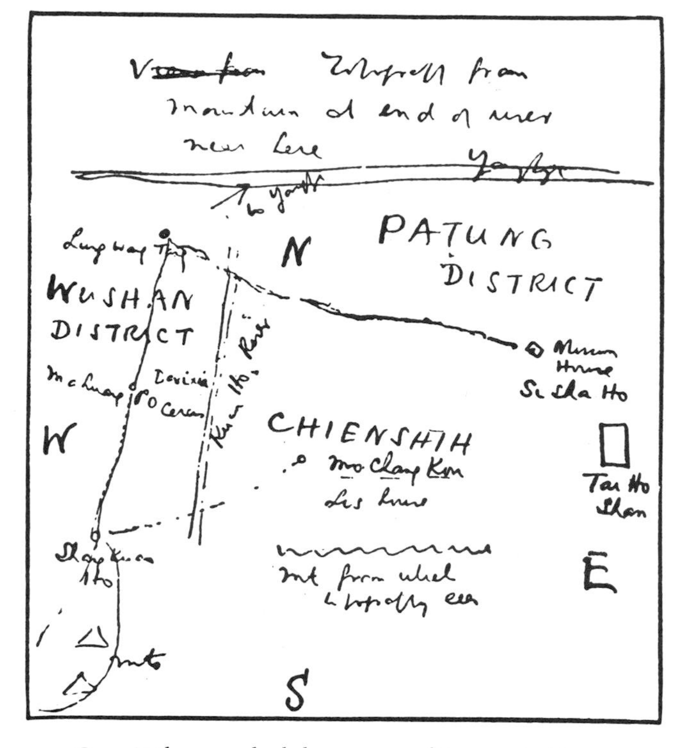

154. Once Wilson reached the outpost of Szemao in Yunnan Province where Augustine Henry was stationed, Henry provided Wilson with this sketch map showing the location of the dove tree he had seen years earlier. The territory shown is roughly equivalent in size to upstate New York.

location of the tree. The map covered an area roughly the size of upstate New York; Wilson's job, or so it seems in retrospect, was akin to finding the proverbial needle in a haystack.

Wilson retraced his steps from Szemao to Mengtse and eventually reached Hong Kong on November 26. From there, encumbered by the equipment he had brought from England, he traveled by steamer to Shanghai and early in 1900 went by boat up the Yangtze River to Ichang, arriving at his jumping-off point on February 24. There, Wilson purchased a native river boat and hired a Chinese cook, interpreter, and men to serve as porters and collectors. A sedan chair for his use was also included in the expedition paraphernalia, although Wilson later related that it was required more as a sign of his status than as a useful piece of traveling equipment. After a few short excursions in the vicinity of Ichang, which were intended to test his men and to provide Wilson with an opportunity to become superficially acquainted with the region's flora, Veitch's agent was ready to risk the turbulent waters of the Yangtze and head upstream in search of the dove tree.

On the morning of Sunday, April 15, 1900—one year and four days after Wilson had left England—his riverboat left its moorings at Ichang to begin the ascent of the Yangtze. A week later, on April 21, Wilson had reached the town of Patung (now Badong) and had become thoroughly aware of the hazards of river navigation; his boat had all but capsized in the swift current in one near disaster, and the rapids and exposed and submerged rocks offered a continuous, life-threatening obstacle course. At Patung another peril to the expedition surfaced that could have brought it to an abrupt end. The town's chief Chinese official frightened Wilson's men with stories of rioting, hostilities toward foreigners, and the prospect of bandits in the regions that lay to the south and west, where Wilson intended to explore. Fearful for Wilson's life, the Chinese functionary begged that the expedition be abandoned. Firm in his resolve to continue, Wilson scoffed at the official's doom-saying but readily accepted the offer of an armed guard of six men. With their misgivings slightly allayed, the coolies in Wilson's employ consented to continue in his service, and the expedition resumed as planned.

Not knowing what misadventures might lie ahead, Wilson and his party left Patung and the Yangtze River behind when they headed south on the morning of April 22. On the evening of the 23rd the party reached the Catholic Mission at Hsi-sha-ho, where Wilson was welcomed as the first foreigner to visit the Christian outpost since Henry had lodged there twelve years earlier. Two years before, one of the priests of the mission had been brutally murdered, and Wilson was given a detailed account of the tragedy and shown photographs of the dead man's remains. Wilson noted that they were "gruesome, nauseating, and horrible to look upon" (Wilson, 1917, p. 288). The Belgian priest who shared his recollections with the plant collector ended by repeating the warnings of imminent danger, stating that he feared further trouble at any time.

With that disquieting prospect in mind, Wilson continued on his journey and arrived at the hamlet of Mahuang-po on the afternoon of April 25. Delighted to learn that some of the residents remembered Henry's visit, he made further inquiries about the *K'ung-tung*, the native Chinese name for the dove tree. He was told that the tree was near at hand, and with local residents as guides, Wilson set out to find the tree. He wrote later that "after walking about two miles we came to a house rather new in appearance. Near by was the stump of

155. This pen and ink drawing of a flowering branchlet of the dove tree was drawn by Blanche Ames Ames, wife of Harvard botanist and orchidologist Oakes Ames, who became supervisor of the Arnold Arboretum in 1930.

156. On his return to England in April of 1902 after the successful completion of his first Chinese expedition, Wilson was presented with this gold watch by James Herbert Veitch, who expressed his sentiments "Well done!" on the inner surface of the watch lid.

Henry's Davidia. The tree had been cut down a year before and the trunk and branches formed the beams and posts of the house! I did not sleep during the night of April 25, 1900" (Wilson, 1917, p. 289).

Despite this discouraging setback, Wilson decided to continue his explorations of western Hupeh Province and to collect any other plants he could find in that region. He also resolved to travel to western Szechwan Province in the late winter of 1900 to visit the region where Abbé David had first discovered the dove tree. This resolve, however, proved unnecessary. On May 19, while collecting near the hamlet of Ta-wan, about five days distant from Ichang, Wilson came upon a fifty-foot specimen of the dove tree in full flower. He located ten additional trees within a one-hundred-mile radius, and in the fall of 1900 he returned to these trees to collect a large quantity of seeds for dispatch to England. These seeds arrived at the Veitch nursery in the early spring of 1901, almost two years after Wilson had agreed to travel to China with the expressed purpose of introducing the dove tree into cultivation in Great Britain.

Wilson continued to collect botanical specimens and seeds in western Hupeh for another year, returning to England in April 1902. Once home, he promptly visited the Veitch Coombe Wood Nursery at Kingston Hill, where his davidia seeds had been sown in the spring of 1901. By April 1902, not one of the seeds had germinated, and Wilson was fearful that they might not be alive. Anxious to determine if disappointment was inevitable, Wilson dug down to examine the seeds; some showed no change since they had been collected in China, while others were beginning to split longitudinally, and still others showed the tip of a protruding root. By May, thousands of young plants had germinated, and with the help of an assistant, Wilson personally potted-up more than 13,000 young plants. In May 1911, the first of these plants flowered in the Veitch nursery.

Irony, however, surrounded Wilson's introduction of the dove tree into cultivation in Europe and North America. As he himself freely admitted, "After my successful introduction of the Davidia in 1901, and its free germination in 1902, I had yet one little cup of bitterness to drain" (Wilson, 1917, p. 294). Unbeknown to Wilson until after his return to England in 1902, seeds of the dove tree had been received in France by Maurice de Vilmorin (1849–1918) in 1897—two years before Wilson left England in their quest—and in 1898 one plant germinated at Vilmorin's arboretum at Les Barres. Père Paul Guillaume Farges, another French missionary stationed in China, sent the seed from which Vilmorin's plant germinated, and he could claim to be the first to introduce the tree into cultivation. Wilson, writing about the dove tree sixteen years after his first trip to China, stated that he had been responsible for the "introduction of *every seedling plant but one* of this remarkable tree" (Wilson, 1917, p. 295).

Two old specimens of the dove tree (*Davidia involucrata* Baillion) grow in the Arnold Arboretum, and each spring avid plantsmen come from all over New England to view the spectacle of these trees in flower. Visiting botanists from the People's Republic of China never fail to request to see the Arboretum specimens, whether in flower or not. Because of the rarity and scattered distribution of the dove tree in nature, the specimens growing in the Arboretum are frequently the first living individuals of this famous Chinese tree that botanists from China have had the opportunity to see.

One of the Arboretum's dove trees originated from a rooted layer of the plant raised from Père Farges's seed and was planted in 1904. The other was grown from one of Wilson's seeds and was received from the Veitch nursery as a sapling in 1911. Both are multiple-stemmed trees, and both grow along Chinese Path on the slope of Bussey Hill. One of the trees is partially obscured by surrounding stewartias, but in the middle of May when the trees are in flower, they are easily seen.

The beauty of these plants is derived from the two large, white bracts associated with the flowers, and also from the glossy, dark green leaves that are reminiscent in size and shape of the leaves of a linden or basswood tree (species of the genus *Tilia*). The rounded or egg-shaped fruits mature in late summer and fall, and the trunks have scaly, brownish bark. The multiple-trunk habit of the Arboretum specimens is attributed to the winter of 1933–1934, when severe cold killed the plants to the ground. Subsequent new growth resulted in the several trunks. Plants growing in their native habitats in China are usually single-stemmed trees.

Once he had located the dove tree and had dispatched a large quantity of its seeds to England, Wilson continued to collect and prepare dried specimens, as well as to gather quantities of seeds of additional Chinese plants. His technique in the field consisted of locating plants in the spring and early summer, when flowering specimens could be collected. Retracing his route in the late summer and fall, he prepared fruiting specimens from trees located earlier in the year, and gathered seeds for germination in England. His winters were spent in Ichang, where dried specimens could be sorted and consignments

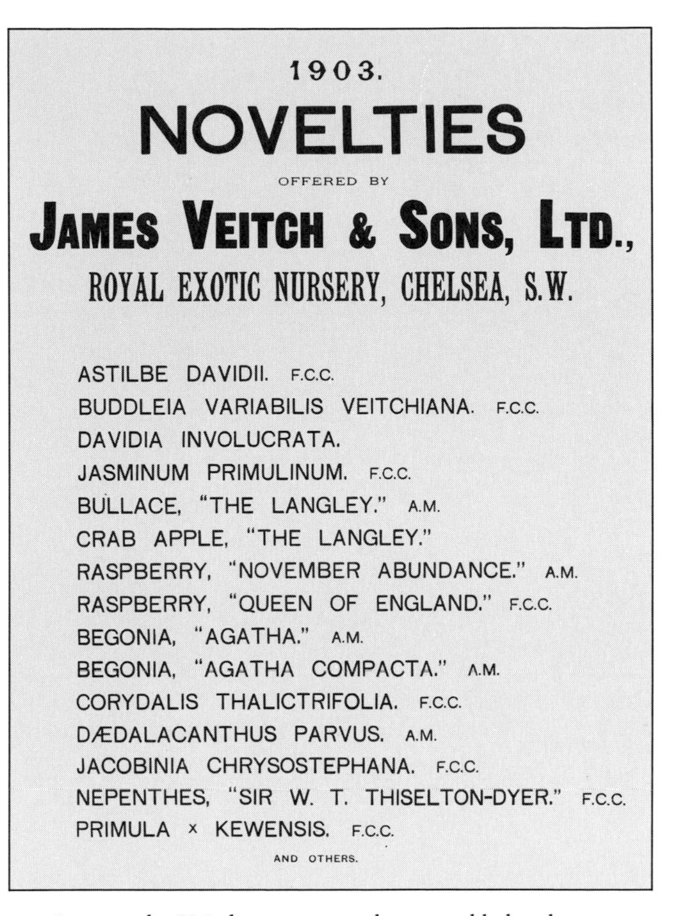
1903.

NOVELTIES

OFFERED BY

JAMES VEITCH & SONS, LTD.,

ROYAL EXOTIC NURSERY, CHELSEA, S.W.

ASTILBE DAVIDII. F.C.C.
BUDDLEIA VARIABILIS VEITCHIANA. F.C.C.
DAVIDIA INVOLUCRATA.
JASMINUM PRIMULINUM. F.C.C.
BULLACE, "THE LANGLEY." A.M.
CRAB APPLE, "THE LANGLEY."
RASPBERRY, "NOVEMBER ABUNDANCE." A.M.
RASPBERRY, "QUEEN OF ENGLAND." F.C.C.
BEGONIA, "AGATHA." A.M.
BEGONIA, "AGATHA COMPACTA." A.M.
CORYDALIS THALICTRIFOLIA. F.C.C.
DÆDALACANTHUS PARVUS. A.M.
JACOBINIA CHRYSOSTEPHANA. F.C.C.
NEPENTHES, "SIR W. T. THISELTON-DYER." F.C.C.
PRIMULA × KEWENSIS. F.C.C.

AND OTHERS.

157. In 1903 the Veitch nursery catalogue could already announce the availability of the dove tree (*Davidia involucrata*) to the gardening public. Since the dove tree did not flower in cultivation until 1911, a photograph of a dried herbarium specimen was used showing the large bracts associated with the flowers to entice prospective buyers. Other Wilson introductions from China were also featured.

158. This photograph shows one of the curious fruits of the dove tree, the prize Wilson went to seek on his first expedition to China for the Veitch nursery firm.

of seeds readied for shipment to England. If the dove tree had been the only plant acquired during his expedition, the adventure would have been judged a success; but a virtual cornucopia of new, horticulturally significant plants resulted from Wilson's field activities. Many of these plants had been collected first by Augustine Henry or Abbé David as dried botanical specimens, but credit is due Ernest Henry Wilson for the introduction of living plants into cultivation.

Perhaps the best known is the paperbark maple (*Acer griseum* (Franchet) Pax), which ranks as one of the finest small trees available—albeit in limited quantities—in the nursery trade. Wilson himself claimed that the paperbark maple (Plate 23) is "easily the most outstanding of all Chinese maples and one of the most distinguished members of [its] large and varied tribe" (AAA, unpublished Wilson MS). The most obvious ornamental attribute of the tree is the paper-thin layers of bark that peel and exfoliate from the trunk and limbs to expose ever deeper sheets. The bark ranges from orangish brown to a lustrous cinnamon color, and when a leafless tree is encountered it is sometimes difficult to recognize the plant as a maple. Attaining heights between fifteen and fifty feet, well-grown specimens develop open, oval-to-rounded crowns, and the main limbs are ascending to somewhat spreading from the trunk. A member of the trifolioliate group of maples that includes the Nikko maple from Japan, the paperbark maple produces leaves with three dark green leaflets, which are grayish green on the under surfaces owing to a covering of grayish hairs. In fall, the leaves assume bright colors ranging from salmon red to crimson and contribute generously to the palette of fall colors in American landscapes.

In its natural range the paperbark maple is confined to the mountains of central China, where it grows on forested slopes between four and six thousand feet in altitude. Nowhere is it a common tree, and even in cultivation it has remained a sought-after rarity. Its seeds, which are frequently produced in abundance, are usually devoid of viable embryos because the flowers are generally either male or female and require cross-pollination for viable seed production.

A venerable old specimen of the paperbark maple grows in the Arnold Arboretum on the slope of Bussey Hill near the two dove trees. Wilson sent this plant from China as a potted sapling in 1907. Another individual, now growing in the maple collection, was received from the Veitch nursery in the same year and was grown from Wilson's seed sent to England in 1901. Other more youthful specimens have been included in various Arboretum locations, and the specimen to the left of the gate into the nursery area at the Dana Greenhouses is one of the most splendid examples of this unusual tree. Yet another superior example grows adjacent to the tree peony collection near the Center Street gate and is worthy of a special trip, in the Japanese tradition, for "maple viewing."

Of the many Chinese shrubs that entered the horticultural marketplace early in this century, two of the most remarkable were sent from Ichang by Ernest Wilson to the Veitch nursery in Chelsea. The rosy dipelta (*Dipelta floribunda* Maximowicz) was first encountered by Wilson in the spring of 1900, as he was searching for Henry's dove tree. On the dry, stony roadside south of Patung, Wilson encountered an amazing number of these large shrubs in full flower, and in the fall of the same year a large quantity of fruit was collected from the same population. Unfortunately, only one or two seeds were viable, and in the fall of 1901 living plants were forwarded to England in an attempt to establish the species in cultivation. The rosy dipelta has likewise failed to produce viable seeds in cultivation, and asexual means of propagation—chiefly layering—were resorted to in the effort to produce enough plants to satisfy the initial market demand. Today, despite its beauty, the rosy dipelta is only rarely offered by commercial growers, undoubtedly because of the difficulties surrounding its propagation.

An extremely floriferous shrub that develops an arching habit and reaches a height of upwards of fifteen feet, the rosy dipelta produces its abundant corymbs of fragrant, tubular flowers in early May. Each flower is subtended by several leafy, shieldlike, green or purplish bracts, which persist and enlarge to surround the developing, albeit seedless, fruits. The tubular corollas are a delicate, pale pink and five-lobed at their mouths, and the inner throat surfaces are suffused a bright yellowish orange in a reticulated pattern. In fall, when the shieldlike bracts have enlarged and turned pale brown, the clusters of fruits held at the ends of the leafy shoots provide another period of seasonal interest.

Closely related to the dipelta but producing smaller flowers that lack associated bracts, the beautybush (*Kolkwitzia amabilis* Graebner) is as common in gardens today as the dipelta is uncommon. Ironically, in nature the beautybush is extremely rare, and during his travels in China Wilson located but one population of the plant on a windswept ridge between nine and ten thousand feet above sea level in the Fang Hsien (Fang Xian) region of Hupeh Province northwest of Ichang. There, in the

159. The beautybush quickly became an extremely popular shrub in American gardens in the early part of this century and was widely promoted by nurserymen. Its easy propagation, coupled with its extremely floriferous habit and dependable flowering each spring, have combined to ensure its continued popularity and use where a large, arching shrub is desired in landscape plantings.

fall of 1901, Wilson gathered the hairy, ripe fruits of the shrubs for the first and last time, dispatching them to England later in the season. Unlike the seeds of the common and widespread rosy dipelta, the majority of the rare beautybush's seeds proved viable, and a large number of plants were raised in the Veitch firm's Coombe Wood Nursery. One of the seedlings was sent from England to the Arnold Arboretum in 1907, and when the plant flowered it was immediately recognized as one of the most important shrubs ever introduced from China. Perfectly adapted to the New England climate and of easy culture throughout much of the East and Midwest, the beautybush soon entered the stock and trade of American nurserymen and was sold by the thousands and widely planted.

The beautybush annually produces an abundance of delicate pink flowers that literally obscure the leaves and twigs of the arching branches of these medium-sized, upright shrubs. The five-lobed corollas of individual flowers, like those of the rosy dipelta, are broadly tubular and essentially bell-like in shape, and the throats exhibit a clear orange reticulated pattern on their inner surfaces. The united sepals form a five-lobed calyx, which is covered by lustrous gray hairs. The calyces are persistent in fruit, and the bristle-like hairs give these dry, indehiscent, seed-filled structures a distinctive appearance. In summarizing his own assessment of the beautybush, Wilson wrote, "For the colder parts of America I esteem this one of the best of my introductions" (AAA, unpublished Wilson MS).

Other plants too numerous to list were grown from Wilson's collections made between 1900 and 1902. These include the tea crabapple (*Malus hupehensis* (Pampanini) Rehder, sometimes incorrectly listed as *Malus theifera* Rehder), an extremely beautiful and floriferous crabapple, the leaves of which were used by the Chinese to brew a substitute for tea; and Veitch spiraea (*Spiraea veitchii* Hemsley), a vigorous bridal wreath with arching stems that are annually covered by corymbs of pure white flowers. The first plants of the Chinese gooseberry (*Actinidea deliciosa* (A. Chevalier) C. F. Liang & A. R. Ferguson, formerly *Actinidia chinensis* Planchon) to be cultivated in Europe resulted from Wilson's seed collection, and plants grown from Wilson's seed were eventually distributed to growers in New Zealand. In that country they became the foundation on which the kiwi fruit industry was based. Kiwi fruits—a name denoting a New Zealand origin that has largely supplanted the earlier name of Chinese gooseberries—are now imported into the United States from New Zealand, are also grown commercially in California, and are becoming commonplace in American supermarkets.

Wilson's expedition to China was so successful in introducing new plants that the Veitch firm endeavored to sponsor the collector on a second collecting trip to the Middle Kingdom. Despite the protestations of his bride, Wilson agreed to undertake another mission for the Veitch nursery, and in this venture he was given another specific goal. The object of his second expedition, which commenced on January 23, 1903, and ended with Wilson's return to England in 1905, was to procure a quantity of seed of the golden poppywort (*Meconopsis integrifolia* (Maximowicz) Franchet), then the most highly regarded of alpine flowers. To accomplish his task Wilson traveled nearly 13,000 miles within the space of five and one-half months from England to the alpine regions of the Chinese–Tibetan borderland.

Wilson had become a veteran of the Chinese outback, familiar with the flora and fauna, the geology, geography, natural resources, and customs of its diverse ethnic cultures. Wilson was equally at home in the herbarium at Kew, the Veitch nursery at Coombe Wood, the foreign compound at Ichang, or a disheveled, bedbug-infested inn on a remote mountainside in Hupeh or Szechwan Province. As Augustine Henry had predicted concerning Wilson in a letter written after he had first met the former Kew student in Szemao, "He is a self-made man, knows botany thoroughly, is young and will get on" (quoted by E. C. Nelson, 1983, p. 33).

160. The yellow, applelike fruits of the tea crab (*Malus hupehensis*) provide a second period of ornamental beauty for this crabapple in the fall. The leaves of this species are sometimes used as a substitute in brewing a tealike beverage in China, thereby giving rise to its common name.

407

161. The so-called kiwi fruit (*Actinidia deliciosa*) is generally associated with New Zealand, where large numbers of the gooseberrylike fruits are produced on a commercial basis for export to American and European markets. This twining vine is actually native to China, and E. H. Wilson was the first horticultural explorer to introduce the plant into cultivation in the West. Stock from Wilson's introduction growing in England constituted the basis on which the kiwi industry was initially established in New Zealand and is but one example of how useful plants have been moved around the globe. This photograph of the fruits was taken by Wilson in China, where the fruits have long been esteemed by the Chinese.

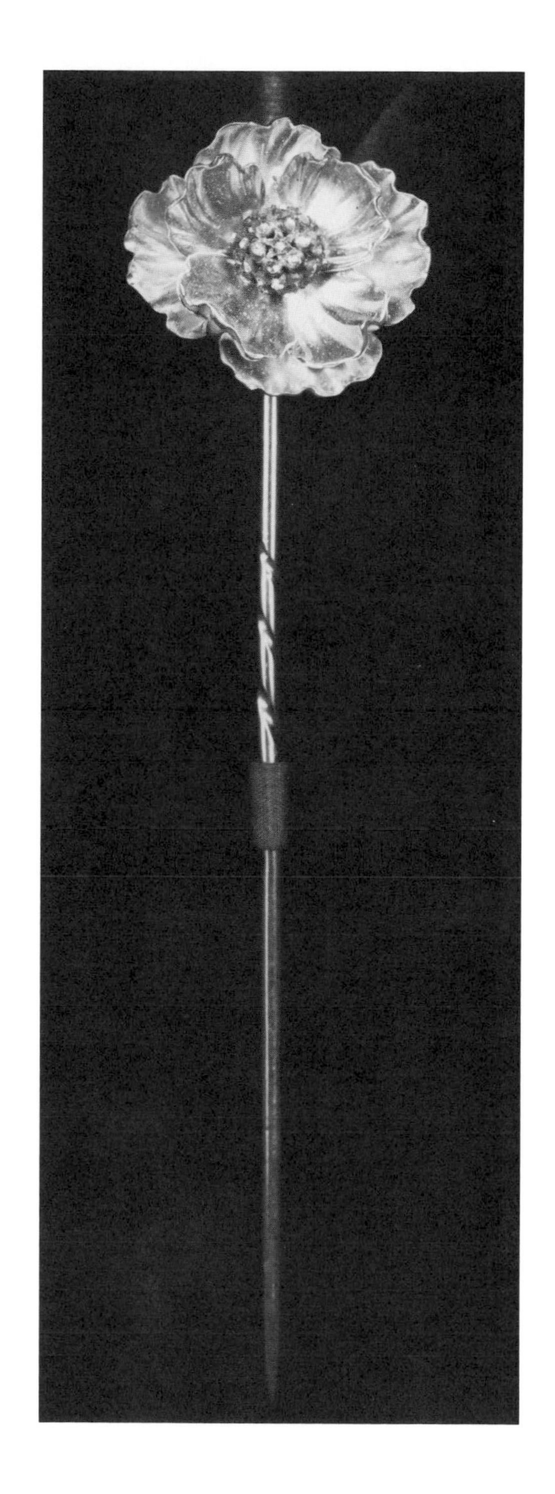

162. In appreciation of his efforts on his second expedition to China for the Veitch firm, James Herbert Veitch presented this gold stickpin to E. H. Wilson. The head of the pin is fashioned to represent a golden poppywort flower, and the central boss of stamens is encrusted with small diamonds.

CHAPTER SIX

The Arboretum in Asia

Faithfully and Obediently Yours

In 1922, when Charles Sprague Sargent wrote his fifty-year review summarizing the progress that had been made at the Arnold Arboretum, he was able to record the following concerning the institution's role in botanical exploration and plant introduction during his forty-nine-year tenure as director.

> During the last forty years the Arboretum has lost no opportunity to increase the number of species of plants cultivated in the United States and Europe. Its officers and agents have continued to explore the forests of North America; they have visited every country in Europe, the Caucasus, eastern Siberia and Korea, and have studied every species of tree growing in the forests of the Japanese Empire from Saghalin to the mountains of Formosa . . . Agents of the Arboretum in pursuit of knowledge and material have visited the Malay Peninsula, Java, the Himalayas, the high mountains of east tropical Africa, southern Africa, Australia, Mexico, Peru, Chile southward to Tierra del Fuego, and the Falkland Islands. (C. S. Sargent, 1922, p. 143)

In no region where agents of the Arnold Arboretum had traveled had they been more successful in locating new species and in collecting specimens for the herbarium and seeds for trial in the living collections than in eastern Asia. Sargent himself had explored Japan in 1892 and returned to the island nation on a round-the-world tour undertaken in 1903. Accompanied by his eldest son, Andrew Robeson, and his friend and energetic California colleague John Muir, Sargent left Boston on his global trip toward the end of May. England was their first destination, and after visiting gardens and museums throughout Europe and undertaking field work in the Caucasus, the threesome boarded a coach on the Trans-Siberian Railroad in Moscow. From Harbin in Manchuria they crossed into China and traveled overland to Peking. From China the globetrotters crossed the Yellow Sea to Korea, botanically a virgin territory, and finally sailed to Japan.

This trip offered Sargent his first opportunity to become briefly familiar with China, and not having satisfied his curiosity, Sargent—with Robeson in tow—went by boat from Japan to Shanghai and meandered further down the Chinese coast, calling at other port cities before sailing for San Francisco. While his personal observations and collections were limited to the region around Peking and in gardens in southern coastal cities, Sargent's ambitions with regard to the botanical and horticultural exploration of central and western China were only whetted by his fleeting exposure to the country that Augustine Henry had compared with the United States. In his sixty-third year, Sargent recognized that he would not be able to undertake botanical and horticultural exploration of the forest resources of China, despite his generally good health. He was still seeking a man to do the work for him and to represent the Arnold Arboretum in central and western China.

In 1905, following in Sargent's footsteps, John G. Jack—Sargent's right-hand man on the Arboretum staff—undertook travel in eastern Asia. Touring in Japan, Korea, and northern China, Jack added to his baggage plants and seeds for the Arboretum collections, and many of the plants grown from seeds he brought back to Boston still flourish in the Arboretum's living collections. Among the noteworthy ornamentals Jack collected were the beautiful royal azalea (*Rhododendron schlippenbachii* Maximowicz) from Korea, the low-growing Kirilow indigo (*Indigofera kirilowii* Palibin), the harlequin glorybower (*Clerodendrum trichotomum* Thunberg), and the Japanese beauty berry (*Callicarpa japonica* Thunberg). Jack's sojourn was relatively brief, since funds came out of his own pocket. Nonetheless, he was able to satisfy his curiosity concerning the Asian nations he had visited, and he returned to the Arbore-

163. Charles Sprague Sargent at work on his celebrated *Silva of North America* in the Museum building at the Arnold Arboretum. Veteran of field work across North and South America and in the forests of Japan, Sargent had by 1905 determined to employ E. H. Wilson to conduct field work in China under Arnold Arboretum auspices.

tum content that he had added to his understanding of the Japanese, Korean, and particularly the northern Chinese floras.

Pressing budgetary shortages at the Arboretum probably figured in any explanation offered the would-be collector for the lack of institutional support. Sargent, while not admitting it openly, probably reasoned that any money he could make available for the support of a collector should be reserved until a man willing to undertake a major Chinese expedition had been found. And in March 1905, Ernest Henry Wilson had returned to England at the conclusion of his second successful trip in China for the house of Veitch. Sargent intended to write to Wilson and propose that the collector undertake a third trip to China, this one under Arnold Arboretum auspices. Money would be needed should the young but experienced Englishman agree to Sargent's proposal, and a high salary might be required for Wilson if Sargent was at long last to be successful in sending a man to China under Arboretum sponsorship.

During the winter of 1905–1906 Sargent again left Boston on a collecting and exploratory mission for the Arboretum, although on this trip he headed in a southerly direction. Ostensibly, Sargent intended to investigate the temperate flora of Peru and Chile, yet the prospect of escaping a New England winter may have helped to determine his destination. While the dendrologist enjoyed the mild climate of the Andes and was absorbed with its flora, his mind was also occupied by thoughts of the Arboretum, its future, and the exploration of China. Almost without warning, Sargent changed his itinerary, resolving to return to Boston via England. He was intent upon a face-to-face meeting with Wilson, and when he reappeared at the Arboretum in April of 1906—tanned and refreshed from his winter sojourn—Sargent resumed his duties with renewed vigor. He had met with Wilson and could not help feeling guardedly optimistic concerning the future.

Anxious to have a response to his verbal proposal, Sargent wrote to Wilson on June 28, 1906, urging,

> I hope that you have been giving serious thought to the proposition for another journey to China and that you will come to see your way clear to doing this. I am more than ever convinced of the importance of further botanical investigation in western China and of the fact that you are the only man to undertake this work with the prospect of carrying it out successfully. I think that the financial part of it can be managed here if you are willing to undertake the work, and I shall take the mat-

164. John George Jack (1861–1949) arrived in Boston from his native Canada in 1886 to study under Professor Sargent at the Arnold Arboretum. In 1891 he was appointed lecturer, and in 1908 assistant professor of dendrology at the Arboretum, and in both capacities Jack was responsible for leading field classes, lecturing, and supervising the plantings in the collections. Jack's presence on the Arboretum staff freed Professor Sargent from his teaching obligations and some of his administrative duties. In addition to visiting Asia, Professor Jack also undertook extensive travels in Europe and in western North America, and on all of his trips he never failed to return with plants and seeds for the Arboretum collections. Jack is shown here (left) with one of the first groups of Chinese students to come to Harvard and the Arboretum to pursue advanced training in botany.

165. *Rhododendron schlippenbachii*, the so-called royal azalea, is a common shrub of the Korean peninsula. The flowers of this species are among the largest produced by the azalea group of rhododendrons. John Jack successfully introduced this species to the Arnold Arboretum in 1905 when he traveled to Japan, northern China, and Korea, the "land of the morning calm."

ter up with some friends of the Arboretum as soon as I receive a favorable reply from you on the subject. Do not decide hastily as the matter is too important to be dismissed with a negative without the most careful consideration . . . The trip is worth making alone for the seeds of the fine conifers you have discovered, and you are likely to remain practically unknown unless you revisit China. (AAA, Sargent correspondence)

Sargent's letter was addressed to Wilson at the Imperial Institute in Kensington. After returning to England in 1905, the veteran of two Chinese expeditions had accepted a government post as a botanical assistant at the Institute. Wilson intended to settle down and establish himself in a botanical career and in family life as well. His second stint in China for the Veitchs had commenced six months after his marriage to Helen Ganderton in 1902 and had forced the couple to postpone having children. On May 21, 1906 Wilson's daughter, Muriel Primrose, was born, and her birth had temporarily intensified Wilson's domestic inclinations. Moreover, Wilson's wife, who had been separated from her husband for two long years, was strenuously opposed to Sargent's proposal, remarking that she hoped that her husband would never again return to China.

But Sargent was persistent and also brought pressure to bear on Wilson through the aid of influential friends in England. George Nicholson, curator of the gardens at Kew, Augustine Henry, who had decided to resign his post with the Maritime Customs to remain in England, and strong-willed Ellen Willmott, the wealthy horticulturist and gardener extraordinaire at Warley Place, all intervened on Sargent's behalf and stressed the scientific and horticultural importance of the Chinese work. In their shared opinion, Wilson had left much undone, particularly in fostering an understanding of the taxonomic and phytogeographic relationships of the flora of the Middle Kingdom. Sargent, to achieve his goal, threatened to return to London once again if his presence there might provide the needed assurances and elicit the positive response from Wilson that he sought.

As negotiations with Wilson neared an end, finances emerged as a crucial consideration. Writing to the professor on August 2, 1906, Wilson explained,

> Your letter of July 16th reached me some ten days ago. It is not altogether to be expected that you and I can look on this Chinese business from exactly the same standpoint. There are many difficulties in the way, difficulties that are well-nigh insuperable, but which you cannot be expected to appreciate in the way I do. Putting all sentiments aside, I have to throw up a government appointment which, if at present modest, has possibilities. Secondly, I have to leave behind my wife and child and these have to be provided for. Thirdly, there is a possibility of the pitcher going once too often to the well. There are other things . . . to be considered, the possibilities of obtaining suitable employment on my return do not appear to me brighter in the future than in the past.

The sum and substance of the whole thing, is that unless the remuneration is sufficient to enable me to put a goodly sum aside, so that I shall not be forced to take the first thing that comes along on my return, simply because it means bread and butter, I am not justified in considering any such proposition as the one you have made. (AAA, Wilson correspondence)

At long last, after Sargent had enlisted subscriptions from numerous friends—Ellen Willmott included—to help underwrite the venture, Sargent and Wilson agreed to terms whereby Wilson would receive a salary of £750 a year for two years, the equivalent of $7,500. Other costs of the expedition were estimated by Wilson, including round-trip travel to and from London. The grand total, a sum Wilson estimated sufficient for a two-year expedition in central and western China and including his salary, was £6,200 or $13,000! Without hesitation Sargent agreed to provide the funds, and to sweeten the bargain, he held out the possibility of employment—at least a temporary position—at the Arnold Arboretum on Wilson's return from China. Sargent had snared his man and was in an uncharacteristically expansive mood.

Wilson began his third expedition to China as he had begun his first. He sailed from England to Boston, where he conferred once again with Sargent and Jackson Dawson before leaving for China via San Francisco in late December. He intended to be in Ichang by April 1907, ready to begin collecting as warm spring weather came to the mountains of western Hupeh. While Wilson was still en route to China, Sargent addressed the first of many communications he would write to the Arboretum's new agent, this one care of the Shanghai office of the Hongkong & Shanghai Bank. In his letter, dated December 26, the Arboretum's director recapitulated and summarized the discussions the two men had had in Boston, stressing the scientific aspects of the mission. Sargent reminded Wilson,

The object of this journey is to increase the knowledge of the woody plants of the Empire and to introduce into cultivation as many of them as is practicable. It is therefore desirable to cover as much territory as possible rather than to attempt to gather large quantities of specimens or seeds of any particular plant, and, although we have generally decided on the territory that you will visit, it is desirable as far as it is practicable for you to do so to cover new regions not previously visited by you. (AAA, Sargent correspondence)

Sargent had insisted that Wilson carry a large camera, despite Wilson's professed ignorance of the art of photography. Sargent had written Wilson earlier suggesting, "It would be well, too, to take along a small instrument in case of accident. The large instrument only means another porter, and that is not a very important item. Bring, too, enough plates and films with you as there will certainly be a large amount to photograph" (AAA, Sargent correspondence, letter dated November 6, 1906).

In compliance with Sargent's wishes, Wilson had purchased a Sanderson whole-plate field camera in London, the best he could find, and one which used glass-plate negatives measuring six and one-half by eight and one-half inches. With a cumbersome tripod and crates of heavy, fragile glass plates in his luggage, Wilson had headed to China prepared to expose the plates as Sargent had admonished. Sargent nevertheless prodded him, writing,

It is desirable to photograph as many trees as possible, provided the tree photographed can be named . . . Photographs of flowers like Magnolias, Rhododendrons, etc., made life size if possible, are of very great value, and so are the photographs of the fruiting branches of Picea, Tsuga, and Abies. Photographs on a large scale of the bark of trees are extremely valuable and should be made showing the character of the country inhabited by the different trees. It would be well, too, if time permits, to take views of villages and other striking and interesting objects as the world knows little of the appearance of central and western China. (AAA, Sargent correspondence, letter dated December 26, 1906)

Wilson inaugurated his side of the correspondence on January 7, 1907, by writing Sargent to inform him that he had arrived safely in Oakland, California, but not without incident and delay.

Just outside of Omaha a switch engine ran into us derailing and smashing up the tender of our engine together with the luggage and dining cars . . . The uninjured portion of the train was taken back to Omaha and after considerable delay a fresh engine, luggage and dining cars were attached and we proceeded on our way. Later in the same day we passed two East-bound express trains in collision. The day following we passed two wrecked freight trains. Altogether I can assure you that I have had enough of American railways to last me for a time. (AAA, Wilson correspondence)

166. Following in the tradition established on his first two Chinese expeditions, E. H. Wilson purchased a houseboat at Ichang for river travel on the Yangtze at the beginning of his first Arboretum-sponsored trip. This sturdy craft was christened *The Harvard*. Wilson's "boy" and the boat's captain are standing on the roof of the cabin, while the expedition's handyman stands on the gangplank.

Wilson sailed from San Francisco on board the *Doric* on January 8, arriving in Shanghai on February 4. By the twenty-sixth of the month he had progressed as far as Ichang, where a native river craft—christened *The Harvard*—was purchased and men were employed for the coming season's travels. Word of Wilson's arrival in Ichang had preceded him, and several of the collectors and porters he had employed on his Veitch trips were awaiting him at the docks. On March 31, the day before Wilson was to set out on his first foray, he informed Sargent, "It will interest you to learn that all my old men have turned up—the trouble is that there are too many of them" (AAA, Wilson correspondence).

The 1907 season was largely spent in the rugged, mountainous regions north and south of the Yangtze River in western Hupeh Province, where Henry had explored twenty years before and where Wilson had gone in search of the dove tree on his first expedition for the Veitch firm. By contrast, during the 1908 season Wilson headed west from Ichang and entered Szechwan Province, where he concentrated his efforts for the remainder of the year. In April he explored the countryside in the vicinity of Wan Hsien (Wan Xian) and Feng-tu Hsien (Fengdu Xian) along the Yangtze River in eastern Szechwan, but by early May he had proceeded farther west into the Red Basin and was exploring in the environs of Chengtu, the provincial capital.

Basing himself at Kiating (now Leshan), a town on the banks of the Min River at the confluence of the Ya and Tung rivers, Wilson planned his strategy for the remain-

167. Wilson turned the lens of his Sanderson camera on a group of his Chinese men, many of whom worked for the botanical and horticultural explorer on all of his Chinese expeditions. These men assisted Wilson by collecting specimens, pressing and drying herbarium specimens, and preparing and packing seeds, roots, and bulbs for shipment to Boston. Without their assistance, Wilson's horticultural and botanical haul would have been greatly diminished.

der of the year. On the seventh of May he outlined his plans in a letter to Sargent, enclosing a sketch map of the region with his proposed collecting routes marked in blue. Wilson planned to travel to the mountains northwest of Chengtu, making a circuit westward and southward and returning eastward to Chengtu. Time would also be spent collecting in the vicinity of Mupin, where Abbé David had based his operations when he discovered the dove tree in 1869. And another trip originating from Kiating would take the explorer west to the Mount Omei (Emei Shan) region and further west to Mount Wa, a mountain never before visited by a Westerner. This foray was planned for late in the season, when fruits and seeds would be ripe for collection.

Returning to Kiating at the end of August from his northern excursion, Wilson learned from Sargent's letters that the shipments of seeds, cuttings, scions for grafting, and bulbs he had dispatched from Ichang after the 1907 collecting season had reached Boston in poor condition. The greatest blow concerned enormous losses

168. Three of a series of six postage stamps depicting Mount Omei, which were issued by the People's Republic of China in 1984. One of China's seven sacred mountains, Mount Omei annually attracts thousands of Chinese and foreign tourists, who make the ascent of the mountain by road and, at higher elevations, by foot. Many temples and shrines serve as way-stations on the steep climb. Considering the number of species that are known from the mountain, this region supports one of the richest and most diverse floras in the North Temperate Zone, with perhaps a greater number of species than for any area of comparable size. The stamps pictured here show the 10,145 foot summit of the mountain.

in the thirty-two crates of lily bulbs shipped to the R. & J. Farquhar Nursery in Boston. Of a total of 18,237 bulbs, only 873 were sound when the crates were opened and unpacked in Boston; 17,364 bulbs had rotted en route from Ichang to New England! Wilson was astonished and quickly responded to Sargent's ugly news, writing,

> I need not enter into my feelings of bitter disappointment and vexation on mastering [your letter's] contents. In slang language I was "knocked all of a heap." The living plants, grafts, and cuttings were packed in the same way as Veitch's were that traveled [so] well. That such shipments are of the nature of a gamble one realizes, and failure to a certain degree is only to be expected, but that it should be so disastrously complete one did not expect. However, there is nothing to be gained by entering into explanatory details and excuses, and to remedy the failure is the task before us. To a certain extent this can fortunately be accomplished. The exact localities of the majority of the plants and grafts sent is known and I can on my return to Ichang in January next make a trip and recollect many. Such a trip will not occupy more than three weeks . . .
>
> The failure with Farquhar's lily bulbs was evidently due to not balling them all in clay. All bulbs sent to Veitch were packed in clay and I made a big mistake in departing from my old methods for the purpose of saving money in cases and freight. (AAA, Wilson correspondence, letter dated August 27, 1908)

Returning from his late summer trip to Mount Omei and Mount Wa, Wilson related on the first of October,

> The trip through the wilderness behind Mts. Omei and Wa proved fairly profitable but, owing largely to bad weather, was desperately hard and uncomfortable. I

> never want to hear of, much less see the region again! Above 4,000 ft. alt. the country is of limestone formation and, as in general with this rock, is much broken up, forming steep cliffs and crags presenting very wild and savage scenery. Charcoal burners have destroyed all the forest, leaving in lieu a dense jungle of shrubs; above 7,000 ft. (to 10,000 ft.) bamboo forms one absolutely impenetrable thicket. Nothing is more disheartening to a botanical collector than these bamboo jungles. Practically nothing can grow in them but bamboos, there is no traversing them save by the recognized paths, unless one has time to cut a new track, and if there is any suspicion of rain falling one is drenched through in less than no time by the overhanging culms. (AAA, Wilson correspondence)

In another letter, one dated November 29, Wilson reported the startling news of the demise of the emperor and empress dowager, and informed Sargent of the strictly enforced period of mourning during which no one could bare their head and blue buttons were to be worn on all hats. White lanterns and scrolls were to be displayed in all streets and at the entryways of all official quarters. Wilson cautioned that "secret societies of an anti-dynastic nature abound everywhere and in certain parts are very strong. Now is their opportunity." (AAA, Wilson correspondence)

Wilson's work in western Szechwan Province came to a close as the fall season advanced, and by mid-December he was prepared to return to Ichang in Hupeh Province. Informing Sargent of his intended departure from Kiating, and hoping that nothing would occur to destroy his work before the valuable and irreplaceable collections could be entrusted to the mails at Ichang, Wilson observed,

> The expedition leaves here tomorrow (Dec. 20th) for down river. Another boat in addition to our own has had to be chartered to accommodate the collections. The most anxious part of the year's work is now before us: every precaution will be taken and in all probability we shall reach Ichang safely; nevertheless, I heartily wish we were safely there at the moment of [this] writing. At this low water season some of the rapids are at their worst and if an accident happens and the boat strikes a rock the chances are that everything will be lost. However, to know and appreciate the dangers is always something towards combating them and I trust the good fortune which has attended us heretofore will not desert us now. (AAA, Wilson correspondence, letter dated December 19, 1908)

Wilson was fortunate to reach Ichang safely on January 15, 1909, with both boats and his precious collections intact. A potentially disastrous mishap with one of the boats on December 24 had been avoided, repairs quickly made, and the journey recommenced on Christmas Day. Concerning the accident Wilson wrote from Ichang, "I of course took the precaution to put half the collections in another boat. But whilst this halves the risks it doubles the anxiety. However, here we are safe and sound in every way and the dangers of the upper Yangtze are all behind us" (AAA, Wilson correspondence, letter dated January 16, 1909).

During the latter part of January and the better part of February Wilson undertook another foray southwest of Ichang to replenish the cuttings and scions that had been lost in shipment to Boston. Arrangements had also been made earlier with local Chinese collectors to amass another astounding consignment of lilies, and Wilson could report that "something like 25,000 lily bulbs [are] in hand" (AAA, Wilson correspondence, letter dated January 18, 1909). These were prepared for shipment using a time-honored procedure: each bulb was individually embedded in moist clay, and the coating was allowed to dry before the bulbs were placed in crates and surrounded by pulverized charcoal for transport.

Wilson was also busy preparing seeds and cuttings, which were packed in moist sphagnum and wrapped in oiled paper, for shipment. The herbarium specimens, too, had to be crated, as did the glass-plate negatives that had been exposed using his "large" camera. At the end of February Wilson claimed that he had "taken 57 dozen photos. I hope to manage 3 dozen more and thus complete six gross" (AAA, Wilson correspondence, letter dated February 20, 1909). The exposed but undeveloped plates were addressed to Mr. E. J. Wallis at the Royal Botanic Gardens, Kew, where Wilson would personally supervise their development upon his return to England. Summarizing this aspect of expedition activity, Wilson informed Sargent, "I have worked pretty hard at this photography business and if anything goes wrong with these plates I vow I will never attempt to handle another camera" (AAA, Wilson correspondence, letter dated March 6, 1909).

By April 25, all of the expedition materials had been dispatched either to Boston or London, Wilson was in Peking where the funerals of the emperor and empress dowager were about to occur, and Sargent's agent in China was about to leave for Harbin to board the Trans-Siberian Railroad for Moscow and his eventual return to England. Writing to Sargent from China at the close of

169. At the same time E. H. Wilson was exploring central and western China for the Arnold Arboretum, Frank Nicholas Meyer (1875–1918) was traveling in northern China for the United States Department of Agriculture and searching for plants of potential agricultural significance for introduction into the United States. Through an arrangement agreed upon by C. S. Sargent and USDA chief David Fairchild, Meyer would be on the lookout for trees and shrubs of ornamental value in the region he was exploring, and Wilson would keep an eye out for interesting or novel crop plants for the USDA in the territories in which he traveled. While both Sargent and Wilson were unsatisfied with this relationship and Meyer's results, the native-born Dutchman did send seeds of a new walnut (*Juglans cathayensis* Dode), and plants of this species grow in the collections of the Arnold Arboretum. Meyer was also responsible for sending a packet of seeds of the callery pear (*Pyrus calleryana* Decaisne)—a species first introduced by Wilson—from which the cultivar 'Bradford' was grown and selected years later at the U.S. Plant Introduction Station at Glen Dale, Maryland. This selection has recently found wide use as a street tree in the eastern United States.

the expedition, Wilson offered his impression of the work accomplished.

> The expedition has proved much more comprehensive and has entailed far more work than I anticipated when I agreed in England to undertake it. Your thoughtfulness in many details and your most encouraging letters from time to time have helped enormously, and the two years and odd months have passed quickly and I may add, happily. I have given the best that was in me and I hope the results satisfy . . . all your friends, but most of all, your own good self. (AAA, Wilson correspondence)

Wilson closed his letter as he had closed all of his communications addressed to Sargent from China since the inception of his third Chinese expedition:

> With cordial regards and greetings,
> I am, dear Professor,
> faithfully and obediently yours,
> E. H. Wilson

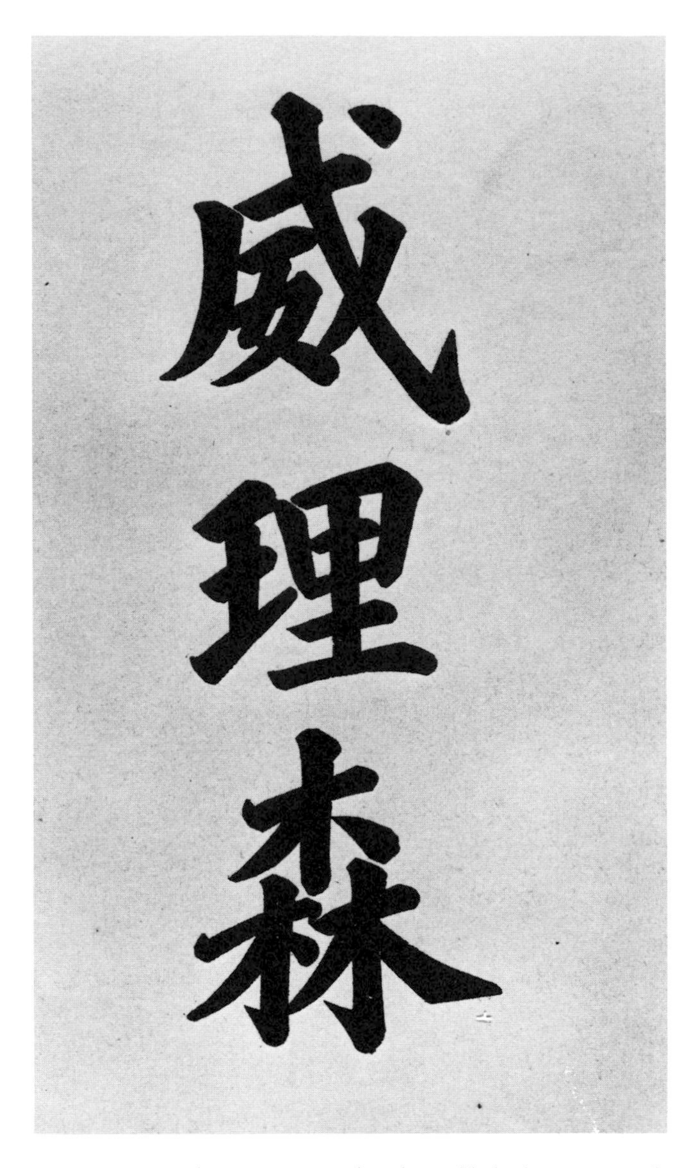

170. Ever a gentleman concerned with established customs and protocol, Wilson carried this calling card with his name in Chinese characters on his Arboretum-sponsored trips to China. Each card has the black Chinese characters printed on crimson paper.

Calamitous Encore

At the conclusion of Wilson's third China expedition, the naturalist returned to England to find that Professor Sargent was willing to stand behind his word. Temporary employment was available at the Arnold Arboretum, where Sargent wanted Wilson to supervise the sorting of his extensive collection of dried specimens that had resulted from the Arboretum expedition. Wilson gladly accepted the offer and agreed to sail for Boston to take up the work in the Arboretum herbarium, where he would be joined in the enterprise by the Arboretum's German-born taxonomist, Alfred Rehder (1863–1949). Wilson's departure for America, however, was postponed until the early fall of 1909 because his daughter had developed a serious case of the whooping cough. But on September 21, Wilson, his wife, and their recuperating daughter sailed on the S.S. *Ivernia,* destined to make a new home, at least temporarily, in Boston.

Personal involvement in the scientific aspects of his Chinese work during the winter of 1909–1910 provided Wilson with an opportunity to sharpen his taxonomic skills and to refresh his knowledge of the intricacies of botanical nomenclature. Working alongside Rehder, a taxonomist of exceptional skill who had encyclopedic knowledge of the ligenous flora of the North Temperate Zone and the pertinent botanical literature, Wilson developed a fuller appreciation of the scientific importance of his own work as a collector. In addition, Jackson Dawson in the Arboretum greenhouse was available to share his enthusiasm and delight as the seeds Wilson had shipped from Ichang began to germinate and grow in Boston. Members of the small Arboretum staff were all delighted with the results of Wilson's Chinese work, and talk centered on the rich Chinese flora. Moreover, Wilson's personal observations and recollections of China and Chinese plants were constantly sought. His splendid photographs of the Chinese landscape and the portraits of many of its trees and shrubs gave perspective to the collector's comments and added significantly to the scientific value of his other collections. Enthusiastic horticulturists in the vicinity of Boston, several of whom had helped to support the Arboretum expedition, were anxious to meet the Chinese explorer, and Wilson was gratified as his new friends and acquaintances began to refer to him affectionately as "Chinese" Wilson.

It is not known when Sargent first broached the subject of another trip to China, or what Wilson's initial reaction had been. One can surmise that Mrs. Wilson—simultaneously attempting to adjust to American culture and striving to establish a new home and make new friends for herself and Muriel Primrose—was vehemently opposed to the idea. Yet Sargent was persuasive and stressed the importance of the introduction into cultivation of the new conifers Wilson had discovered in western China. Cone production and seed set in these evergreen trees had been minimal during the 1908 sea-

Plate 19 (top left). The star magnolia (*Magnolia stellata*) from Japan was the first Asian magnolia to be introduced into North America. Because of its low, mounded, twiggy habit and the profusion of flowers produced each spring, it has become one of the most widely cultivated magnolias in southern New England. Old specimens of this species flank the steps of the Hunnewell Visitors' Center at the Arnold Arboretum. This photograph depicts the form 'Rosea' that produces pink flowers.

Plate 19 (top right). The black seeds of Japanese yew are embedded in a bright red flesh, which makes the seeds attractive to foraging birds. While the red flesh is free of poisonous compounds, the seeds and other plant parts contain taxines, a group of alkaloids that are potentially toxic if ingested by humans.

Plate 19 (bottom). The capsular fruits of the pearlbush (*Exochorda racemosa*) provide ornamental interest during the late summer and fall. In the Arnold Arboretum pearlbushes are cultivated on the slope of Bussey Hill as well as in the Bradley Collection of rosaceous plants.

Plate 20 (top). Short-shoot leaves of the katsura tree (*Cercidiphyllum japonicum*) and clusters of its banana-like fruits, technically follicles. The sexes are separate in the katsura tree, and both carpellate or female and staminate or male individuals grow along Meadow Road in the Arnold Arboretum.

Plate 20 (bottom). The large, three-lobed leaves of the Boston ivy are characteristically a deep emerald green during the growing season but turn shades of brilliant orange and crimson before falling in the fall. Trifoliolate leaves with three distinct leaflets are also produced by the vines on new growth; leaves on older shoots are simple and three-lobed like the ones shown here.

Plate 21. *Syringa villosa* is one of the several Asiatic lilacs that has proven hardy in New England. This species was first made known in Western horticulture from plants grown at the Arnold Arboretum; seeds of this species were sent to Charles Sargent by Emil Bretschneider.

Plate 22. This botanical illustration of the flower of *Davidia involucrata,* the dove tree, accompanied Franchet's account of the tree in *Plantae Davidianae.* Based on pressed and dried botanical specimens, the drawing fails to capture the correct posture of the flowers in relation to the leaves and branches in life.

171. E. H. "Chinese" Wilson is shown here, seated in the center, at the French legation in Chengtu, the capital of Szechwan Province. Wilson sought the company of Americans and Europeans in China when he visited the cities and towns where they were stationed.

son, and while Wilson had returned from China with ample herbarium material, no viable seeds had been collected.

Sargent had great faith in Wilson's abilities, and his proposal for a second Arboretum expedition verified that fact in the mind of the collector. It also eased the sense of resentment Wilson had tried to suppress when he had learned toward the end of his first Arboretum trip that Sargent had entered into an arrangement with Wilson's previous employer, Sir Harry Veitch, to co-sponsor another collector in China. William Purdom, a former Veitch employee at the Coombe Wood Nursery who had achieved the sub-foremanship of the arboretum at Kew, had been hired as the new agent, and his mission was to explore the northern tier of Chinese provinces for horticultural novelties. Wilson, en route to England, had met with Purdom in Peking in April 1909 and had coached the novice to the extent possible before recommencing his homeward journey. The fact that Veitch and particularly Sargent had passed over Wilson in favor of Purdom without so much as a warning had rankled the veteran collector, despite the fact that by 1909 he was anxious to leave China behind.

Despite the impending political and social turmoil that he knew was brewing in China, Wilson left Boston for England with his family in tow in February of 1910. Reestablishing his wife and daughter in England, Sargent's agent made his farewells and traveled to China

172. Although traveling with a heavy camera, cumbersome crates of glass plate negatives, and a sturdy tripod required hiring an additional porter, Wilson put the camera to good use. Not limiting his subjects to plants or vegetation, Wilson frequently captured the Chinese landscape on film, as in this image of the rugged terrain of the Ta-Pao-Shan range on the Chinese-Tibetan border. The snowclad peaks of these mountains attain elevations in excess of 21,000 feet.

again, this time by the overland route through Siberia. He arrived in China in May, and on the first of June he once again disembarked from a Yangtze steamer, which had brought the collector from Shanghai to Ichang.

With many of his Chinese collectors and porters employed for the fourth time, Wilson pushed westward following the overland route, which took his party through the rugged, mountainous district on the Hupeh-Szechwan border. The exhausted group finally arrived in Chengtu on July 29, after a long, hot, wearisome journey. From that metropolis Wilson wrote Sargent informing him that all was well and that he intended to leave "for Sungpan, in the northwest corner of the province, on August 8th" (AAA, Wilson correspondence, letter dated August 3, 1910). Wilson intended to be away about five weeks and would investigate the previously discovered new conifers in the region, hoping they had produced seeds.

"I also have in view seeds of several interesting plants" (AAA, Wilson correspondence, letter dated August 3, 1910), Wilson wrote, but he failed to mention that uppermost in his mind was a new lily, one which he had encountered in June of 1908 growing by the thousands in the semi-arid valley of the Min River south of Sungpan. He had determined to arrange for a large number of bulbs to be dug in October for eventual shipment—after each bulb had been embedded in clay—to the Farquhar Nursery in Boston.

After these arrangements had been made with local tribesmen, Wilson's party headed south on their return trip to Chengtu, following the ribbonlike trail high above the Min River. The River Min runs in a southerly course for more than 250 miles from its headwaters north of the Chinese trading post of Sungpan (now Songpan) in northern Szechwan Province to its confluence with the Yangtze River below Kiating. In its upper reaches the river flows through a high mountainous region that once formed the natural boundary between China and Tibet, a no-man's-land sparsely populated by tribesmen of unknown origin. In this remote region, the river's turbulent waters course through a deep and precipitous, semi-arid valley, which over the millennia has been carved and eroded through the underlying bedrock of mud shales and granites. Each year, fed by hundreds of small tributaries that carry the melt water of the eternal snows capping the mountain peaks within its watershed, the river deepens its gorge as it races southward to join the mighty Yangtze.

Early in this century, a narrow, ribbonlike trail followed the river, etched high on the steep wall of the valley above the high water mark of vernal floods. Linking Sungpan in the north with Chengtu and the fertile and populous Red Basin in the south, the trail constituted the main trading route through the region. Caravans transporting brick tea and laden with medicinal herbs and other articles of commerce were forced to follow this treacherous track, as no other direct route linked the two regions.

On September 3, 1910, Wilson's party—small compared with most of the commercial caravans—was slowly making its way southward along the narrow track high above the Min. Men carrying two wicker sedan chairs led the way, followed by porters balancing their suspended loads from the ends of split bamboo poles slung slant-wise over their bare, calloused shoulders. Their pace was determined by the muscular, bronzed men shouldering the two chairs, but the coolies were accustomed to adjusting their bouncing and rhythmic but constant pace to suit the grade and terrain; a few lagged behind, setting a leisurely pace of their own. Far in advance of the party a small, overly energetic black spaniel ran down the trail, stopped, turned, and retraced his steps at full speed over the rocky road. Darting in and out between the porters's sandaled feet, the dog went back and then forward again along the path, seemingly checking to see that no one fell too far behind and urging the party onward. Suddenly the spaniel stopped short in his tracks and then dashed forward as a small rock hit the hard-packed trail from above, rebounding and plummeting into the river three hundred feet below.

For the next few moments the head of the procession was thrown into chaos as boulders, rocks, soil, and uprooted shrubs slipped almost silently down the side of the steep valley into the path of the caravan, taking with them the lead chair. The porters near at hand lurched forward to avoid catastrophe, and the quick-thinking occupants of the wicker chairs leaped from their seats, seeking the shelter of protruding outcrops on the lee of the trail. Miraculously, when the dust had settled and the stunned coolies who had brought up the rear of the train raced to their comrades at the head of the line, none were found seriously injured. But a worse fate had befallen Wilson, leader of the small band of expeditionaries. In seeking the shelter of the outcrop, he had been hit by a large boulder that had broken his right leg in two places below the knee. Had he remained in his chair, the broken fragments of which his men retrieved from the river's edge, Wilson might not have lived to write Charles Sargent in Boston concerning the calamity that effectively brought Wilson's fourth Chinese expedition to an end.

173. While this suspension bridge constructed from plaited bamboo may appear hazardous, Wilson nonetheless welcomed any easy means of crossing the numerous rivers and streams—many of them raging torrents—he encountered in western China. Perhaps as a consequence, bridges were one of Wilson's favorite photographic subjects.

In the wake of the accident, the tripod of the bulky Sanderson camera was sacrificed to provide splints for Wilson's leg, and the distance to Chengtu, normally a four-day march, was covered in three. Eight miles from the city, Wilson's party met a group of missionaries, who escorted the wounded botanist to the home of a Dr. and Mrs. Davidson in the compound of the Friends' Presbyterian Mission. There Dr. Davidson attempted to set the bones, but in six weeks the leg had become infected and the bones had failed to knit. Amputation of the limb below the knee seemed inevitable, but with the assistance of a French surgeon, a Dr. Mouillac, another operation was performed and the leg saved. Within three months of the accident Wilson was hobbling around on crutches, and by the end of January 1911 he had returned by boat to Ichang.

In spite of Wilson's incapacity, his faithful, devoted, and well-trained Chinese assistants continued to collect specimens and gather seeds throughout the fall and into the winter months. At Ichang Wilson personally supervised the packing of seeds and specimens for shipment to Boston, including the cones and seeds of many of the conifers he had been sent half way around the world to harvest. Also prepared for shipment were thousands of bulbs of the regal lily (*Lilium regale* Wilson) that had

174. The opening portion of E. H. Wilson's typed letter to C. S. Sargent informing the Professor of the calamitous conclusion of the naturalist's fourth and final Chinese expedition.

Chengtu, West China, Sept.10,1910.

Dear Professor Sargent:-

I returned to Chengtu on the 6th inst. from a round trip to Sung P'an. The trip was highly successful up to within a few days of the finish, when it ended in catastrophe. Three days' journey from Chengtu, whilst descending the lower reaches of the Min valley by the main highway we were overtaken by a landslide, and I escaped only with a badly broken leg. Both bones are broken about a foot above the ankle, and there is a nasty wound on the outside of the calf. We improvised some crude splints, and I was carried to Chengtu, spending two nights at Chinese inns on the way. Eight miles before reaching Chengtu I overtook some missionaries who very kindly escorted me to the house of Dr.Davidson of the Friends' Mission in Chengtu. Dr.Davidson and his wife, assisted by other friends, immediately took me in hand and rendered every possible assistance. Drs.Davidson and Sheridan set the limb. As over sixty hours had elapsed since the accident, the leg was much swollen, and the operation was therefore a long one, requiring more than an hour under chloroform. Dr.Davidson most kindly placed at my disposal a room on the ground floor, and Mrs.Davidson has taken over the duties of nurse, and everything possible for my comfort etc. is being done. The limb has now been set four days, and is apparently taking a normal course.

175. Dr. and Mrs. Davidson posed for Wilson on the veranda of their home in the compound of the Friends' Mission in Chengtu before Wilson departed on his homeward journey.

been collected on the steep slopes above the Min River and its tributaries south of Sungpan. These were transported to Boston via the Pacific route and across the North American continent on the Canadian Pacific Railroad. Most were sound when the crates were unpacked at the Farquhar nursery in Roslindale, Massachusetts, and the plant was destined to become "Chinese" Wilson's single most famous introduction from China.

Hundreds of other Chinese plants that today enrich and embellish North American and European gardens—as well as those in other regions of the world where climatic conditions favorable to their culture prevail—owe their introduction as cultivated subjects to Ernest Henry Wilson. Little-known species from Japan, Sakhalin, Korea, Formosa, and the Liukiu and Bonin Islands joined their Chinese congeners in cultivation as a result of collecting expeditions Wilson undertook in those regions during the years between 1914 and 1919. These trips, like his last two China expeditions, were under the auspices of the Arnold Arboretum, where employment was once again offered Wilson shortly after his return from China in March 1911.

Of the many woody plants introduced into North America as a result of Wilson's two Arboretum expeditions to China, the sand pear (*Pyrus pyrifolia* (Burman f.) Nakai; Plate 24, top) can be singled out as one of the most beautiful spring-flowering trees. The large, round-headed individual that grows on the south-facing slope of Bussey Hill in the Arnold Arboretum is catalogued in the Arboretum records under accession number 7272 and was grown from Wilson's seed lot 395, collected near Ichang. This tree never fails to produce an abundance of white flowers in late April and early May, and in late summer and fall a crop of small, brownish-green, pear-shaped fruits hide among the leaves along the branches. The flesh of these pears is hard, not particularly sweet, and filled with abundant stone cells—the thick-walled cellular components that give the flesh of the common, edible pear its slightly gritty texture. Because of the abundance of stone cells in its flesh, the sand pear is not edible; the common name was probably given to the tree because eating its fruit is akin to eating a mouthful of sand. The sand pear is nevertheless an outstanding flowering ornamental and worthy of wide cultivation, even if its fruits prove disappointing.

The significant and majestic collection of beeches in the Arnold Arboretum was augmented and made nearly complete in species representation through Wilson's efforts in China. He had initially confused their identity in the field on his earlier expeditions and had repeatedly failed to introduce the Chinese species into cultivation in the West. Nevertheless, Wilson determined to make a final try in 1910. In an unpublished Wilson manuscript, the author recalled his plan to accomplish their introduction, writing,

> I made a special journey into northwestern Hupeh and in different localities affixed wire marks on about a dozen seedlings of each species. This done, I arranged to leave in the district until the autumn a man whose sole duty was to dig up these plants and bring them to me on my return to Ichang from western Szechuan. He fell sick and failed to carry out the trust. Winter had set in when I returned to Ichang, but I determined to make a final effort. I offered two of my collectors, who were with me when the plants were marked, fifty

176. Wilson's Chinese assistants preparing lily bulbs for shipment to the Farquhar Nursery Company in Boston. Each bulb was embedded in clay, allowed to dry, and then packed in heavy wooden crates that were filled with pulverized charcoal to prevent damage to the precious bulbs during shipment.

177. Regal lilies in full flower in the growing fields of the Farquhar Nursery Company in Roslindale, Massachusetts, in 1920. This lovely lily was destined to become E. H. Wilson's most famous plant introduction from China.

ounces of silver if they would journey forth and bring in the plants. This they accomplished for the prize to them was great.

The plants, wrapped carefully in sphagnum moss and packed in a ventilated trunk, accompanied me to the Arnold Arboretum where they arrived in excellent condition. Today all three are fairly well established in cultivation. There was no special reason why the identity or introduction of these Beeches should have proved so tantalizing, but the conquest, even today, affords me greater pleasure than that of any other group of trees. (AAA, unpublished Wilson manuscript)

Eight species of beech occupy wide to narrow ranges across the North Temperate Zone and serve as an excellent example of species of a genus that together occupy a broken and discontinuous range in the northern hemisphere today. Prior to the successive glacial epochs of the Pleistocene, the genus had a wider and more continuous distribution across northern latitudes. Today the center of generic diversity occurs in eastern Asia, the region least affected by the continental ice sheets. On the Asian continent three species are known from China (*Fagus engleriana* Seemen, *Fagus longipetiolata* Seemen, and *Fagus lucida* Rehder & Wilson), and two (*Fagus japonica* Maximowicz and *Fagus crenata* Blume) are represented

178. The genus *Fagus* is characterized by its many scaled, elongate winter buds, its simple, toothed leaves, and its pendulous staminate inflorescences and terminal carpellate flowers. It is also well known for its nuts that are partially or completely enclosed in four-valved husks, which split open at maturity. Shown here are the fruits of *Fagus sylvatica* from Europe, *Fagus orientalis* from the Caucusus region, *Fagus engleriana* from China, and *Fagus crenata* and *Fagus japonica*, both from Japan. The curved prickles of the husks aid in dispersal by attaching the fruits to the furry coats of animals. This mode of attachment is reminiscent of the principal employed by modern-day velcro fasteners.

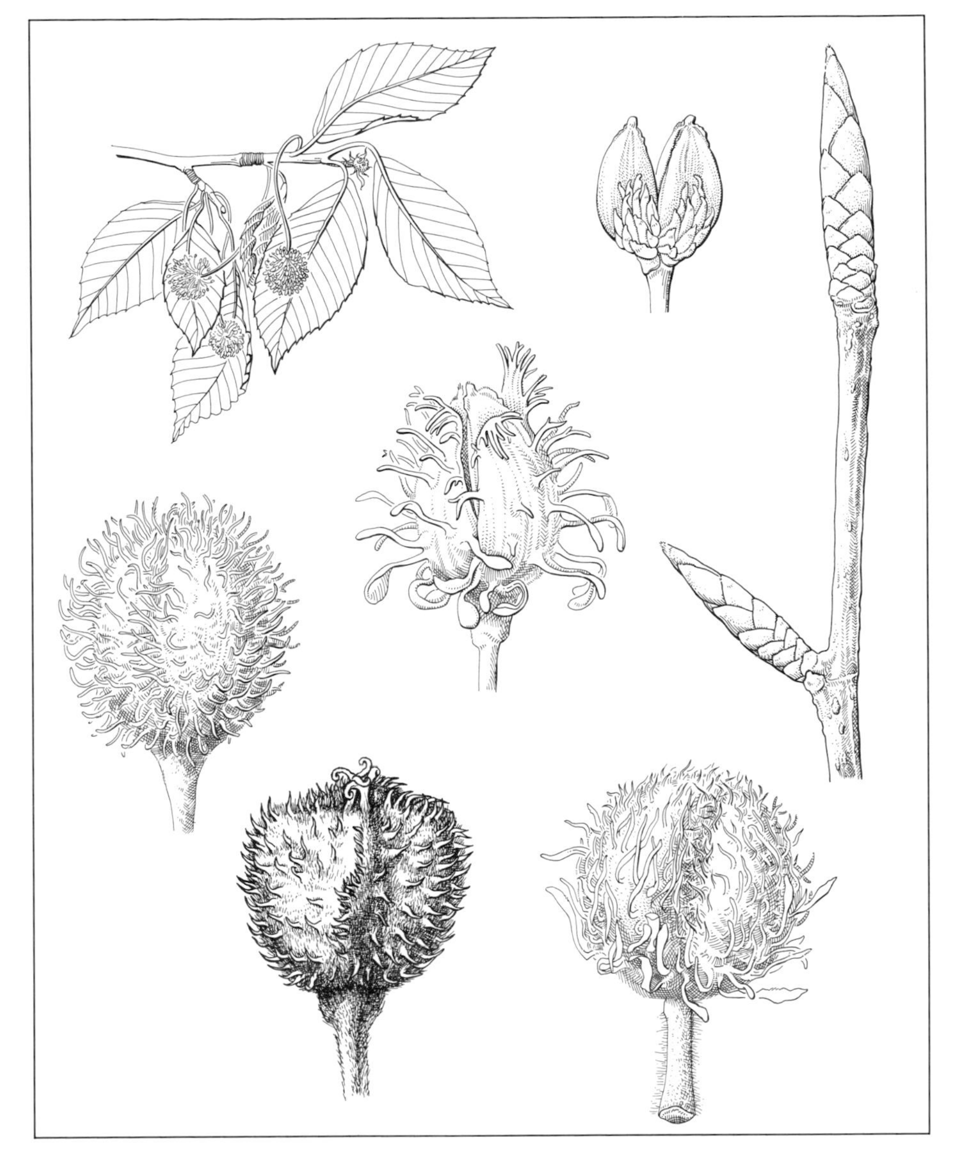

in the Japanese flora. One species (*Fagus orientalis* Lipsky) occurs in the Caucasus region of Asia Minor, an area that, like eastern North America and eastern Asia, supports a flora comprised of numerous relict species that have survived from pre-Pleistocene times or have evolved from pre-Pleistocene precursors. Only one beech (*Fagus sylvatica* L.) is widespread in Europe, and a solitary species (*Fagus grandifolia* Ehrhart) occupies a wide range in the postglacial forests of eastern North America.

Two additional species, both of extremely limited distribution, bring the total number of species in the genus *Fagus* to ten. Of these last two, one (*Fagus hayatae* Palibin) is known from a single mountain in Taiwan, while the second (*Fagus multinervis* Nakai) is restricted to the small island of Ullung-do off the coast of Korea in the Sea of Japan. Neither of these species is currently cultivated in the Arnold Arboretum, but all except one (*Fagus longipetiolata*) of the eight Chinese species currently grow in the beech collection of the Arnold Arboretum. While the American and particularly the elephantine European beeches dominate the collection—located on the slope of Bussey Hill facing Hemlock Hill—the Caucasian and Asiatic species have been planted in scattered locations among the numerous variants of the European species. Higher up the slope of

Bussey Hill a solitary tree of one of the Chinese species (*Fagus engleriana*) appears to grow as a tree belonging to a remnant of the native Arboretum woodland. In fact, this tree has grown from one of Wilson's saplings dug in western Hupeh Province in the winter of 1910. It has grown to maturity in this sheltered Arboretum location, half way around the world from the forest in which it grew as a seedling plant.

Like the beeches, the Chinese zelkova (*Zelkova sinica* Schneider) and the Chinese cork tree (*Phellodendron chinense* Schneider)—both introduced by Wilson on his Arboretum expeditions to China—are prized primarily as shade trees. Seeds of a third, unique shade tree were also obtained by Wilson in 1910, and large specimens of the Chinese eucommia (*Eucommia ulmoides* Oliver) are now growing in the Arnold Arboretum along the boundary wall near the Center Street gate. Wilson discovered that seed of this species proved difficult to obtain because reproductively mature individuals were rarely encountered in nature. Moreover, young trees planted and cultivated by the Chinese rarely produced seed. This situation exists for two reasons. First, the bark of the tree is highly esteemed in China as a tonic and as an ingredient in herbal remedies. Consequently, both naturally occurring and cultivated trees were felled and stripped of their bark before they reached maturity, a practice that continues to the present day. Second, the sexes are differentiated between different individuals, and carpellate and staminate trees must grow in close proximity for pollination and seed set to be assured. In the Arnold Arboretum, however, seeds are annually produced on the old carpellate tree that grows near the Center Street gate, inasmuch as a staminate individual grows near by and shares the microclimate favorable for their growth below the Center Street wall.

The reputed medicinal value of the eucommia is associated with one of its unique characteristics, which can be demonstrated if a leaf is carefully torn in two. Thin threads of an elastic, white latex or rubber stretch between the two halves as the leaf is torn and maintain tenuous links between the separated portions. Should a small piece of bark be torn, a copious amount of the white, gummy rubber connects the two halves, and its tough, elastic quality makes separation of the two pieces somewhat difficult. Although the rubber synthesized by the plant is of an inferior quality, the tree is the only species hardy in New England that produces this valuable vegetable substance.

From a botanical standpoint, the eucommia—sometimes referred to as the hardy rubber tree—is amply distinct from its nearest relatives and has been classified as constituting its own, monotypic family, the Eucommiaceae. Its closest relatives are found in the elm family, the Ulmaceae, and the specific epithet *ulmoides* of the solitary species of *Eucommia* signifies this relationship. The similarities of the eucommia with the elms are best reflected by the morphology of its glossy green leaves and its narrowly winged, elongated fruits. Fossil leaves of a now-extinct eucommia have been found in brown coal deposits of Tertiary age in Germany. Despite their age, these fossils produce a smoke that has the rank, acrid odor of burning rubber when they are exposed to an open flame. Ancient eucommias also grew in North American forests, and fossil fruits are known from Eocene strata in Tennessee and Oligocene or early Miocene strata in Montana. Today, the solitary, remaining species of eucommia is restricted in nature to central and eastern China. In cultivation, where it is included in collections as an anomaly and a pleasing, pest-free shade tree, the eucommia again grows in regions where it once flourished in the geological past.

Wilson's accident forced him to return to Boston prematurely in 1911, but he had managed to send 1,285 packets of seed from his brief, fourth sojourn in the Middle Kingdom. Compared with the 304 packets sent by Purdom during an equal time period, Wilson had once again more than exceeded expectations and proved his abilities as plant hunter extraordinaire. Alfred Rehder claimed that over one thousand species were introduced into cultivation in the West by "Chinese" Wilson, and that his results in that endeavor exceeded those of any other collector. From a scientific perspective, Wilson and his Chinese assistants prepared upwards of 50,000 herbarium specimens. From these dried collections, 4 genera, 382 species, and 323 varieties or forms were described as new to science and added to the catalogue of the woody plants of China. New taxa had already resulted from Wilson's travels for the Veitch firm, and others would be based on his collections from other regions of Asia. While some of these taxa have not withstood the test of time and are no longer considered valid, other new taxa have been described, as succeeding generations of taxonomists have studied and scrutinized Wilson's valuable collections.

North American and European horticulture shared equally in Wilson's harvest. Through the skill of Jackson Dawson as a plant propagator, the seeds Wilson collected were coaxed to germinate, and the grafts, cuttings, and seedling plants sent and carried back from China were provided growing conditions that satisfied the initial

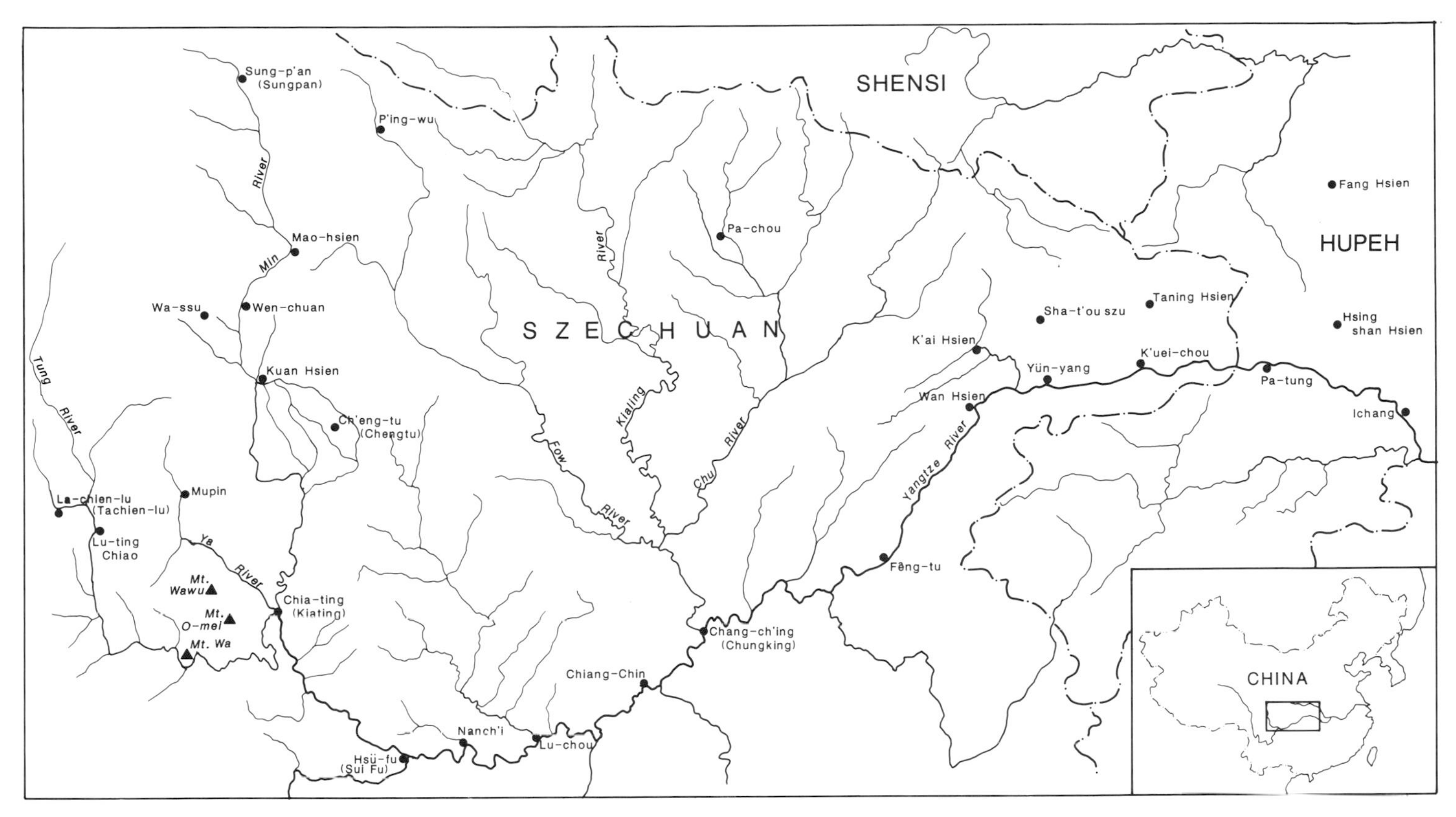

179. Sketch map of central and western China with placenames of important collecting localities visited by Ernest Henry Wilson on his four expeditions to the region. Wilson's itineraries have proved very difficult to trace with any accuracy; consequently, no attempt is made here to show Wilson's exact routes.

requirements for their establishment in American soil. Sargent could report in 1916 that 918 species and varieties raised from Wilson's collections were growing in the Arnold Arboretum and that 445 species not hardy in the Boston region had been distributed to cultivators in more temperate regions, where their survival was more likely. It was due to the munificence of Charles Sargent that Wilson's introductions were shared widely with the co-sponsors of Wilson's travels both in America and England, with other friends, and with nurserymen in America and abroad. Wisely, he sent to gardens in milder climes those species that promised not to fare well in the climate of Boston, to assure their establishment in cultivation. Thus, Rehder could write years later that Wilson's plants "have spread to all the gardens of temperate and subtropical regions" (Rehder, 1930, p. 185). For example, many of the magnolias Wilson introduced from China were shared with the Chenault Nursery in the south of France, where they were propagated and widely distributed.

The End of an Era

On his return to Boston in 1911 E. H. Wilson entered a Boston hospital, where the broken bone in his right leg was reset and his foot was fitted with a special boot to compensate for the shortened limb. After a well-deserved three-month vacation with his family in England, Wilson returned to the Arnold Arboretum to begin work in the herbarium on a descriptive treatise of the scientific results of his expeditions in China. Published between 1913 and 1917, *Plantae Wilsonianae* grew into a three-volume compendium, which C. S. Sargent signed as editor. But the work of preparing the detailed studies that summarized taxonomic knowledge of the Chinese genera of trees and shrubs Wilson had collected fell largely to Alfred Rehder. Wilson personally authored twelve of the taxonomic treatments and collaborated with Rehder on forty-seven. But Rehder masterminded the work, single-handedly authoring forty-four articles and coordinating the contributions of

180. One of Wilson's beautiful photographs of Japan that has captured the quintessential atmosphere of the Japanese landscape. Using a glass plate negative that has survived to the present day in the photographic archives of the Arnold Arboretum, Wilson took this photograph at Lake Towada; it shows a small island grown with Japanese white pines.

181. Arnold Arboretum botanists took time from work for this photograph in August of 1916. From the left are Alfred Rehder, Ernest Wilson, Charles Sargent, Charles Faxon, and Camello Schneider.

other world authorities. These included the treatments of the German-born but Austrian-trained dendrologist Camillo Schneider, who had been forced to abandon field work in China and to seek asylum and employment at the Arboretum while World War I raged in Europe.

One reason Wilson did not become more involved in the ambitious undertaking was his simultaneous involvement in authoring numerous articles concerning his Chinese travels that popularized many of the plants he had been able to introduce into Western horticulture. A two-volume work, *A Naturalist in Western China,* appeared in 1913 and was illustrated with many of the author's photographs. But another, more obvious, reason Wilson contributed fewer articles to *Plantae Wilsonianae* than Rehder rested in the fact that Sargent recognized Wilson's true value to the Arboretum. While Wilson knew his way around the herbarium and library and could produce sound taxonomic work, his most important contributions were made as an extraordinarily gifted naturalist who combined the skills of botanist with those of horticultural collector. Sargent was no fool; he intended to capitalize once again on Wilson's field abilities.

Sargent's plan was to send Wilson to Japan for an extended visit. Travel there was easy by comparison with travel in China, and despite the work of previous horticultural explorers, Sargent himself included, the aging director knew much remained to be done. Complete data on Japanese conifers and on the myriad forms of Japanese cherries alone would prove of scientific interest and horticultural significance. Wilson needed little coaxing to agree to another, more relaxed trip, and his wife offered few if any objections, as she and seven-year-old Muriel Primrose would accompany the Arboretum agent on his Japanese tour.

The Wilsons arrived in Japan on February 3, 1914, and the next year was spent exploring the forests of the island nation, with particular emphasis being placed on a complete study of the conifers. A rich haul of collections resulted, and numerous seed lots were forwarded to Sargent in Boston. Visits were also made to the horticultural centers of Japan, where nurseries were surveyed for novelties that Japanese plantsmen had brought into cultivation. The Sanderson camera was put to good use once again, and portraits of many notable Japanese trees as well as astonishingly beautiful views of the Japanese countryside were recorded on film.

By the time the Wilsons returned to Boston in February of 1915, the war was well under way in Europe and additional travel plans had to be shelved. Wilson began work at once on two monographs—one on the conifers and the other on the cherries of Japan—that had been made possible by his extended personal observations and work in the field. These careful studies, illustrated with photographs by the author, were published in 1916, and by January 1917, after American doughboys had joined the Allied forces in the trenches in Europe, Wilson set out again in pursuit of new plants for the Arnold Arboretum. On this journey—his sixth and final trip to eastern Asia, and his fourth expedition for the Arnold Arboretum—Wilson explored the forests of the Ryukyu Islands and Okinawa, the Bonin Islands, the Korean peninsula, and the offshore islands of Quelpaert (now Cheju Do) and Warrior Island (Ooryongtō), as well as Japan. Further exploration was undertaken in Formosa, where Mount Morrison, the highest peak on the island, was climbed and where conditions in the field were reminiscent of those in central and western China.

Wilson returned to the Arboretum in March of 1919, having forwarded quantities of seed of totally new species as well as plants not previously introduced directly into North America. Crates of documenting specimens also reached the Arboretum, where Rehder welcomed their addition to the herbarium. Once again Wilson had exceeded expectations, and despite the troublesome, newly enforced regulations of the United States Department of Agriculture, which required that all incoming plant materials be inspected for diseases and insect pests by agents in Washington, D.C., the greenhouse staff was once more overwhelmed with the extra workload that they had come to expect when Wilson returned home at the end of an expedition.

On this last, extensive Asian trip, the little-explored flora of the Korean peninsula provided Wilson with the greatest number of new introductions to prove hardy in the Arnold Arboretum and New England. The Korean forsythia (*Forsythia ovata* Nakai) was located in the Diamond Mountains, an extremely scenic corner of Korea where waterfalls cascade over high cliffs and the forested slopes are robed with native maples, birches, and oaks. While the 1918 season was generally a poor one for seed production, Wilson managed to locate a sufficient number of shrubs of the forsythia to obtain a quantity of seed large enough to ensure successful germination. In cultivation, the Korean forsythia has proven to be the hardiest of all the forsythias, and it—or hybrids involving it—now grows in northern New England, in the northern tier of the midwestern states, and in southern Canada, where the severity of the winter cold frequently freezes the preformed flower buds of other species, rendering them undependable as flowering ornamentals.

Also from the forests of Korea came three maples that have proven to be distinguished ornamentals in cultivation. The shrubby Korean maple (*Acer pseudosieboldianum* Komarov), a member of the Japanese maple group, matures to a height of about twenty-five feet, and its bluish-green, deeply lobed leaves are dramatic both in coloration and outline. But the greatest beauty of this small tree is achieved in the fall, when its leaves turn as bright a red as can be conceived, although they sometimes flaunt shades of deep, rich orange. When planted as a specimen tree against a backdrop of conifers or broad-leaved evergreens, the year-round effect of this tree is pleasing and ever changing but most spectacular in late September and October.

Wilson wrote to Sargent from Seoul concerning the second Korean species of maple, the so-called three-flowered maple (*Acer triflorum* Komarov), shortly after he had harvested its seed in the field. He extolled its beauty, reporting that "perhaps the best capture was *Acer triflorum*. This [maple] makes a large tree and has loose, papery, reddish-gray bark and wonderful autumn tints of yellow and orange. Please have a good lot of this seed sown for, like that of all its section, a large percentage of the seed is infertile" (AAA, Wilson correspondence, letter dated October 22, 1917). In mentioning its section, Wilson referred to the trifoliolate group of maples, of which the paperbark maple—introduced earlier by Wilson from China—is also a member. The third and last maple Wilson introduced from Korea, the Manchurian maple (*Acer mandshuricum* Maximowicz), also belongs to the trifoliolate group and it, too, provides spectacular fall color. Unlike the brilliant crimson leaves of *Acer pseudosieboldianum* or the golden yellows and pumpkin shades of *Acer triflorum*, the leaves of the Manchurian maple assume surprising hues of pink. When seen in fall the warm pink coloration of this maple seems to impress itself on the memory of all who behold it, and the identity of the tree is assured by the color alone, regardless of where the tree is encountered.

Another maple, the Japanese red maple (*Acer pycnanthum* K. Koch), first grew as a cultivated tree in the Western hemisphere as a result of Wilson's successful

182. In response to E. H. Wilson's request for information concerning the distribution of conifers on the southern Japanese island of Kyushu, a Japanese colleague, T. Miyoshi, sent this sketch map of the island on which he pinpointed the localities of many of its notable trees. With friendly cooperation and assistance such as this, Wilson was able to study all of the Japanese conifers in their natural habitats.

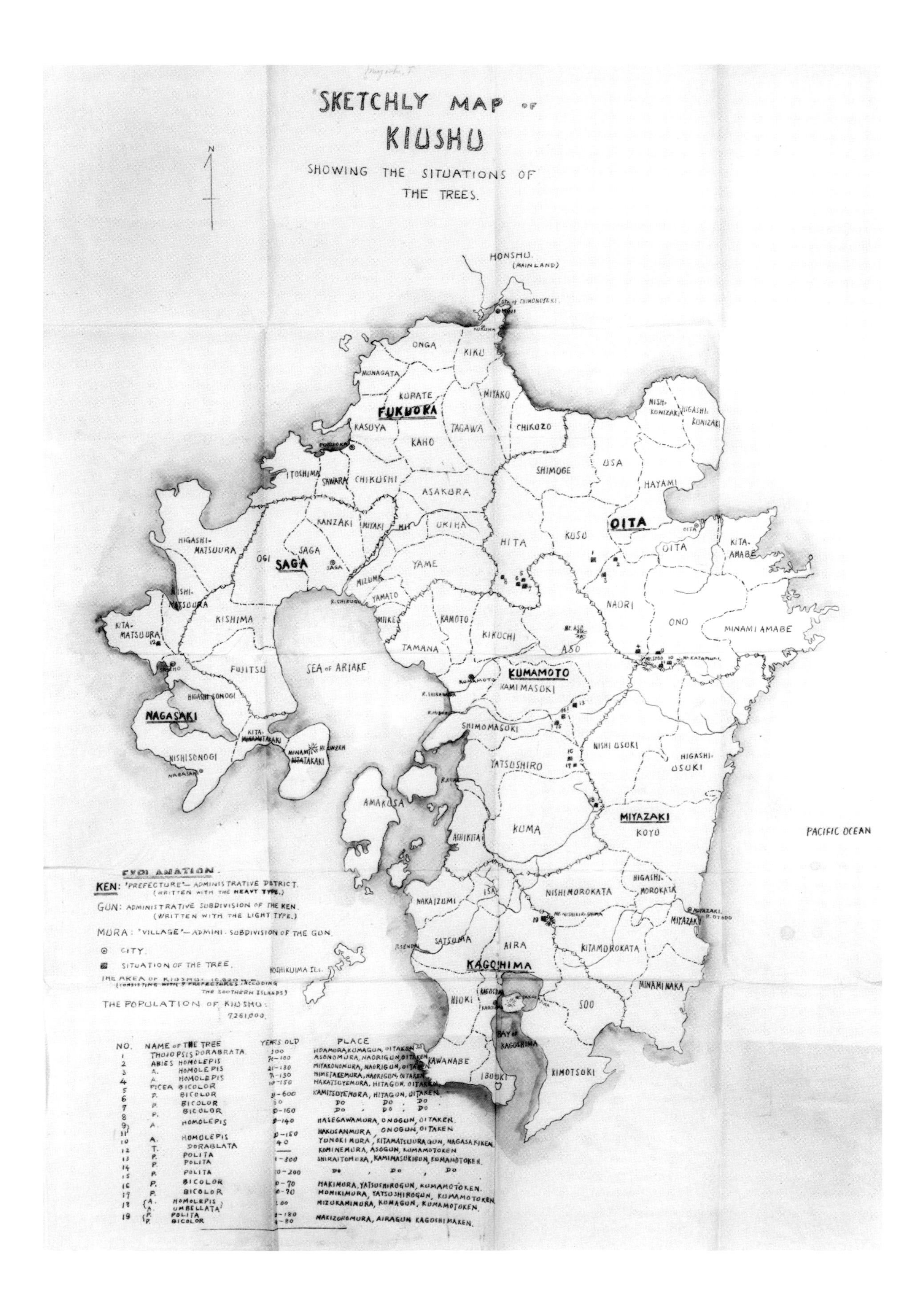
SKETCHLY MAP OF
KIUSHU
SHOWING THE SITUATIONS OF
THE TREES.
N
HONSHU.
(MAINLAND)
PACIFIC OCEAN
SEA OF ARIAKE
FUKUOKA
SAGA
NAGASAKI
OITA
KUMAMOTO
MIYAZAKI
KAGOSHIMA
ONGA
KIKU
MUNAGATA
KURATE
MIYAKO
KASUYA
TAGAWA
CHIKUZO
KAHO
ITOSHIMA
SAWARA
CHIKUSHI
ASAKURA
SHIMOGE
USA
HAYAMI
KANZAKI
MIYAKI
UKIHA
HIGASHI-MATSUURA
OGI
HITA
KUSU
OITA
KITA-AMABE
YAME
YAMATO
MIZUMA
NAORI
KISHIMA
MIIKE
KAMOTO
KIKUCHI
ONO
MINAMI AMABE
KITA-MATSUURA
TAMANA
ASO
FUJITSU
KAMI MASOKI
HIGASHI SONOGI
SHIMOMASOKI
NISHI USUKI
HIGASHI-USUKI
NISHISONOGI
YATSUSHIRO
AMAKUSA
ASHIKITA
KUMA
KOYU
NAKAIZUMI
ISA
NISHIMOROKATA
HIGASHI-MOROKATA
MIYAZAKI
SATSUMA
AIRA
KITAMOROKATA
HOSHIKIJIMA ILs.
HIOKI
SOO
MINAMI NAKA
BAY OF KAGOSHIMA
KAWANABE
IBUSUKI
KIMOTSUKI
EXPLANATION.
KEN: 'PREFECTURE'— ADMINISTRATIVE DISTRICT.
(WRITTEN WITH THE HEAVY TYPE.)
GUN: ADMINISTRATIVE SUBDIVISION OF THE KEN.
(WRITTEN WITH THE LIGHT TYPE.)
MURA: 'VILLAGE'— ADMINI. SUBDIVISION OF THE GUN.
CITY.
SITUATION OF THE TREE.
THE POPULATION OF KIUSHU:
7,261,000.
NO. NAME OF THE TREE YEARS OLD PLACE
1 THUJOPSIS DORABRATA. 100 HIDAMURA, KUMAGUN, OITAKEN
2 ABIES HOMOLEPIS 71–100 ASONOMURA, NAORIGUN, OITAKEN.
3 A. HOMOLEPIS 21–130 MIYAKONOMURA, NAORIGUN, OITAKEN.
4 A. HOMOLEPIS 71–130 HIMETAKEMURA, NAORIGUN, OITAKEN.
5 PICEA BICOLOR 10–150 NAKATSUYEMURA, HITAGUN, OITAKEN.
6 P. BICOLOR –600 KAMITSUYEMURA, HITAGUN, OITAKEN.
7 P. BICOLOR DO, DO, DO
8 P. BICOLOR –160 DO, DO, DO
9 A. HOMOLEPIS –140 HASEGAWAMURA, ONOGUN, OITAKEN.
11 –150 HAKUSANMURA, ONOGUN, OITAKEN
10 A. HOMOLEPIS 40 YUNOKIMURA, KITAMATSUURAGUN, NAGASAKIKEN.
12 T. DORABLATA KOMINEMURA, ASOGUN, KUMAMOTOKEN
13 P. POLITA –300 SHIRAITOMURA, KAMIMASUKIGUN, KUMAMOTOKEN.
14 P. POLITA
15 P. POLITA –200 DO, DO, DO
16 P. BICOLOR –70 MAKIMURA, YATSUSHIROGUN, KUMAMOTOKEN.
17 P. BICOLOR –70 MOMIKIMURA, YATSUSHIROGUN, KUMAMOTOKEN.
18 A. HOMOLEPIS, A. UMBELLATA 100 MIZUKAMIMURA, KOMAGUN, KUMAMOTOKEN.
19 P. POLITA, P. BICOLOR –180, –80 NAKIZONOMURA, AIRAGUN KAGOSHIMAKEN.

efforts to obtain the tree in Japan in 1914. In October of that year Wilson made a special trip to Sakamoto, a small town south of Sendai on the Pacific coast of Honshu. Locating the tree inland from the town, Wilson took photographs and secured herbarium specimens, as well as several young saplings that were packed and sent to the Arnold Arboretum. Wilson noted that the trees grew in a sphagnum swamp and looked for all the world like the red maple (*Acer rubrum* L.) of eastern North America. Indeed, the relationship of the two species is very close, and some botanists claim the Japanese plant is only varietally distinct from its North American congener. Wilson admitted that the two were most easily distinguished from one another if one could use geographic origin as a criterion in their determination. Surprisingly, while the red maple of eastern North America ranks as the most abundant tree in terms of numbers of individuals in the forests throughout its wide range, the Japanese red maple is a very rare species restricted to a small region of central Honshu.

Another tree closely related to an eastern North American species was sent by Wilson from the forests of Japan and Korea. The Korean birch (*Betula schmidtii* Regel) is uncommon in Japan, but in Korea the tree is commonplace in the mountain forests and particularly abundant in the Diamond Mountains. Concerning that rugged, picturesque district, which lies inland along the coast of the Sea of Japan, just north of the boundary that now separates North and South Korea, Wilson commented, "The wealth of autumn color is supplied almost entirely by three kinds of trees—the orange, red and crimson by *Acer pseudosieboldianum*, bright yellow by *Betula schmidtii*, [and] yellow to leather brown by *Quercus mongolica*" (AAA, Wilson correspondence, letter dated October 22, 1917).

In habit and morphology, the Korean birch is closely allied to the river birch (*Betula nigra* L.) of eastern North America, and both are prized ornamentals, which are grown primarily for the beauty of the dark, rugged bark that clothes the trunks of these handsome trees. On mature individuals the bark of the Korean birch consists of large, overlapping, slate-gray, buff, and pewter-hued plates that peel or exfoliate in a manner suggestive of the shagbark hickory (*Carya ovata* (Miller) K. Koch) of eastern North American forests.

183. A view of Mirror Rock in the Kongo-san or Diamond Mountains of Korea, a region that Wilson visited several times during 1917 and 1918.

One last tree collected by E. H. Wilson in Korea must be mentioned, as several individuals grow in various locations in the Arnold Arboretum and add significantly to the beauty of the summer landscape of Boston's "museum without walls." The mimosa or silk tree (*Albizia julibrissin* Durazzini) had been cultivated in Europe since the middle of the eighteenth century, when the cavalier Filippo Albizzi, a Florentine nobleman, introduced the tree into cultivation in Tuscany in 1749. Shortly thereafter the tree spread in cultivation throughout the warmer regions of Europe, and André Michaux brought seeds of the tree to America when he arrived from France in 1785. Michaux shared seeds of the tree, which he may have collected in Persia during his sojourn there prior to representing the King of France in the United States, with William Bartram in Philadelphia. William Hamilton also obtained the plant, and Thomas Jefferson acquired it from both Bartram and Hamilton.

The plants grown from Michaux's seed thrived in the Philadelphia region and prospered at Monticello on the Virginia Blue Ridge. Michaux also cultivated the tree in his South Carolina nursery garden, where it grew at a rapid rate and attracted the attention of visitors from nearby Charleston. The mimosa soon became a commonly cultivated tree in southern gardens, and its seeds were widely shared by gardening friends. The tree was well adapted to the southern climate, and as its popularity increased it began to invade forest margins; today, the mimosa is widely naturalized throughout much of the southeastern United States from Washington, D.C., southward. But attempts to grow the tree north of Philadelphia always ended in failure, since the plants could not endure the colder, longer winters of more northern climes.

The mimosa has a wide natural range in tropical, subtropical, and temperate regions across Asia, extending from Iran in the west to China, Japan, and southern Korea in the east. Michaux's seeds probably originated from trees in Iran at the western extreme of the native range, where the indigenous populations are adapted to the warm, dry climate of the region. In 1918 Ernest Wilson encountered the mimosa growing at the eastern, more northerly limits of its range in Korea, and he also saw the tree cultivated in the garden of the Chosen Hotel in Seoul. The Chosen Hotel plant was growing north of the natural limits of the species in Korea, and one plant grown from seeds it produced has displayed remarkable hardiness. It has persisted in the Arnold Arboretum

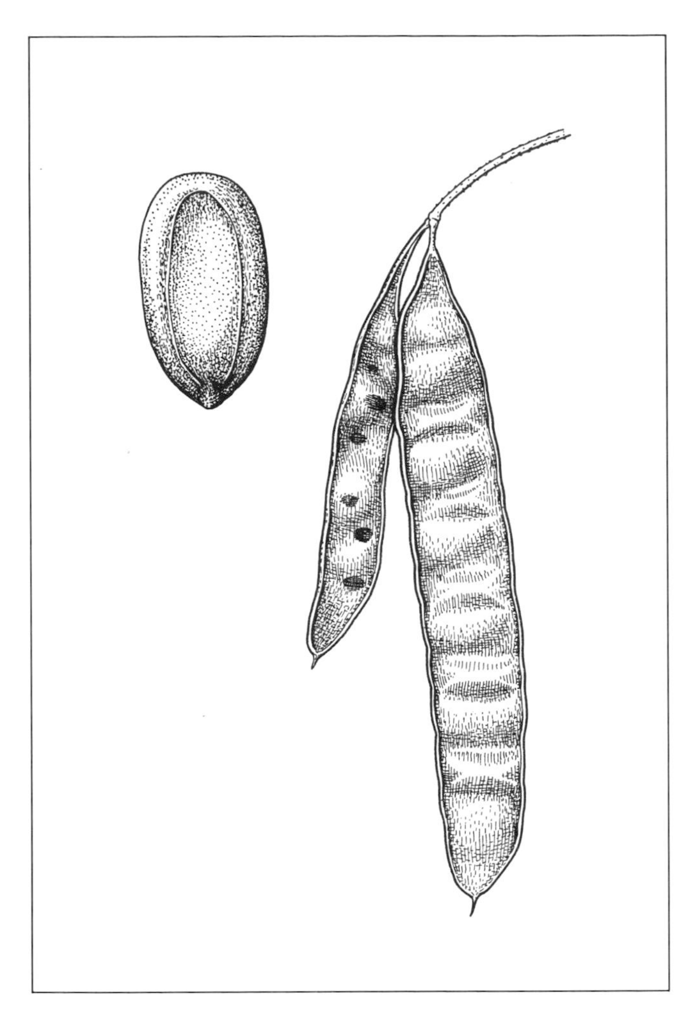

184. The long, tan, seed pods of the silk tree persist on the branches of the tree long after the leaves have fallen, and their familiar pea-podlike structure is indicative of the fact that the tree is a member of the pea or legume family. The seeds, one of which is shown here, have very hard seed coats.

since Wilson's time and has subsequently been given the cultivar name 'Ernest Wilson' to note its hardiness and to pay tribute to the great botanical and horticultural explorer responsible for bringing it to America. Seeds from this plant, moreover, have produced plants that can withstand New England winters, and through asexual propagation techniques the clone 'Ernest Wilson' is now successfully cultivated in the region between Philadelphia and Boston.

Wilson's mimosa or silk tree, frequently referred to as the hardy form, has the added ornamental attribute of producing highly colorful flower clusters (Plate 24, bottom). Close observation will reveal that the beauty of the abundantly produced inflorescences resides in the twenty to fifty long stamen filaments comprising the androecium of each otherwise inconspicuous flower. The filaments are white toward their bases but rosy pink distally, and each is terminated by a small, yellow, peltate anther. Because the flowers are produced in tight clusters that are aggregated together to comprise compound inflorescences, the filaments of the numerous flowers combine to yield the pompon or powder-pufflike appearance characteristic of the tree. The flowering period, moreover, is a long one, and in the Arnold Arboretum the first flowers usually appear in early July, but flowering continues unabated into the first weeks of September.

In habit the mimosa is a deciduous tree to about thirty feet, frequently with more than one main trunk, and with a wide-spreading, flat-topped crown that achieves a layered structure as the trees mature. The fernlike leaves emerge from the buds very late in the spring, and many people fear their prized specimens have died over the winter when no growth is evident by the middle of May. Once growth has finally recommenced in late May or early June, the large, doubly compound leaves expand rapidly. Comprised of hundreds of small leaflets, the leaves can measure up to twenty inches in length, and each is held in a more-or-less horizontal plane, adding to the layered appearance characteristic of the trees. The leaflets exhibit the curious sleep movements, whereby the leaflets fold together from the horizontal into a vertical plane, typical of many tropical species of the pea family, and a small, glandular nectary, which serves no readily apparent purpose, occurs near the base of each leaf stalk.

The hardy mimosa was one of the very last plants Ernest Henry Wilson brought back to the Arnold Arboretum from eastern Asia, and the return of the collector from that region to Boston in 1919 signaled the end of an era. Only one shipment of seeds of Chinese plants had been received at the Arboretum in the interim since Wilson returned from that country in 1911, and further shipments of seeds and plants from foreign lands seemed unlikely, given the restrictions on plant and seed importation that had been imposed by the federal government in 1912. These regulations were strengthened in 1919, and enforcement by federal inspectors promised to become more rigorous.

Joseph Hers, a Belgian railroad official stationed in northern China, had sent welcome specimens collected in Honan (now Henan) Province, but only one shipment of seeds was forwarded from the Department of Agriculture in Washington, D.C., to the Arboretum in Boston. While further consignments of seed and specimens were received from Hers between 1920 and 1926, by comparison with the influx of materials that had been received earlier in the century, incoming shipments were slowing to a trickle.

While Sargent quickly joined the opposition to the new federal regulations that threatened the Arboretum's plant introduction program, officials of the Department of Agriculture had congressional support for the legislation, and foresters, agriculturists, and orchardists had ample reason for concern. The virulent chestnut blight from Asia, which was destined to decimate the majestic chestnut trees of eastern forests, had been identified in New York in 1904, and in 1909 the white pine blister rust was also recognized in New York state on nursery stock that had been imported from Germany. The days of free traffic in plants and seeds and the sometimes indiscriminate importation of diseased plants and their insect pests into the United States—even if part of a scientifically based program of the nation's leading arboretum—had to come under governmental scrutiny that frequently required inspection and quarantine. Charles Sargent bristled under the regulations but was chagrined when federal inspectors found four species of foreign insects never before recorded from North America thriving in the Arnold Arboretum!

Nevertheless, Sargent was optimistic and could not restrain himself when Joseph Rock visited the Arnold Arboretum during the summer of 1924. Rock, a thoroughly versed Austrian botanist, explorer, and cartographer with remarkable linguistic abilities, had recently returned from a highly successful expedition in China on which he had gathered over 80,000 herbarium specimens, seeds of hundreds of plants, and over 1,600 ornithological specimens. Before he left Boston, Rock agreed to undertake another Chinese expedition—this

185. Joseph Hers (1884–1965), a Belgian railroad official stationed in northern China, developed a great interest in botany and sent shipments of both seeds and specimens of Chinese plants to the Arnold Arboretum for about a decade beginning in 1919. Hers is shown here in official dress at the Belgian consulate in Shanghai in 1910.

time under Arboretum sponsorship—that would explore the mountainous districts of northern Tibet and northwestern China. In September Rock started again for China.

The regions pinpointed for Rock's exploration proved woefully disappointing with regard to its ligneous flora. Political upheaval in China added to the difficulties the collector faced, and circumstances involving disputes and uprisings between the tribes of the Tibetan borderland forced circuitous routes, restricted movement, and the constant threat of open hostility by the Chinese in the border regions Rock was to explore. Sargent was surprisingly understanding of the difficulties Rock faced, and the elderly director repeatedly reassured the Austrian collector in his letters. With regard to the dearth of trees in the region, Sargent wrote from Boston, "No one certainly has a better reason not to be disheartened than you. Do not forget that it is as important to discover that no plants grow in a country as it is to find what grows in it, and for this reason I consider your Tibet journey a great success" (AAA, Sargent correspondence, letter dated October 27, 1926).

As the political situation throughout China worsened, Sargent and Wilson, who had been appointed assistant director of the Arboretum in April of 1919, feared for Rock's safety. Regretfully, but in light of the circumstances, they decided that the work of the expedition should be brought to a premature close. In December of 1926 Sargent wrote to Rock, urging him to abandon his plans and return to the United States as quickly as possible. Receiving this summons, Rock made preparations to leave the Tibetan Marches and began his long trek to Shanghai in March of 1927. As Rock was traveling homeward, Charles Sprague Sargent died in his sleep at Holm Lea on March 22, just a few weeks before his eighty-sixth birthday. E. H. Wilson, forced to assume administrative responsibilities at the Arboretum, quickly wrote Rock with the lamentable news of Sargent's death and reiterated that his return to Boston should be made with all possible speed. The red-letter field days of Arnold Arboretum agents in Asia had reached a hiatus.

In the wake of Sargent's death, uncertainty surrounded the Arboretum staff, and the course and direction of the institution he had created almost single-handedly was temporarily questioned. Wilson, however, was placed in charge and, in the English tradition, given the title of Keeper. In that capacity the veteran naturalist brought continuity to the institution, and the established routine and emphasis on sound taxonomic studies emanating from the herbarium and library and the pre-

186. Joseph Rock (1884–1962), Viennese-born botanist, ethnologist, and linguist, undertook field work in western China for the Arnold Arboretum between 1924 and 1927. Rock also collected for the United States Department of Agriculture and conducted explorations under the auspices of the National Geographic Society in southwestern China and Tibet. In addition to his contributions to botany and geography, Rock made significant contributions to linguistic studies of the languages of the minority nationalities populating the areas in which he collected. Rock is shown here in Tibetan dress.

eminence of the living collections was continued. But Wilson's tenure, unlike the 54-year directorship of Sargent, was brief. Tragically, on October 15, 1930, three and a half years after Sargent's death, Ernest Henry Wilson and his wife, Helen, were both killed in an automobile accident near Worcester, Massachusetts, as they were returning to Boston after visiting their daughter. With Wilson's premature death, the old order at the Arnold Arboretum came to an abrupt end.

A Million-Dollar Tree

Shortly after Christmas, winter tightened its grip on New England, and gusty, northwesterly winds blew cold over the snow-encrusted landscape of the Arnold Arboretum. The greenhouses were nonetheless warm, as the furnace was kept well stoked; and excitement and anticipation filled the air during the first short days of the new year, 1948. Dr. Merrill had received a package from war-torn China and let it be known that additional help would be required at the greenhouse. A taxonomist of world repute who specialized in the floras of eastern and southeast Asia, Merrill had also distinguished himself as an extremely able and energetic administrator and an active participant in the politics of the worldwide botanical community. Merrill had resigned his post as director of the New York Botanical Garden to come to Harvard in 1936 as administrator of the university's botanical collections and to assume the Arnold Professorship as director of the Arnold Arboretum. Slight of build, Merrill was a dynamo, who promptly became known to the Arboretum staff as the "Chief." Recently arrived in Boston from China to begin her graduate studies in botany under Merrill's direction, Shu-ying Hu was quickly given marching orders during her first interview with the Chief. Merrill suggested that as soon as she was settled and could see her way clear, that she report to the greenhouse and lend a hand in sorting the seed. It would be of enormous interest to her as it came from a living fossil recently discovered in a remote corner of her homeland.

With the helping hands of staff members from the library and herbarium, the greenhouse staff quickly and efficiently filled coin envelopes with quantities of the precious seed from the shipment that had been received from China early in January. These packets, in turn, were placed in envelopes, which were addressed, stamped, and mailed to botanical gardens, arboreta, and Merrill's professional colleagues across North America

187. Elmer Drew Merrill (1876–1956) accepted the position of Arnold Professor of Botany and Director of the Arnold Arboretum in 1935.

and in England. Another, far larger shipment of seed was received from China on March 16, 1948, and the distribution process was repeated, although by that date requests for a share of the seed had been received at the Arboretum from correspondents around the world. Additional allotments from this batch of the seed, which Merrill likened to rolled, dried kernels of corn, were shared with other botanical and horticultural colleagues and with nurserymen and interested amateur gardeners from among the general public as well. In total, something in excess of six hundred packets of seed were sent from the Arboretum to institutions and individuals across the United States, in all of the European countries, in Canada and Mexico, and to many destinations in subtropical countries in both hemispheres. The receipt of the seeds at the Arboretum had been widely reported in the press, and the romantic notoriety surrounding the discovery of a tree that was reported to constitute a "living fossil" from the age when dinosaurs roamed the earth helped to fuel the demand.

Despite the generosity of the Arboretum in attempting to send a few seed to all who were interested in obtaining it, a quantity of seed sufficient for experimentation was retained for germination studies in the Arboretum greenhouses. Fortunately, germination was rapid, thereby quelling questions concerning the viability of the seed, and by the end of January young seedlings were soaking up the rays of the winter sun in New England.

188. Chi-ju Hsueh, the Chinese forester who collected the specimens of the dawn redwood in 1946 that were studied by Hu and Cheng and provided the basis for their description of *Metasequoia glyptostroboides*. This photograph was taken in October of 1984, some thirty-eight years after Hsueh made his historic collection.

Surprisingly, as the mails carried the seeds around the world from the Arboretum clearinghouse and as the young seedlings were germinating in the Jamaica Plain greenhouse, the tree from which the seed had been gathered in China lacked a scientific name. But by the time the seed had been collected for shipment to the Arboretum it was known that the tree was a living representative of the genus *Metasequoia*, which had been described only seven years earlier. In 1941 Shigeru Miki, a Japanese paleobotanist, founded the new genus to accommodate fossils of Pliocene age—from deposits roughly five million years old—that had previously been confused with the fossil remains of bald cypresses (*Taxodium* species) and redwoods (*Sequoia* species).

Unbeknown to Miki in wartime Japan, a Chinese forester, T. Kan, from the National Central University, chanced on a strange new coniferous tree in 1941 on the outskirts of the hamlet of Mo-tao-chi (now Modaoqi) in eastern Szechwan Province. Kan's discovery led to a series of events that would result in the worldwide distribution of the seed of the tree, which proved to be an extant species of the genus that Miki had established to accommodate fossil remains millions of years old.

Kan failed to collect specimens from the tree he discovered, as it was in a leafless condition when he encountered it in the early winter of 1941. The following year, however, Kan asked the principal of a local school, Lung-hsin Yang, to gather specimens during the growing season and forward them to his attention. These collections were duly made, yet the specimens were never identified and may have been lost. Probably perturbed that his efforts had come to naught, Yang mentioned the strange tree to T. Wang of the Central Bureau of Forest Research, who was conducting forestry

189. Professor Hsen-hsu Hu (1894–1968) with his wife and daughter. Hu was director of the Fan Memorial Institute of Biology in Peking and a close friend, colleague, and correspondent of E. D. Merrill. Professor Hu was the first of many Chinese students of botany who received their graduate training at Harvard. In 1924 he earned a master's degree, followed by a doctor of science degree in 1925. Returning to China, Hu became one of his country's leading botanists and was instrumental in arranging for the special expedition to collect seeds of the dawn redwood in the fall of 1947.

reconnaissance in the region during 1944. As a result, Wang visited Mo-tao-chi to examine the tree and gather fragmentary specimens—a few leafy branchlets and some cones picked up from the ground beneath the large tree. Two of the cones and one of the branchlets were passed on to Chung-lung Wu, an assistant in the Department of Forestry of the National Central University, who, in turn, gave the incomplete material to Professor W. C. Cheng of the same academic department for identification.

After his initial examination of the samples, Cheng thought they might possibly represent the Chinese swamp cypress (*Glyptostrobus lineatus* (Poiret) Druce), but the botanist was frustrated by Wang's incomplete specimens. Wanting to finalize the tree's identity, Professor Cheng sent C. J. Hsueh, a recent graduate of his department, to the remote village in 1946 to gather a complete suite of specimens for further study. Dutifully, Hsueh visited Mo-tao-chi in February and again in May of that year. Cheng's messenger returned with adequate herbarium material to suggest that the tree was not, in fact, a Chinese swamp cypress.

Finally convinced that the tree represented an undescribed species and constituted a new genus, Professor Cheng sought the opinion of a colleague. Fortunately, Cheng forwarded some of Hsueh's specimens to Dr.

190. This map, published by professors Hu and Cheng in 1948, shows the metasequoia area in the border region of Szechwan and Hupeh provinces.

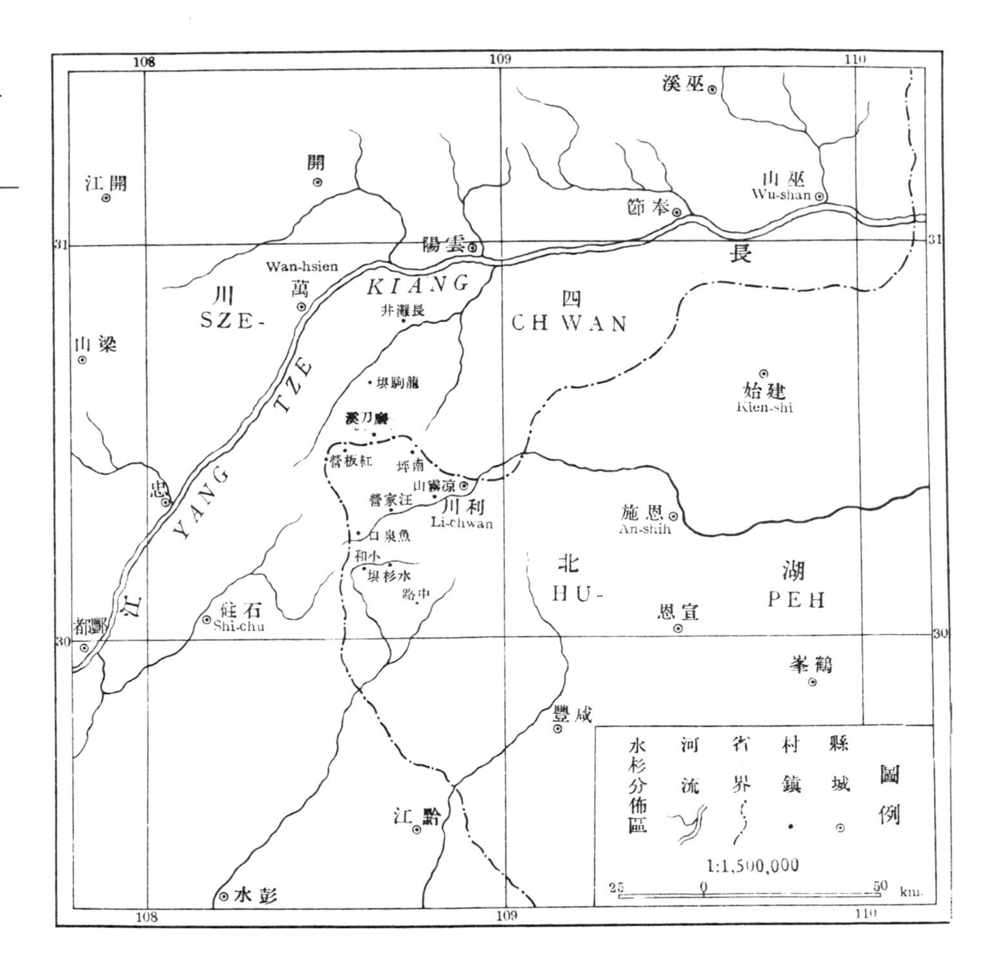

191. One of the many newspaper stories announcing the arrival of dawn redwood seeds at the Arnold Arboretum.

'Extinct' Tree Yields Seeds for America

CAMBRIDGE, Mass., Feb. 9—(AP)—Seeds of a tree, believed extinct for 100,000,000 years, have arrived at Harvard's Arnold Arboretum and soon will be shipped to botanical gardens in the United States and England.

Up to two years ago the metasequoi, a type of fir tree, was known only from fossil remains. The first living specimen was found in February, 1946, by Chinese explorers in a remote valley of central China.

The seeds were obtained later by an Arboretum expedition which located 100 more metasequoias, described as more than 100 feet high with 7½-foot trunks. Although cone-bearing, the tree loses its leaves in winter.

Dr. E. D. Merrill, Arnold professor of botany at Harvard, said today that the seeds, resembling rolled, dried corn kernels, may not germinate in this climate.

H. H. Hu in the fall of 1946. Hu, China's leading dendrologist and director of the Fan Memorial Institute of Biology in Peking, was aware of Miki's article and was overwhelmed by the similarity between Miki's fossils and Hsueh's specimens. While it hardly seemed possible, it nevertheless became obvious that the tree growing at Mo-tao-chi belonged to the genus the Japanese botanist had based on fossil specimens. Some of Hsueh's herbarium material was also forwarded to Elmer D. Merrill at the Arnold Arboretum, and Professor Merrill, realizing the significance of the new discovery, immediately corresponded with both Professor Cheng and Dr. Hu and requested seed. In response to Merrill's request, Hu pointed to the sorry state of the Chinese economy and his woefully inadequate institutional budget; funds to support a collector and finance the trip to Mo-tao-chi were simply not available. Anxious to obtain the seed, Merrill immediately provided $250 from a special Arboretum endowment fund for a seed-collecting expedition, which was arranged by Professor Cheng for the late summer and fall of 1947. Due to the extreme inflation that the Chinese economy was experiencing in the aftermath of the Sino-Japanese war, Merrill's $250 yielded the astounding equivalent of almost a million dollars once it was converted into Chinese currency! The mission to Mo-tao-chi proved successful, and two consignments of seeds of the living fossil were forwarded from China to the Arnold Arboretum. These were received in early January and March of 1948 and were immediately shared with institutions and individuals around the world.

Later in 1948, as seeds of the Mo-tao-chi tree were germinating in greenhouses and nurseries around the globe, Professors Hu and Cheng collaborated in formally describing the new conifer. Their publication appeared in the pages of the *Bulletin of the Fan Memorial Institute of Biology,* where the tree was given the name *Metasequoia glyptostroboides* Hu & Cheng. In the American press, however, the living fossil was quickly given the romantic moniker of dawn redwood at the suggestion of Ralph W. Chaney, a professor of paleobotany at the Berkeley campus of the University of California. In part, the name dawn redwood was derived from the generic name, which had been based on the Greek, *meta,* akin to, and *Sequoia,* the generic name of the coast redwoods of California, which in turn commemorates the half-breed Cherokee Indian, Sequoya (1770–1843), who recorded the Cherokee language. Because of its early fossil record, the epithet "dawn" seemed appropriate and served to romanticize the name. The specific epithet, however, alluded to the genus *Glyptostrobus,* the Chinese swamp cypress with which the dawn redwood had been initially confused.

Because of the paleobotanical significance of metasequoia, Chaney became deeply interested in the occurrence of the extant tree at Mo-tao-chi. Wanting to satisfy his curiosity about the ecology of the region, Chaney hastily made plans to visit the area in March of 1948. Accompanied by Milton Silverman, science editor of the *San Francisco Chronicle,* Chaney boarded a trans-Pacific flight for China. After a harrowing journey across seemingly never-ending mountain ranges, the paleobotanist and his journalist companion finally arrived in Mo-tao-chi, and Chaney and Silverman could claim to be the first Westerners to pay homage to the living fossil in its native habitat. The visit of the California travelers astonished the local residents as well as the populace of a neighboring valley, where upwards of a thousand dawn redwoods had been discovered in 1947 by the team of seed collectors financed by the modest grant from the Arnold Arboretum.

During the late summer of 1948 the local populace was again dumbfounded as a larger team of expeditionaries filed along the narrow road that entered their secluded valley from the north. On this occasion the scientists, their Chinese colleagues, and their escorts made arrangements to remain in the area for an extended period, and two farmhouses were rented for a base camp. This time, too, the local inhabitants from one end of the valley to the other would have ample opportunity to satisfy their curiosity concerning J. Linsley Gressitt, the American in their midst who represented the California Academy of Sciences. Gressitt, for his part, intended to satisfy his curiosity concerning the flora and fauna—particularly the insects—of the metasequoia valley.

The California Academy–Lingnan University Dawn Redwood Expedition, which Linsley Gressitt headed, constituted the last effort an American institution would be allowed to make toward a greater understanding of Chinese biota before the bamboo curtain quietly fell during 1949. The dawn redwood had achieved international notoriety in a matter of months and aroused renewed interest in the rich and varied Chinese flora. Chinese botanists, moreover, anxious to reestablish their professional activities following the devastating hiatus wrought by the Sino-Japanese war, were anxious to cooperate and collaborate with botanical colleagues in America and Europe. The era of institutional cooperation, which the introduction into cultivation and worldwide distribution of the dawn redwood had made clearly manifest, could be international in scope and involve interested parties in both hemispheres. Yet the political turmoil in mainland China in the late 1940s and the founding of the People's Republic in 1949 dictated missed opportunities in the era that lay ahead. The bamboo curtain effectively sealed the Middle Kingdom from the outside world once again. For a brief 110-year period in her long history China had been open to world travelers. Robert Fortune first explored that country for plants at the end of the Opium War in the nineteenth century, but similar opportunities for twentieth-century botanists from the majority of Western nations came to a temporary close in 1949, as China once again became isolationist.

While communications between the Arnold Arboretum and its botanical colleagues in China came to a halt in 1949, Donald Wyman, a Cornell-trained horticulturist whom Merrill had appointed to the Arboretum staff in 1936 to take charge of the living collections, astutely planted the young metasequoia saplings in a variety of Arboretum locations to determine the cultural requirements of this living fossil. Some individuals were planted high up on the slope above Bussey Brook in the conifer collection, where soil drainage would be good and the surrounding trees would provide shelter from

192. Milton Silverman, science editor of the *San Francisco Chronicle*, stands among the three dawn redwoods that were growing at Mo-tai-qui in March 1948. Silverman accompanyed Ralph Chaney of the University of California, Berkeley, on a brief visit to the metasequoia region, and this snapshot is one of the first pictures taken of the "living fossils." In 1980 only the large tree in the center was still growing in this location, and the small shrine or temple at its base had been removed.

193. The beautiful needlelike leaves of the dawn redwood are closely spaced along its branches, and like those of the bald cypress they fall to the ground in the fall. Also depicted here is a winter twig with dormant buds, as well as an individual bud magnified several times. An unopened and opened cone of the dawn redwood along with two of its seeds are also shown; the notch at one end of the seed marks the point at which it was attached to a cone scale. Seed production has not been commonplace in cultivated individuals. Fortunately, the dawn redwood is easily propagated from cuttings, and this has been the chief means by which the species has been propagated in North America and Europe.

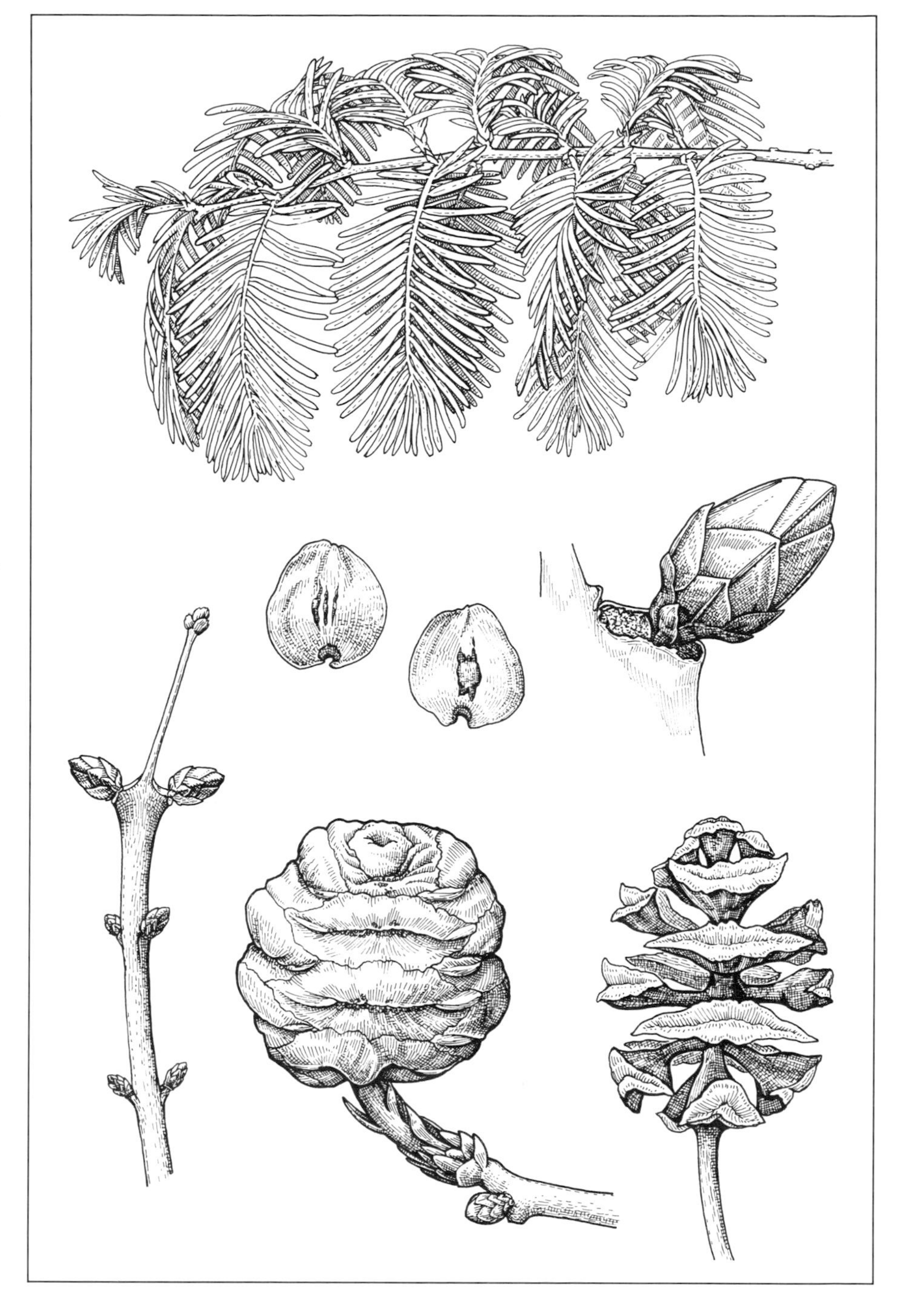

strong winds. Others were sited along the brook itself, where soil moisture would be less subject to summer drought but cold air drainage during the winter months and in early spring might prove to be a stressful environmental factor. A solitary specimen was planted at the edge of the meadow across from the Hunnewell building near the Arborway gate. If it survived, it would be one of the first trees visitors would encounter as they entered the Arboretum by its most frequented gate. Still other individuals were planted in a small grove just inside the gate in association with a bald cypress (*Taxodium distichum* (L.) Richard) from the southeastern United States, a species that has numbered as a cultivated plant from the time the Tradescants grew the tree in their London garden over three hundred years earlier.

Ironically, the bald cypress was one of the first North American trees to be cultivated, while the dawn redwood ranks as one of the last-discovered Chinese trees to be grown as an ornamental. The dawn redwood has an extremely limited natural distribution in central China, while the bald cypress is widespread on the Coastal Plain of eastern North America. Both are deciduous in habit, and despite the generic and common names of the Chinese tree—which allude to an alliance with the redwoods or sequoias of California—the dawn redwood and the bald cypress are very closely related. The similarities between the two also extend to their native habitats, where both are riverine species and stand as ecological equivalents in the natural communities in which they occur.

The ancestors of both, moreover, have a shared history in the geological record that attests to their once widespread distribution across the Northern Hemisphere and their subsequent reduction in range. Metasequoia endured the glacial epochs, to survive to the present day—albeit on the brink of extinction—in one small corner of China, while the bald cypress persisted to reinvade the swamp forests of eastern North America, which now extend along the Atlantic and Gulf Coastal Plains. Yet at one time during the Cenozoic Era the dawn redwood grew in Greenland and Spitzbergen, and representatives of both genera were widespread across Eurasia and particularly well represented in the forests of western North America.

A New Beginning

Subsequent to the introduction of the metasequoia into cultivation, no active field work in the temperate regions of China was possible by foreigners, and the Korean War of the 1950s offered little opportunity for horticultural exploration in that Asian country. With the advent of the Cold War, moreover, the Soviet Union, like the People's Republic, became off limits to botanically inclined travelers from the Arnold Arboretum and other horticultural and botanical institutions in the United States. The introduction of new plants into North American gardens continued, but it centered most frequently on the importation of new, named selections from pre-existing garden stock or the hybrid progeny resulting from breeding programs in European horticultural centers. Many of the new plants given fancy vernacular—or cultivar—names were also of American origin, resulting from the selection, trial, and propagation of outstanding clones by nurserymen and horticulturists.

By mid-century the Arnold Arboretum was certainly not alone in seeking new trees and shrubs for American landscapes, and many new plants were raised from seeds received from cooperating botanical gardens around the world. Venerable sister institutions like the New York Botanical Garden, the Missouri Botanical Garden, the Fairchild Tropical Garden, the Morton Arboretum, the Morris Arboretum, the botanical garden network of the University of California at its several campuses, and the University of Washington Arboretum, among others, all catered to and continue to serve the horticultural and botanical communities in their regions. This work included the search for new and improved plants for urban settings and their introduction into the nursery industry, as well as maintenance of diverse living collections for use by professional botanists, nurserymen, and amateur plantsmen alike.

Yet the botanists and horticulturists at these and other gardens were frequently frustrated by the problems associated with plant exploration in, and introduction from, foreign lands. The expense and disruption of the war effort during the 1940s, government regulations and quarantine requirements, and the high costs of travel and field work seemed to conspire to bring the halcyon, golden era of plant exploration to an end. In the early 1950s, however, Dr. Russell Seibert, then director of Longwood Gardens in Kennet Square, Pennsylvania, advanced a proposal that would link the considerable financial resources of the Longwood Foundation with the

194. Arnold Arboretum plant propagator Jack Alexander explains softwood cutting techniques employed at the Arboretum to members of the delegation of Chinese botanists who visited the Dana Greenhouses in 1979.

skill and governmental status of the United States Department of Agriculture. Through the joint efforts of the two organizations, which were formalized in 1956, the United States Department of Agriculture–Longwood Ornamental Plant Exploration Program was established, and a new phase of plant exploration was inaugurated.

As a consequence of this joint agreement, the Longwood Foundation funded thirteen expeditions between 1956 and 1971 that used U.S. Department of Agriculture personnel and facilities. A series of exploratory trips to Japan marked the considerable efforts of Dr. John Creech, who was to become the director of the National Arboretum, to introduce horticulturally meritorious plants suitable for culture in the southeastern United States and northward along the eastern seaboard. Frederick G. Meyer likewise surveyed the native flora, gardens, and nurseries of much of Western Europe for valuable new ornamental plants for introduction into the United States under the auspices of the program. All told, over 10,000 plants were introduced into the United States from regions as diverse as New Guinea, Japan, and Scotland as a result of the joint program.

Specific programs of the United States National Arboretum in tree breeding and specialized programs in horticulture to provide ornamentals undertaken by Longwood Gardens as a direct consequence of the joint program added to the dynamic undertakings of horticultural institutions in North America. A seed-collecting trip sponsored by the Arnold Arboretum in the late summer and fall of 1977 to Japan and Korea provided the opportunity for the Arboretum to renew and revitalize its collections from those regions, to add genetic diversity to the existing collections, and to reintroduce species that had been lost over the years to disease, insect pests, old age, and natural calamity. Another important aspect of the 1977 mission—the first Arboretum-sponsored undertaking in Asia by its staff since the days of E. H. Wilson—was the attempt to gather sufficient quantities of seed from species that had previously failed to prove hardy in the Boston region. If these seeds could be collected from specimens growing in northern latitudes, hardy strains might be selected for successful cultivation in New England.

But the real opportunity came in 1978, when Dr. Richard A. Howard, at that time director of the Arboretum, was invited to represent that Harvard department as a member of a delegation of American botanists invited to tour botanical gardens and institutions in the

People's Republic of China. For one month during the late spring of 1978 Dr. Howard and his colleagues from other American institutions were fêted by Chinese colleagues and given the opportunity to reforge old ties that had been broken three decades before. In response to the warm hospitality accorded the American group, the Botanical Society of America extended an invitation to botanical colleagues in China to visit the United States.

As a consequence, a delegation of Chinese botanists headed by Professor P. S. Tang—a Harvard-trained plant physiologist and director of the Institute of Botany in Peking—visited the United States in the late spring and early summer of 1979. The delegation was welcomed at the Arnold Arboretum in early June by the staff and Professor Peter S. Ashton, the Arboretum's recently appointed fifth director. While visiting the Arboretum, the Chinese botanists delighted in inspecting so many Chinese plants that had reached their maturity growing in Boston. The dove trees and the paperbark maple, which some of the delegation had never seen growing before, drew appreciative responses, while the many dawn redwoods, which have proven perfectly hardy in southern New England, were among the first individuals of this species the Chinese visitors had seen growing outside China. At the conclusion of their U.S. tour, a meeting of interested individuals and representatives from leading American botanical institutions was held at the University of California in Berkeley. The Arnold Arboretum was represented at this meeting, where the primary topic of discussion centered on the potential for future collaboration between Chinese and American botanists.

As a result of discussions there, early in 1980 five American botanists received coveted invitations from the Chinese Academy of Sciences to participate in a field expedition to western Hupeh Province during the late summer and fall. The field work would be undertaken in the Shennongjia Forest District, a rugged region north of the Yangtze River near the Szechwan border where both Augustine Henry and Ernest Henry Wilson had collected. For me, participation in the expedition was the opportunity of a lifetime, combining as it did the chance to visit a region explored by Wilson, the ability to study well-known cultivated trees and shrubs in their native habitats, and the possibility of obtaining once again herbarium specimens and seeds of Chinese plants for trial in the Arnold Arboretum. David E. Boufford, at the time employed by the Carnegie Museum of Natural History in Pittsburgh but currently on the Arboretum staff, was another enthusiastic member of the American team, as were Dr. James Luteyn of the New York Botanical Garden, Dr. Theodore Dudley of the United States National Arboretum, and Dr. Bruce Bartholomew, then of the University of California Botanical Garden at Berkeley and the leader of the American team.

With the sponsorship of the Botanical Society of America, travel support from the National Geographic Society, and additional financial assistance from the American Association of Botanical Gardens and Arboreta, the five-man American team gathered in Tokyo in early August of 1980 before boarding a flight for Peking. Leaving the Chinese capital behind, the group traveled by air with Chinese colleagues to Wuhan on the Yangtze River and then overland by jeep and minibus caravan to the mountainous districts of western Hupeh. The field work concluded with a cruise up the Yangtze through its famous gorge region to Wan Hsien in Szechwan Province. From that city the expeditionaries went overland by jeep to Lichuan in Hupeh Province, and for three days field work was conducted in the fabled metasequoia valley. The five American botanists were the first Westerners to revisit that area since the establishment of the People's Republic of China, and it was with botanical reverence that they could gather data and specimens from the dawn redwood growing at Mo-tao-chi.

When the American team returned to the United States in November of 1980, they carried with them the seeds of many species previously introduced into cultivation in the West by the plant hunters of earlier days, but they also brought with them seeds of plants never before tested in cultivation. Seeds of species not previously known to science were also carried to the United States. A small grove of a recently described Chinese sweet gum (*Liquidambar acalycina* Chang) now grows in the Arnold Arboretum near trees of its American congener (*Liquidambar styraciflua* L.), which was first cultivated in England by Bishop Compton in the mid-seventeenth century. And a new simple-leaved mountain ash (*Sorbus yuana* Spongberg), named to recognize the tireless and sincere efforts of Professor T. T. Yü in making the 1980 expedition possible and in forging a cooperative spirit between Chinese and American botanists in the era following ping-pong diplomacy, promises to be a new ornamental that will rank high among trees introduced to the New World from the Old. Another potentially versatile ornamental shrub—previously introduced but subsequently lost in Western gardens—the so-called seven-son flower (*Heptacodium miconioides* Rehder), was brought back to the gardens of North America by the 1980 Sino-American Botanical Expedition team. And

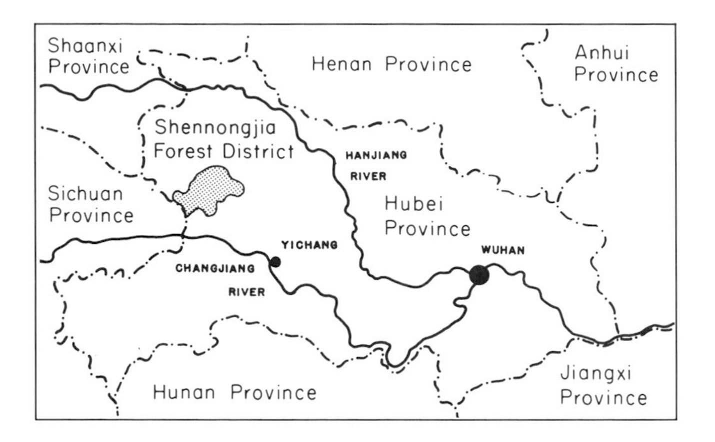

195. This map, which utilizes Pinyin spellings for place names, indicates the location of the Shennongjia Forest District.

196. Despite his advanced age, Professor S. C. Sun, director of the Wuhan Institute of Botany and chairman of the Department of Biology, Wuhan University, served as expedition leader for the 1980 Sino-American team.

197. Chinese members of the 1980 Sino-American expedition team prepare herbarium specimens at the base camp in the Shennongjia Forest District. More than 1,700 collections were made, each represented by numerous sheets. Half of the materials collected remained in China, while the other half was shared by the participating American institutions.

198. This group of Chinese youth had assembled and was waiting for the arrival of the 1980 Sino-American Botanical team in Mo-tai-qui, the small hamlet in southwestern Hupeh Province where the "type" tree of the dawn redwood still grows.

a gift of seeds of an elusive Chinese magnolia (*Magnolia zenii* W. C. Cheng), never before cultivated in the West, was presented to the American botanists by colleagues at the Nanking Botanical Garden. One of the several trees raised from these seeds at the Arnold Arboretum flowered for the first time during the early spring of 1988, and the species has been added to the long list of cultivated magnolias of Chinese origin.

Ernest Wilson, when asked by Charles Sargent to undertake his first Arboretum expedition to China, was fearful of "going once too often to the well" (AAA, Wilson correspondence, letter dated August 2, 1906). While the prospect of diminishing returns might occasionally pass through the mind of the field botanist and naturalist, the urge to travel and experience plants in their natural habitats soon gains the upper hand. As long as there are those who are born with a consuming interest in and love of plants, exploration will continue. For the Arnold Arboretum and many of its sister institutions, their renewed relationships with colleagues in China constitutes a new beginning and a promise for the future.

Today, in urban Boston and in gardens and parklands across North America and Eurasia, trees, shrubs, and other plants unique to each region and species of lineages long separated from one another as a consequence of the earth's geological history mingle in common landscapes. This phenomenon constitutes living testimony to the combined efforts of generations of botanical and horticultural explorers and keen cultivators of plants. As a consequence of these human endeavors and the tireless efforts of its staff, the collection of trees and shrubs that grows on the glacier-scarred landscape of the Arnold Arboretum is unique among all the parklands of North America. Open for the edification and enjoyment of all, this première collection constitutes a dynamic living museum where the woody plants of the Northern Hemisphere have held a century-long reunion, one which continues to enlarge in scope.

Plate 23. The most distinctive attribute of the paperbark maple, one of E. H. Wilson's most famous introductions from China, consists of its paper-thin, cinnamon-colored bark that peels from its trunk and branches in thin sheets. Coupled with its neat habit and the beautiful fall color of its leaves, the tree has year-round appeal.

Plate 24 (top). The characteristic fruits of the sand pear (*Pyrus pyrifolia*) are russet in color and filled with an abundance of stone or grit cells that render them inedible. Their size is similar to that of a ping-pong ball.

Plate 24 (bottom). The silk tree or hardy mimosa produces a profusion of pink, powder-pufflike inflorescences through the summer months. Most of the trees hardy in New England comprise the cultivar 'Ernest Wilson' which Wilson brought from Korea to the Arnold Arboretum in 1918.

199. The American participants on the 1980 Sino-American Botanical Expedition gather at the base of the "type" tree of the dawn redwood on the outskirts of Mo-tai-qui. The American team consisted of (from left to right) David E. Boufford, James L. Luteyn, Bruce Bartholomew, Stephen A. Spongberg, and Theodore R. Dudley. A large crowd of curious onlookers followed the American team and observed their every action.

200. This painting of *Heptacodium miconioides* (formerly known as *Heptacodium jasminoides*), the so-called seven-son flower, was featured on a cover of *Arnoldia*, the magazine of the Arnold Arboretum, and an article discussed its history and ornamental attributes.

BIBLIOGRAPHY

The titles in the bibliography include repeatedly consulted references and those from which information was gleaned that was incorporated into *A Reunion of Trees*. In quoting from the correspondence of C. S. Sargent and E. H. Wilson, the abbreviation AAA has been used in the text to indicate that the original letters or copies thereof are housed in the Archives of the Arnold Arboretum in Jamaica Plain, Massachusetts.

Aiton, W. T. 1811. *Hortus Kewensis; or, a Catalogue of the Plants Cultivated in the Royal Botanic Garden at Kew.* 2nd ed. 5 vols. London.

Alden, R. H., and J. D. Ifft. 1943. *Early Naturalists in the Far West.* Occasional Papers, California Academic of Sciences, 20. San Francisco: California Academy of Sciences.

Allan, M. 1964. *The Tradescants, Their Plants, Gardens, and Museum, 1570–1662.* London: Michael Joseph.

Anderson, M. P. 1907. "Early European botanists in Japan." *Journal of the New York Botanical Garden* 8: 99–110.

Anonymous. 1896. "Plant notes: *Stuartia Pseudo-Camellia.*" *Garden and Forest* 9: 34, 35.

Anonymous. 1900. *A Short Account of the Big Trees of California.* United States Department of Agriculture, Division of Forestry, Bulletin No. 28. Washington, D.C.

Anonymous. 1901. "Dr. Emil Bretschneider." *Kew Bulletin of Miscellaneous Information,* pp. 201, 202.

Anonymous. 1942. *Centennial Year, 1792–1892: The Massachusetts Society for Promoting Agriculture.* Boston: The Meador Press.

Anonymous. 1986. "Professors Gray and Sargent pursue *Shortia.*" *Arnoldia* 46(3): 26–32.

Anonymous. 1987. "The bicentenary of the Botanical Magazine." *Newsletter, Royal Horticultural Society,* No. 6.

Bartholomew, B., D. E. Boufford, A. L. Chang, Z. Cheng, T. R. Dudley, S. A. He, Y. X. Jin, Q. Y. Li, J. L. Luteyn, S. A. Spongberg, S. C. Sun, Y. C. Tang, J. X. Wan, and T. S. Ying. 1983. "The 1980 Sino-American botanical expedition to western Hubei Province, People's Republic of China." *Journal of the Arnold Arboretum* 64: 1–103.

Bartholomew, B., D. E. Boufford, and S. A. Spongberg. 1983. "*Metasequoia glyptostroboides*—Its present status in central China." *Journal of the Arnold Arboretum* 64: 105–128.

Bartram, J. 1942. "Diary of a Journey through the Carolinas, Georgia, and Florida from July 1, 1765 to April 10, 1766." [Annotated by Francis Harper.] *Transactions of the American Philosophical Society,* New Series 33:1–120. Philadelphia: American Philosophical Society.

Bartram, W. 1943. "Travels in Georgia and Florida, 1773–74: a report to Dr. John Fothergill." [Annotated by Francis Harper.] *Transactions of the American Philosophical Society,* New Series 33: 121–242.

Bean, W. J. 1970–1980. *Trees and Shrubs Hardy in the British Isles.* 8th ed. D. L. Clarke and Sir G. Taylor, eds. 4 vols. London: John Murray.

Becker, L. E., and G. B. Skipworth. 1975. "Ginkgo-tree dermatitis, stomatitis, and proctitis." *Journal of the American Medical Association* 231(11): 1162.

Berkeley, E., and D. S. Berkeley. 1969. *Dr. Alexander Garden of Charles Town.* Chapel Hill: University of North Carolina Press.

——— 1974. *Dr. John Mitchell, the Man Who Made the Map of North America.* Chapel Hill: University of North Carolina Press.

———. 1982. *The Life and Travels of John Bartram: From Lake Ontario to the River St. John.* Tallahassee: University Presses of Florida.

Berry, E. W. 1923. *Tree Ancestors: A Glimpse into the Past.* Baltimore: Williams & Wilkins Company.

Betts, E. M., ed. 1944. *Thomas Jefferson's Garden Book.* Philadelphia: American Philosophical Society.

Blunt, W. 1950. *Tulipomania.* Harmondsworth, Middlesex, England: Penguin Books, Ltd.

Böhmer, M. L. 1876. "General observations on the flora of Hokkaido." *Gardener's Monthly and Horticulturist* 18: 180–183, 211, 212, 244, 245, 275–277.

Born, W. 1949. "Early botanical gardens." *CIBA Symposium* 11: 1099–1109.

Bowers, J. Z. 1970. *Western Medical Pioneers in Feudal Japan.* Baltimore and London: Johns Hopkins Press.

Bretschneider, E. 1898. *History of European Botanical Discoveries in China.* London: Sampson Low Marston & Company, Ltd.

Brett-James, N. G. n.d. *The Life of Peter Collinson, F.R.S., F.S.A.* London: Edgar G. Dunstan & Co.

Britten, J. 1899. "Bibliographical notes: Frasers' catalogues." *Journal of Botany, British & Foreign* 37: 481–487.

———. 1905. "Bibliographical notes: Frasers' catalogues." *Journal of Botany, British & Foreign* 43: 329–331.

Browne, D. J. 1851. *The Trees of America; Native and Foreign, Pictorially and Botanically Deliniated, and Scientifically and Popularly Described.* New York: Harper & Brothers.

Buel, J. 1828. "On the horticulture of the United States of America." *Gardener's Magazine* 4: 193–196.

Butler, J. R. 1938. "America—a hunting ground for eighteenth century naturalists, with special reference to their publications about trees." *Papers of the Bibliographical Society of America* 32: 1–16.

Catesby, M. 1731–1743. *The Natural History of Carolina, Florida and the Bahama Islands.* 1st ed. 2 vols. London.
———. 1747. *The Natural History of Carolina, Florida and the Bahama Islands: Appendix.* London.
———. 1763. *Hortus Britanno-Americanus: or, A Curious Collection of Trees and Shrubs, the Produce of the British Colonies in America; Adapted to the Soil and Climate of England.* London.
Chaney, R. W. 1950. "A revision of fossil *Sequoia* and *Taxodium* in western North America based on the recent discovery of *Metasequoia.*" *Transactions of the American Philosophical Society,* New Series 40(3): 171–239.
———. 1979. *Redwoods of the Past.* San Francisco: Save-the-Redwoods League.
Chu, K. L., and W. S. Cooper. 1950. "An ecological reconnaissance in the native home of *Metasequoia glyptostroboides.*" *Ecology* 31: 260–278.
Chvany, P. J. 1976. "E. H. Wilson, photographer." *Arnoldia* 36: 181–236.
Clausen, K., and S. Y. Hu. 1980. "Mapping the collecting localities of E. H. Wilson in China." *Arnoldia* 40: 139–145.
Cleveland, H. W. S. 1888. "Letter to the editor." *Garden and Forest* 1: 227.
Coats, A. M. 1963. *Garden Shrubs and Their Histories.* London: Vista Books.
———. 1970. *The Plant Hunters.* New York: McGraw-Hill Book Company.
———. 1975. *Lord Bute, an Illustrated Life of John Stuart, Third Earl of Bute, 1713–1792.* Aylesbury, Buckinghamshire, England: Shire Publications, Ltd.
———. 1976. "The Hon. and Rev. Henry Compton, Lord Bishop of London." *Garden History* 4(3): 14–20.
Cohen, I. B., ed. 1980. *The Life and Scientific and Medical Career of Benjamin Waterhouse.* 2 vols. New York: Arno Press.
Coker, W. C. 1911. "The garden of André Michaux." *Journal of the Elisha Mitchell Scientific Society* 27: 65–72.
Cole, A. B., ed. 1947. *A Scientist with Perry in Japan: The Journal of Dr. James Morrow.* Chapel Hill: University of North Carolina Press.
———. 1947. *Yankee Surveyors in the Shogun's Seas: Records of the United States Surveying Expedition to the North Pacific Ocean, 1853–1856.* Princeton: Princeton University Press.
Corner, B. C. 1958. "Dr. Fothergill's friendship with Benjamin Franklin." *Proceedings of the American Philosophical Society* 102: 413–419.
Corner, B. C., and C. C. Booth, eds. 1971. *Chain of Friendship: Selected Letters of Dr. John Fothergill of London, 1735–1780.* Cambridge: Belknap Press of Harvard University Press.
Coville, F. V. 1897. "The itinerary of John Jeffrey, an early botanical explorer of western North America." *Proceedings of the Biological Society of Washington* 11: 57–60.
Cox, E. H. M. 1943. "Robert Fortune." *Journal of the Royal Horticultural Society* 68: 161–171.
———. 1944. "The Hon'ble East India Company and China." *Proceedings of the Linnean Society of London* 156: 5–8.
———. 1945. *Plant-hunting in China: A History of Botanical Exploration in China and the Tibetan Marches.* London: Collins.
Creech, J. L. 1957. *Plant Explorations: Ornamentals in Southern Japan.* Washington, D.C.: Agricultural Research Service, U.S.D.A., in cooperation with Longwood Gardens of the Longwood Foundation, Inc.
———. 1963. "The greatest service." In A. Stefferud, ed., *After a Hundred Years: The Yearbook of Agriculture 1962,* pp. 100–105. Washington, D.C.: United States Department of Agriculture.
———. 1966. *Ornamental Plant Explorations—Japan, 1961.* Washington, D.C.: Agricultural Research Service, U.S.D.A., in cooperation with Longwood Gardens of the Longwood Foundation, Inc.
———. 1973. "Ornamental plant introduction—building on the past." *Arnoldia* 33: 13–25.
———. 1974. "Searching the world to obtain new and better plants." In J. Hayes, ed., *Landscape for Living: The Yearbook of Agriculture, 1972.* Washington, D.C.: United States Department of Agriculture.
———. 1974. "Highlights of ornamental plant introduction in the United States." *The Longwood Programs Seminars 1974* (The University of Delaware, Newark) 6: 21–25.
———. 1988. "Pioneer plantsmen in Japan." *Garden, Journal of the Royal Horticultural Society* 113: 380–383.
Cronquist, A. 1977. "Editor's note on *Metasequoia.*" *Botanical Review* 43: 281–284.
Cunningham, I. S. 1987–1988. "I. The U.S. National Arboretum: leader in ornamental plant germplasm collection." *Diversity* No. 12: 20, 21. "II. Exploration in Korea." *Diversity* No. 13: 23, 24. "III. Exploration in Japan." *Diversity* No. 14: 28, 29. "IV. Exploration in the People's Republic of China." *Diversity* No. 15: 33, 34.
Cutler, M. 1783. "An account of the vegetable productions, naturally growing in this part of America, botanically arranged." *Memoirs of the American Academy of Arts and Sciences* 1: 396–493.

Darlington, W. 1849. *Memorials of John Bartram and Humphrey Marshall with Notices of Their Botanical Contemporaries.* Philadelphia: Lindsay & Blakiston.
Darwin, C. 1859. *On the Origin of Species by Means of Natural Selection, or the Preservation of Favoured Races in the Struggle for Life.* London: John Murray.
Davies, J. 1980. *Douglas of the Forests: The North American*

Journals of David Douglas. Seattle: University of Washington Press.

Dawes, C. B. 1972. "Manasseh Cutler (1742–1823): forefather of American botany and American botanical gardens." Unpublished MS. Library of the Arnold Arboretum of Harvard University.

Deleuze, M. 1805. "Memoirs of the life and botanical travels of André Michaux." *Annals of Botany (London)* 1: 321–355.

Del Tredici, P. 1983. "Resurrecting Gardiner Greene's ginkgo." *Horticulture,* New Series 61(11): 12–16.

———. 1986. "The great catalpa craze." *Arnoldia* 46(2): 2–10.

Desmond, R. 1977. *Dictionary of British and Irish Botanists and Horticulturists Including Plant Collectors and Botanical Artists.* London: Taylor & Francis, Ltd.

De Vos, F. 1967 [1968]. "Early plant introductions from China and Japan." *Plants and Gardens* 23: 46–49.

Dillwyn, L. W. 1843. *Hortus Collinsonianus.* Swansea, England: Murray and D. Rees, Printers.

Dirr, M. A. 1977. *Manual of Woody Landscape Plants.* Rev. ed. Champaign, Illinois: Stipes Publishing Company.

Duncan, W. H., H. Venard, and G. W. McDowell. 1950. "*Shortia galacifolia* from Georgia." *Rhodora* 52: 229–232.

Dupree, A. H. 1953. "Science vs. the military: Dr. James Morrow and the Perry expedition." *Pacific Historical Review* 22: 29–37.

———. 1956. "Asa Gray and Andrew Jackson Downing: a bibliographical note." *Rhodora* 58: 243–245.

———. 1959. *Asa Gray, 1810–1888.* Cambridge: Belknap Press of Harvard University Press.

———. 1976. "The national pattern of American learned societies, 1769–1863." In A. Oleson and S. C. Brown, eds., *The Pursuit of Knowledge in the Early American Republic: American scientific and Learned Societies from Colonial Times to the Civil War,* pp. 21–32. Baltimore and London: Johns Hopkins University Press.

Eastwood, A. 1911. "Explorations of William Lobb." *Muhlenbergia* 7: 100–103.

———. 1939. "Early botanical explorers on the Pacific Coast and the trees they found there." *Quarterly of the California Historical Society* 18: 335–346.

Ellwanger, W. D. 1909. *A Snuff-Box Full of Trees and Some Apocryphal Essays.* New York: Dodd, Mead & Company.

Elwes, H. J., and A. Henry. 1906–1913. *The Trees of Great Britain and Ireland.* 8 vols. Edinburgh: privately printed.

Ewan, J. 1955. "San Francisco as a mecca for nineteenth century naturalists." In E. L. Kessel, ed., *A Century of Progress in the Natural Sciences, 1853–1953,* pp. 1–63. San Francisco: California Academy of Sciences.

———. 1960. "Bernard M'Mahon (c. 1775–1816), pioneer Philadelphia nurseryman, and his *American Gardener's Calendar.*" *Journal of the Society for the Bibliography of Natural History* 3: 363–380.

———. ed. 1968. *William Bartram: botanical and zoological drawings, 1756–1788.* Memoirs of the American Philosophical Society, vol. 74. Philadelphia: American Philosophical Society.

———. 1969. "Calendar of events and early history." In J. Ewan, ed., *A Short History of Botany in the United States,* pp. 1–48. New York and London: Hafner Publishing Company.

———. 1970. "Plant collectors in America: backgrounds for Linnaeus." In P. Smit and R. J. Ch. V. ter Laage, eds., *Essays in Biohistory,* pp. 19–54. Regnum Vegetabile, vol. 71. Leiden: International Association of Plant Taxonomists.

———. 1973. "William Lobb, plant hunter for Veitch and messenger of the big tree." *University of California Publications in Botany* 67: 1–36.

———. 1979. "History of exploring for rhododendrons in southeastern United States." *Quarterly Bulletin of the American Rhododendron Society* 33: 206–213.

———. 1979. "Pursh's flora in American botany." In F. Pursh, *Flora Americae Septentrionalis,* pp. 7–117. Reprint ed., Historiae Naturalis Classica, vol. 104. Vaduz: J. Cramer.

Ewan, J., and N. Ewan. 1963. "John Lyon, nurseryman and plant hunter, and his journal, 1799–1814." *Transactions of the American Philosophical Society,* New Series 53(2): 1–69.

———. 1970. *John Banister and His Natural History of Virginia, 1678–1692.* Urbana: University of Illinois Press.

Eyde, R. H. 1985. "The case for monkey-mediated evolution in big-bracted dogwoods." *Arnoldia* 45(4): 2–9.

Farquhar, F. P. 1948. *Yosemite, the Big Trees, and the High Sierra: A Selective Bibliography.* Berkeley and Los Angeles: University of California Press.

Farrington, E. I. 1931. *Ernest H. Wilson, Plant Hunter.* Boston: The Stratford Company.

Favretti, R. 1986. "Restoring Bartram's garden." *Green Scene* 15: 8–13.

Fisher, J. 1982. *The Origins of Garden Plants.* London: Constable.

Fletcher, H. R. 1969. *The Story of the Royal Horticultural Society, 1804–1968.* London: Oxford University Press for the Royal Horticultural Society.

Florin, R. 1952. "On *Metasequoia,* living and fossil." *Botaniska Notiser* 1: 1–29.

———. 1963. "The distribution of conifer and taxad genera in time and space." *Acta Horti Bergiani* 20(4): 121–312.

Fogg, J. M., Jr. 1960. "*Chionanthus* in the Philadelphia area." *Morris Arboretum Bulletin* 11: 3–6.

Foley, D. J. 1969. *The Flowering World of "Chinese" Wilson.* London: Macmillan Company.

Fortune, R. 1846. "Sketch of a visit to China." *Journal of the Horticultural Society of London* 1: 208–224.

———. 1847. *Three Years' Wanderings in the Northern Prov-*

inces of China, Including a Visit to the Tea, Silk, and Cotton Countries: With an Account of the Agriculture and Horticulture of the Chinese, New Plants, etc. London: John Murray.

———. 1847. "Experience in the transmission of living plants to and from distant countries by sea." *Journal of the Horticultural Society of London* 2:115–121.

———. 1848. "Observations upon the best methods of packing seeds for a voyage to India or China." *Journal of the Horticultural Society of London* 3: 41–44.

———. 1857. *A Residence among the Chinese: Inland, on the Coast, and at Sea.* London: John Murray.

———. 1880. "Mr. Fortune's introductions." *Gardeners' Chronicle,* New Series 13: 11.

———. 1880. "Notes on plants introduced from China and Japan." *Gardeners' Chronicle,* New Series 13: 234.

Fox, H. M., ed. 1949. *Abbé David's Diary.* Cambridge: Harvard University Press.

Fox, R. H. 1919. *Dr. John Fothergill and His Friends: Chapters in Eighteenth Century Life.* London: Macmillan and Co., Ltd.

Francis, J. W. 1858. *Old New York; Or Reminiscences of the Past Sixty Years.* New York: Charles Roe.

Frick, G. F., and R. P. Stearns. 1961. *Mark Catesby, the Colonial Audubon.* Urbana: University of Illinois Press. [The Appendix includes (pp. 114–119) a section entitled "Thomas More and his expedition to New England."]

Fuller, K. A. P., and J. M. Langdon. 1973. "The house of Veitch." In *International Dendrology Society Year Book 1972,* pp. 63–69. London: International Dendrology Society.

Fulling, E. H. 1976. "*Metasequoia*—fossil and living." *Botanical Review* 42: 215–315.

Gardener, R. 1971. "Robert Fortune and the cultivation of tea in the United States." *Arnoldia* 31: 1–18.

Gardener, W. 1972. "E. H. Wilson's first trip to China." *Arnoldia* 32: 103–114.

Gifford, G. E., Jr. 1972. "George Rogers Hall: pills and plants." *Harvard Medical Alumini Bulletin* 47(2): 22–24.

Gilmore, J. 1946. *British Botanists.* London: Collins.

Gordon, A. 1832. "Notices of some of the principal nurseries and private gardens in the United States of America, made during a tour through the country, in the summer of 1831; with some hints on emigration." *Gardener's Magazine* 8: 277–289.

Gordon, W. J. 1899. "Robert Fortune, plant collector." *Journal of Horticulture and Cottage Gardener,* New Series 38: 411, 412, 434, 435.

Gray, A. 1842. "Notes of a botanical excursion to the mountains of North Carolina, etc.; with some remarks on the botany of the higher Alleghany Mountains (in a letter to Sir Wm. J. Hooker)." *American Journal of Science and Arts* 42: 1–49.

———. 1856. "List of dried plants collected in Japan, by S. Wells Williams, Esq., and Dr. James Morrow." In *Narrative of the Expeditions of an American Squadron to the China Seas and Japan, Performed in the Years, 1852, 1853, and 1854, Under the Command of Commodore M. C. Perry, United States Navy, by Order of the Government of the United States,* vol. 2, pp. 305–332. Washington, D.C.

———. 1859. "Diagnostic characters of new species of phaenogamous plants, collected in Japan by Charles Wright, botanist of the U.S. North Pacific Exploring Expedition. (Published by request of Captain John Rogers, Commander of the Expedition.) With observations upon the relations of the Japanese flora to that of North America, and other parts of the Northern Temperate Zone." *Memoirs of the American Academy of Arts and Sciences II* 6: 377–452.

———. 1873. "Address of ex-president of the Association." *Proceedings of the American Association for the Advancement of Science* 21: 1–31. [Reprinted as "*Sequoia* and its history." In C. S. Sargent, ed., *Scientific Papers of Asa Gray* 2: 142–173. Boston and New York: Houghton, Mifflin and Company, 1889.]

———. 1878. "Forest geography and archeology." *American Journal of Science* 16: 85–94, 183–196. [Reprinted in C. S. Sargent, ed., *Scientific Papers of Asa Gray* 2: 204–233. Boston and New York: Houghton, Mifflin and Company, 1889.]

Green, C. H. 1939. *Trees of the South.* Chapel Hill: University of North Carolina Press.

Greene, J. C. 1984. *American Science in the Age of Jefferson.* Ames: Iowa State University Press.

Gressitt, J. L. 1953. "The California Academy–Lingnan dawn-redwood expedition." *Proceedings of the California Academy of Sciences IV* 28(2): 25–58.

Gunther, R. T. 1922. *Early British Botanists and Their Gardens, Based on Unpublished Writings of Goodyer, Tradescant, and Others.* Oxford: Oxford University Press.

Hadfield, M. 1955. *Pioneers in Gardening.* London: Routledge & Kegan Paul.

———. 1976. "Trees and their periods." *Garden History* 4(2): 23–29.

Hammond, C. A. 1982. *"Where the Arts and the Virtues Unite": Country Life Near Boston, 1637–1864.* Unpubl. Ph.D. dissertation, Boston University.

———. 1987. "The botanic garden in Cambridge, Massachusetts, 1805–1834." *Herbarist* 53: 45–73.

Hardt, R. A. 1986. "Japanese honeysuckle: from 'one of the best' to ruthless pest." *Arnoldia* 46(2): 27–34.

Harper, F., ed. 1958. *The Travels of William Bartram: Naturalist's Edition.* New Haven: Yale University Press.

Harper, F., and A. N. Leeds. 1937. "A supplementary chapter on *Franklinia alatamaha.*" *Bartonia* 19: 1–13.

Hedrick, U. P. 1950. *A History of Horticulture in America to 1860.* New York: Oxford University Press.

Henry, A. 1889–1898. "Letters from China." *Kew Bulletin of Miscellaneous Information* 1889: 225–227; 1897: 99–101, 407–414; 1898: 289–297.

———. 1986. *Notes on the Economic Botany of China.* Reprint

ed. with an Introduction by C. Nelson. Clarabricken, Kilkenny, Ireland: Boethius Press.

Henry, A., and M. G. Flood. 1919. "The history of the London Plane." *Proceedings of the Royal Irish Academy* 35B: 9–28.

Hindle, B. 1956. *The Pursuit of Science in Revolutionary America, 1735–1789*. Chapel Hill: University of North Carolina Press.

Hogg, T. 1863, 1864. "Letters from Japan." *The Horticulturist and Journal of Rural Art and Rural Taste* 18: 66–68, 110–112, 235–237; 19: 12–14.

———. 1879. "History of *Sciadopitys*, and other Japan trees." *Gardener's Monthly and Horticulturist* 21: 53, 54.

———. 1879. "Introduction of the *Cercidiphyllum*." *Gardener's Monthly and Horticulturist* 21: 155.

Holeman, M. M. 1985. "The history of the *Rhododendron*." *Rosebay* (Massachusetts Chapter of the American Rhododendron Society) 14(2): 10–15.

Hooker, W. J. 1826. "On the botany of America." *Edinburgh Journal of Science* 2: 108–129.

———. 1837. "Biographical sketch of John Fraser, the botanical collector." *Companion to the Botanical Magazine* 2: 300–305.

Horsfall, F., Jr. 1969. "Horticulture in eighteenth-century America." *Agricultural History* 43: 159–167.

Howard, R.A. 1980. "E. H. Wilson as a botanist." *Arnoldia* 40: 102–138, 154–193.

Howard, R. A., and G. W. Staples. 1983. "The modern names for Catesby's plants." *Journal of the Arnold Arboretum* 64: 511–546.

Howe, J. M., Jr. 1923. "George Rogers Hall, lover of plants." *Journal of the Arnold Arboretum* 4: 91–98.

Hsueh, C. J. 1985. "Reminiscences of collecting the type specimens of *Metasequoia glyptostroboides* H. H. Hu & Cheng." *Arnoldia* 45(4): 10–18.

Hu, H. H. 1948. "How *Metasequoia*, the 'living fossil,' was discovered in China." *Journal of the New York Botanical Garden* 49: 201–207.

Hu, H. H., and W. C. Cheng. 1948. "On the new family Metasequoiaceae and on *Metasequoia glyptostroboides*, a living species of the genus *Metasequoia* found in Szechuan and Hupeh." *Bulletin of the Fan Memorial Institute of Biology II* 1: 153–166.

Hu, S. Y. 1979. "*Ailanthus*." *Arnoldia* 39: 29–50.

Hughes, K. W. 1939. "A contribution toward a bibliography of Oregon botany with notes on the botanical explorers of the state." *Oregon State College Thesis Series* No. 14.

Hunkin, J. W. 1942. "William and Thomas Lobb: two Cornish plant collectors." *Journal of the Royal Horticultural Society* 67: 48–51.

J[ack, J. G.] 1889. "Notes from the Arnold Arboretum: *Syringa villosa*." *Garden and Forest* 2: 308–309.

Jack, J. G. 1949. "The Arnold Arboretum—some personal notes." *Chronica Botanica* 12: 184–200.

Jarvis, P. J. 1973. "North American plants and horticultural innovation in England, 1550–1700." *Geographical Review* 63: 477–499.

———. 1979. "The introduced trees and shrubs cultivated by the Tradescants at South Lambeth, 1629–1679." *Journal of the Society for the Bibliography of Natural History* 9: 223–250.

Jenkins, C. F. 1943. "Franklin's tree." *National Horticultural Magazine* 22: 119–127.

Johnson, K. 1985. "Sycamore, 110 million years in the making." *Morris Arboretum Newsletter* 14(2): 4, 5.

Johnstone, J. T. 1939. "John Jeffrey and the Oregon Expedition." *Notes from the Royal Botanic Garden Edinburgh* 20: 1–53.

Koller, G. L. 1986. "Seven-son flower from Zhejiang: introducing the versatile ornamental shrub *Heptacodium jasminoides* Airy Shaw." *Arnoldia* 46: 2–14.

Latourette, K. S. 1917. "The history of early relations between the United States and China, 1784–1844." *Transactions of the Connecticut Academy of Arts and Sciences* 22: 1–209.

Lawrence, G. H. M. 1969. "Horticulture." In J. Ewan, ed., *A Short History of Botany in the United States*, pp. 132–145. New York and London: Hafner Publishing Company.

Leet, J. 1987. *The Botanical Paintings of Esther Heins*. New York: Harry N. Abrams, Inc.

Lehmer, M. 1961. "The Walter Street 'berrying' ground." *Arnoldia* 21: 75–82.

Leith-Ross, P. 1984. *The John Tradescants, Gardeners to the Rose and Lily Queen*. London: Peter Owen.

Le Rougetel, H. 1971. "Gardener extraordinary—Philip Miller of Chelsea, 1691–1771." *Journal of the Royal Horticultural Society* 96: 556–563.

———. 1986. "Philip Miller/John Bartram botanical exchange." *Garden History* 14: 32–39.

Li, H. L. 1956. "A horticultural and botanical history of the ginkgo." *Morris Arboretum Bulletin* 7: 3–12.

———. 1957. "The origin and history of the cultivated plane-trees." *Morris Arboretum Bulletin* 8: 3–9, 26–31.

———. 1963. *The Origin and Cultivation of Shade and Ornamental Trees*. Philadelphia: University of Pennsylvania Press.

Lindley, J., ed. 1853. [Untitled article.] *Gardeners' Chronicle*, pp. 819, 820.

———. 1853. "New plants. 33. *Wellingtonia gigantea*." *Gardeners' Chronicle*, p. 823.

Loudon, J. C., ed. 1827. "Foreign notices: North America. Linnean Botanic Garden, Flushing, near New York." *Gardener's Magazine* 2: 90.

———. 1831. "Bartram's botanic garden on the Schuylkill, near Philadelphia." *Gardener's Magazine* 7: 665–671.

———. 1838. *Arboretum et Fruticetum Britannicum; Or, the Trees and Shrubs of Britain, Native and Foreign, Hardy and Half-Hardy*. 8 vols. London.

Manning, R. 1880. "Sketch of the history of horticulture in the United States up to the year 1829." In R. Manning, ed., *History of the Massachusetts Horticultural Society, 1829–1878,* pp. 1–54. Boston.

Marlor, C. S. 1972. "James Hogg." Unpublished ms. Library of the Gray Herbarium, Harvard University.

———. 1972. "Thomas Hogg, Sr." Unpublished ms. Library of the Gray Herbarium, Harvard University.

———. 1972. "Thomas Hogg, Jr." Unpublished ms. Library of the Gray Herbarium, Harvard University.

Marshall, H. 1967. *Arbustrum Americanum: The American Grove (Facsimile of the Edition of 1785), and Catalogue Alphabétique des Arbes et Arbrisseaux. (Facsimile of the Edition of 1788).* J. Ewan, ed. Classica Botanica Americana, vol. 1. New York and London: Hafner Publishing Company.

McGourty, F., Jr. 1967 [1968]. "Hall—of Hall's honeysuckle." *Plants and Gardens* 23: 85.

McKelvey, S. D. 1955. *Botanical Exploration of the Trans-Mississippi West, 1790–1850.* Jamaica Plain, Mass.: Arnold Arboretum of Harvard University.

Merrill, E. D. 1948. "*Metasequoia,* another 'living fossil.'" *Arnoldia* 8: 1–8.

———. 1948. "A living *Metasequoia* in China." *Science* 107: 140.

Meyer, F. G. 1959. *Plant Explorations: Ornamentals in Italy, Southern France, Spain, Portugal, England, and Scotland.* Washington, D.C.: Agricultural Research Service, U.S.D.A., in cooperation with Longwood Gardens of the Longwood Foundation, Inc.

———. 1963. *Plant Explorations: Ornamentals in the Netherlands, West Germany, and Belgium.* Washington, D.C.: Agricultural Research Service, U.S.D.A., in cooperation with Longwood Gardens of the Longwood Foundation, Inc.

———. 1976. "A revision of the genus *Koelreuteria* (Sapindaceae)." *Journal of the Arnold Arboretum* 57: 129–166.

Miki, S. 1941. "On the change of flora in eastern Asia since Tertiary Period. I. The clay or lignite beds flora in Japan with special reference to the *Pinus trifolia* beds in central Hondo." *Japanese Journal of Botany* 11: 237–303.

Miller, P. 1768. *Gardeners Dictionary.* 8th ed. London.

Miller, W. H. 1948. "Mark Catesby, an eighteenth century naturalist." *Tyler's Quarterly Historical and Genealological Magazine* 29: 167–180.

Morley, B. D. 1979. "Augustine Henry: his botanical activities in China, 1882–1890." *Glasra* 3: 21–81.

Morong, T. 1893. "Thomas Hogg." *Bulletin of the Torrey Botanical Club* 20: 217, 218.

Morison, S. E. 1967. *"Old Bruin": Commodore Matthew C. Perry, 1794–1858.* Boston and Toronto: Little, Brown and Company.

Morwood, W. 1973. *Traveler in a Vanished Landscape: The Life and Times of David Douglas.* New York: Clarkson N. Potter, Inc.

Mower, D. R., Jr. 1984. "Bartram's garden in Philadelphia." *Magazine Antiques* 125: 630–636.

Muir, J. 1961. *The Mountains of California.* Garden City, N.Y.: Anchor Books, Doubleday & Co., Inc.

Nelmes, E. 1944. "Robert Fortune, pioneer collector, the centenary of whose departure for China has fallen in this year." *Proceedings of the Linnean Society of London* 156: 8–16.

Nelson, E. C. 1983. "Augustine Henry and the exploration of the Chinese flora." *Arnoldia* 43(1): 21–38.

———. 1984. "The garden history of Augustine Henry's plants." In S. Pim, *The Wood and the Trees,* pp. 217–236. 2nd ed. Kilkenny, Ireland: Boethius Press.

Oldschool, O. 1809. "American scenery for the Port Folio: The Woodlands." *Port Folio,* New Series 2(6): 505–507.

Oliver, D. 1891. "*Davidia involucrata* Baill." *Hooker's Icones Plantarum* 20: *t. 1961.*

O'Neill, J. 1984. "Peter Collinson and the American garden." *Garden,* July/August, pp. 20–23, 31.

Parkinson, J. 1629. *Paradisi in Sole: Paradisus Terrestris.* London.

Parsons and Co. 1862. "Japanese trees." *The Horticulturist or Journal of Rural Art and Rural Taste* 17: 186, 187.

Paterson, A. 1986. "Philip Miller: a portrait." *Garden History* 14: 40, 41. [Rather, the lack of a portrait. The only likeness thought to be of Miller is not.]

Peattie, D. C. 1953. *A Natural History of Western Trees.* New York: Bonanza Books.

———. 1964. *A Natural History of Trees of Eastern and Central North America.* 2nd ed. New York: Bonanza Books.

Peck, W. D. 1818. *A Catalogue of American and Foreign Plants, Cultivated in the Botanic Garden, Cambridge, Massachusetts.* Cambridge.

Pim, S. 1966. *The Wood and the Trees: A Biography of Augustine Henry.* London: Macdonald & Co., Ltd. [2nd ed., Kilkenny, Ireland: Boethius Press, 1984.]

Pineau, R., ed. 1968. *The Japan Expedition, 1852–1854: The Personal Journal of Commodore Matthew C. Perry.* Washington, D.C.: Smithsonian Institution Press.

Piper, C. V. 1906. "Flora of the state of Washington." *Contributions from the United States National Herbarium,* vol. 11. Washington, D.C.: Government Printing Office.

Poor, J. M., ed. 1984. *Plants That Merit Attention.* Vol. 1: *Trees.* Portland, Oregon: Timber Press.

Pulteney, R. 1790. *History and Biographical Sketches of the Progress of Botany in England.* 2 vols. London.

Pyle, H. 1880. "Bartram and his garden." *Harper's New Monthly Magazine* 60: 321–330.

Raven, C. E. 1950. *John Ray, Naturalist: His Life and Works.* 2nd ed. Cambridge: Cambridge University Press.

Rehder, A. 1927. "Charles Sprague Sargent." *Journal of the Arnold Arboretum* 8: 69–86.
———. 1930. "Ernest Henry Wilson." *Journal of the Arnold Arboretum* 11: 181–192.
———. 1940. *Manual of Cultivated Trees and Shrubs Hardy in North America Exclusive of the Subtropical and Warmer Temperature Regions.* 2nd ed. New York: Macmillan Publishing Co., Inc.
———. 1946. "On the history of the introduction of woody plants into North America." *Arnoldia* 6: 13–23. [Reprinted from *National Horticultural Magazine* 16: 245–257.]
Robbins, W. J., and M. C. Howson. 1958. "André Michaux's New Jersey garden and Pierre Paul Saunier, journeyman gardener." *Proceedings of the American Philosophical Society* 102: 351–370.
Rusby, H. H. 1884. "Michaux's New Jersey garden." *Bulletin of the Torrey Botanical Club* 11: 88–90.

Sargent, C. S. 1879. "The climbing hydrangea." *Gardener's Monthly and Horticulturist* 21: 2.
———. 1879. "*Sciadopitys* and other Japan plants." *Gardener's Monthly and Horticulturist* 21: 121.
———. 1890–1902. *The Silva of North America.* 14 vols. Boston: Houghton Mifflin.
———. 1891. "*Syringa pubescens.*" *Garden and Forest* 4: 262.
———. ed. 1913–1917. *Plantae Wilsonianae.* 3 vols. Publication of the Arnold Arboretum No. 4. Cambridge: Harvard University Press.
———. 1915. "Washington and Michaux." *Rhodora* 17: 49, 50.
———. 1922. "The first fifty years of the Arnold Arboretum." *Journal of the Arnold Arboretum* 3: 127–171.
———. Correspondence. Archives of the Arnold Arboretum of Harvard University.
Savage, H., Jr., and E. J. Savage. 1986. *André and François André Michaux.* Charlottesville: University Press of Virginia.
Schilling, T. 1984. "David Douglas (1798–1834)." *Garden, Journal of the Royal Horticultural Society* 109: 276–280.
Senn, T. L. 1969. "Farm and garden: landscape architecture and horticulture in eighteenth-century America." *Agricultural History* 43: 149–157.
Severin, T. 1976. *The Oriental Adventure: Explorers of the East.* Boston: Little, Brown & Co.
Siebold, P. F. von. 1863. *Catalogue Raisonné et Prix-courant des Plantes et Graines du Japon et de la Chine, Cultivés dans le Jardin D'acclimation de Ph. F. von Siebold a Leide.* Amsterdam.
Sims, J. 1806. "*Hydrangea quercifolia:* Oak-leaved hydrangea." *Curtis's Botanical Magazine* 25: *t. 975.*
Smith, B. H. 1905. "Some letters from William Hamilton, of the Woodlands, to his private secretary." *Pennsylvania Magazine of History and Biography* 29: 70–78, 143–159, 257–267.
Smith, P. C. F. 1984. *The Empress of China.* Philadelphia: Philadelphia Maritime Museum.
Smith, W. W. 1931. "The contribution of China to European gardens." *Notes from the Royal Botanic Garden Edinburgh* 16: 215–221.
Society of Gardeners. 1730. *A Catalogue of Trees, Shrubs, Plants and Flowers which are Propagated for Sale in the Gardens Near London.* London.
Spongberg, S. A. 1974. "A review of deciduous-leaved species of *Stewartia* (Theaceae)." *Journal of the Arnold Arboretum* 55: 182–214.
———. 1984. "C. S. Sargent: seeing the forest and the trees." *Orion Nature Quarterly* 3(4): 4–11.
Stafleu, F. A. 1971. *Linnaeus and the Linnaeans: The Spreading of Their Ideas in Systematic Botany, 1735–1789.* Utrecht: A. Oosthoek's Uitgeversmaatschappij N.V.
Stearn, W. T. 1958. "Botanical exploration to the time of Linnaeus." *Proceedings of the Linnean Society of London* 169: 173–196.
———. 1971. "Sources of information about botanic gardens and herbaria." *Biological Journal of the Linnean Society of London* 3: 225–233.
Stearns, R. P. 1952. "James Petiver: promoter of natural science, c. 1663–1718." *Proceedings of the American Antiquarian Society* 62: 243–365.
———. 1970. *Science in the British Colonies of America.* Urbana: University of Illinois Press.
Steiger, G. N. 1927. *China and the Occident: The Origin and Development of the Boxer Movement.* New Haven: Yale University Press.
Stern, F. C. 1937. "The flora of China: early investigations." *Journal of the Royal Horticultural Society* 62: 347–351.
———. 1944. "The discoveries of the great French missionaries in central and western China." *Proceedings of the Linnean Society of London* 156: 16–20.
Stetson, S. P. 1949. "Traffic in seeds and plants from England's colonies in North America." *Agricultural History* 23: 45–56.
———. 1949. "William Hamilton and his 'Woodlands.'" *Pennsylvania Magazine of History and Biography* 73: 26–33.
Stewart, W. N. 1983. *Paleobotany and the Evolution of Plants* Cambridge: Cambridge University Press.
Sutton, S. B. 1970. *Charles Sprague Sargent and the Arnold Arboretum.* Cambridge: Harvard University Press.
———. 1974. *In China's Border Provinces: The Turbulent Career of Joseph Rock, Botanist-Explorer.* New York: Hastings House.
Swain, R. B. 1985. *Field days: Journal of an Itinerant Biologist.* New York: Penguin Books.
Swem, E. G. 1948. "Brothers of the spade: correspondence of Peter Collinson, of London, and John Custis, of Williamsburg, Virginia, 1734–1746." *Proceedings of the American Antiquarian Society* 58: 17–190.
Swingle, W. T. 1916. "The early European history and the

botanical name of the tree of heaven, *Ailanthus altissima.*" *Journal of the Washington Academy of Science* 6: 490–498.

Taylor, G. 1980. "The contribution from America to British gardens in the early nineteenth century." In E. B. Macdougall, ed., *John Claudius Loudon and the Early Nineteenth Century in Great Britain,* pp. 105–123. Washington, D.C.: Dumbarton Oaks.

Thomas, J. H. 1979. "Botanical explorations in Washington, Oregon, California and adjacent regions." *Huntia* 3: 5–66.

Thorhaug, A., ed. 1978. *Botany in China: Report of the Botanical Society of America Delegation to the People's Republic, May 20–June 18, 1978.* Stanford, California: U.S.–China Relations Program, Stanford University.

Tschanz, E. N. 1977. "A history: the U.S.D.A.–Longwood ornamental plant exploration program." *The Longwood Program Seminars 1977* (The University of Delaware, Newark) 9: 67–70.

Veitch, J. G. 1860, 1861. "Extracts from Mr. J. G. Veitch's letters on Japan." *Gardeners' Chronicle* 1860: 1104, 1126; 1861: 24, 25, 97, 98.

———. 1860. "Notes on the vegetation of Japan." *Gardeners' Chronicle* 1860: 1126, 1127.

———. 1861. "Diary kept by J. G. Veitch during his trip to Mount Fusi Yama, September, 1860." *Gardeners' Chronicle* 1861: 49, 50.

———. 1861. "A trip to the nurseries and botanic gardens of Yeddo." *Gardeners' Chronicle* 1861: 120, 121.

Vos, R. de. 1967 [1968]. "Early plant introductions from China and Japan." *Plants and Gardens* 23: 46–49.

Wade, M. 1942. *Francis Parkman, Heroic Historian.* New York: Viking Press.

Ward, N. B. 1836. "Letter from N. B. Ward, Esq. to Dr. Hooker, on the subject of his improved method of transporting living plants." *Companion to the Botanical Magazine* 1: 317–320.

Waring, R. H., and J. F. Franklin. 1979. "Evergreen coniferous forests of the Pacific Northwest." *Science* 204: 1380–1386.

Warner, M. F. 1954. "The Morins." *National Horticultural Magazine* 33: 168–176.

Whitehill, W. M. 1973. "Francis Parkman as horticulturist." *Arnoldia* 33: 169–183.

Whittle, T. 1970. *The Plant Hunters.* London: William Heinemann, Ltd., and Philadelphia: Chilton Book Co.

Wick, P. A. 1976. "Gore Place, federal mansion in Waltham, Massachusetts." *Magazine Antiques* 110: 1250–1261.

Wilson, E. H. 1913. *A Naturalist in Western China, with Vasculum, Camera and Gun.* 2 vols. London: Methuen & Co., Ltd.

———. 1916. "The cherries of Japan." *Publications of the Arnold Arboretum* No. 7. Cambridge: Harvard University Press.

———. 1916. "The conifers and taxads of Japan." *Publications of the Arnold Arboretum* No. 8. Cambridge: Harvard University Press.

———. 1917. *Aristocrats of the Garden.* Garden City, N.Y.: Doubleday, Page & Co.

———. 1920. *The Romance of Our Trees.* Garden City, N.Y.: Doubleday, Page & Co.

———. 1927. "Charles Sprague Sargent." *Harvard Graduates Magazine* 35: 605–615.

———. *Wilson's Plants in Cultivation.* Unpublished MS. Archives of the Arnold Arboretum of Harvard University.

———. Correspondence. Archives of the Arnold Arboretum of Harvard University.

Wilson, E. H., and A. Rehder. "A monograph of Azaleas, *Rhododendron* subgenus *Anthodendron.*" *Publications of the Arnold Arboretum,* No. 9. Cambridge: Harvard University Press.

Wright, R. 1949. *Gardener's Tribute.* Philadelphia and New York: J. B. Lippincott Company.

Wyman, D. 1951. "*Metasequoia* brought up to date." *Arnoldia* 11: 25–28.

———. 1959. "These are the forsythias." *Arnoldia* 19: 11–14.

———. 1968. "*Metasequoia* after twenty years in cultivation." *Arnoldia* 28: 113–123.

Wynne, W. 1832. "Some account of the nursery gardens and the state of horticulture in the neighbourhood [sic] of Philadelphia, with remarks on the subject of emigration of British gardeners to the United States." *Gardener's Magazine* 8: 272–276.

ACKNOWLEDGMENTS

Grateful acknowledgment is made to the National Endowment for the Humanities for a grant to the Arnold Arboretum that has made this project a reality. Publication of this book has also been aided by funds provided by the National Endowment for the Humanities.

Many people gave freely of their time and knowledge while I was gathering material for and writing this book, and although I cannot name them all, I extend special thanks to my colleagues at the Arnold Arboretum and at the Harvard University Herbaria. These include Peter S. Ashton, David E. Boufford, Barbara Callahan, Carin Dohlman, Ida Hay, Ethan Johnson, Karen Kane, Geraldine Kaye, Rob Nicholson, Jennifer Quigley, and Elizabeth Shaw. Hollis Bedell, John Creech, my mother, Barbara E. Spongberg, and two anonymous reviewers all willingly read and commented on the manuscript, and I appreciate their advice. I am also grateful to Professor Sam Bass Warner, whose introductory essay provides an added dimension to the book. A. Hunter Dupree made helpful and insightful comments on much of the initial manuscript and helped to shape its final form.

James Lewis, Curator of the Reading Room, and the staff at the Houghton Library gave me unlimited access to the Arboretum's Houghton deposit and permitted me to photograph, on site, illustrations from those volumes for use as artwork in this book. Christopher Burnett, photographer for this project, tirelessly and cheerfully took the many excellent photographs of materials in the Houghton Library and in the libraries and archives of the Arnold Arboretum and the Gray Herbarium. Donald Pfister allowed access to materials in the archives of the Gray Herbarium.

Many of the pen and ink botanical illustrations reproduced in this book are the skillful work of Robin S. Lefberg and Karen Stoutsenberger, who prepared the drawings under a grant from the Stanley Smith Horticultural Trust; the support of the Trust and the help of Sir George Taylor, Director of the Trust, is gratefully acknowledged. Other drawings were prepared by Dorothy H. Marsh under grants to Carroll E. Wood, Jr., from the National Science Foundation for the Generic Flora of the Southeastern United States project, and I am grateful to Dr. Wood for permission to reproduce them here. The map illustrating the routes of David Douglas in the Pacific Northwest, the map of Wilson territory in China, and the map of the Arnold Arboretum are the exacting and careful work of Herb Heidt.

Mrs. George Gifford generously made available the photograph of George Rogers Hall's Shanghai residence, and Dr. H. G. Hers kindly provided the photograph of his uncle, Joseph Hers. Richard Eyde and Alice Tangerini, both of the Department of Botany at the Smithsonian Institution, granted permission to use the illustration of *Cornus florida* and *Cornus kousa* painted by Miss Tangerini.

Several institutions allowed materials in their care to be used as illustrations (see Illustration Credits, below), and to their directors and staffs I express my warm thanks. These include the American Philosophical Society; the British Museum (Natural History); the Gray Herbarium of Harvard University; the Harvard College Library; the Harvard University Art Museums; the Harvard University Portrait Collections; Houghton Library, Harvard University; the Hunt Institute for Botanical Documentation; the Massachusetts Historical Society; the Missouri Botanical Garden; the Morton Arboretum; the National Botanic Garden, Glasnevin; the National Portrait Gallery, London; the Pennsylvania Historical Society; the Royal Botanic Gardens, Kew; the Royal Horticultural Society, London; the Society for the Preservation of New England Antiquities; and the Waltham, Massachusetts, Public Library. Albert Bussewitz, David E. Boufford, and Peter S. Ashton also kindly provided photographs, for which I am grateful.

I wish to extend my appreciation to Gunder Hefta for his encouragement and enthusiasm in the early stages of this project, and to Howard Boyer of Harvard University Press for his continuing support. I am particularly indebted to Susan Wallace, my editor, for her informed influence on the entire enterprise; to Marianne Perlak for her sensitive and artistic approach to the design of this book; and to David Foss, whose technical and managerial expertise set high standards during the book's production. I am also very grateful to Margaretta Fulton, who first suggested that the Press might be interested in publishing a work of this kind.

A special and heartfelt note of thanks is extended to Sheila Connor, who served as principal investigator for the grant from the National Endowment for the Humanities to the Arnold Arboretum. I appreciate all her efforts and, in particular, her friendship and assistance with this volume.

Last, I should like to express in writing my deep appreciation to my wife, Happy, who not only proofread the manuscript but offered useful suggestions for modification, and who brought to this task the opinion and viewpoint of the layperson with an interest in plants. If this volume has value to nonprofessionals, it is largely due to her judgment and influence.

ILLUSTRATION CREDITS

CHAPTER ONE

1 Library of the Arnold Arboretum.

2, 3 Illustrations from Franz Schmidt, *Osterreichs Allgemeine Baumzucht* (Vienna, 1792). Library of the Arnold Arboretum.

4, 5 Library of the Arnold Arboretum.

6 From *Le Jardin du Roy* (Paris, 1608). Library of the Arnold Arboretum.

7 Engraving by N. Langlois (Paris, ca. 1690). Library of the Arnold Arboretum.

8, 9 Reproduced from *Chronica Botanica*, volume 1 (1935). Library of the Arnold Arboretum.

10 Library of the Arnold Arboretum.

11 From *Musaeum Tradescantianum* (London, 1656). By permission of the Houghton Library, Harvard University.

12 From *Gardeners' Chroncile*, New Series, volume 24 (1885). Library of the Arnold Arboretum.

13 Courtesy of the *Annals of the Missouri Botanical Garden.*

14 Photograph by Stephen A. Spongberg.

15 Photograph by Albert Bussewitz.

16 Drawings by Dorothy H. Marsh, courtesy of C. E. Wood, Jr., and the Generic Flora of the Southeastern United States project, Arnold Arboretum.

17 Painting by Sir Godfrey Kneller (ca. 1700), courtesy of the National Portrait Gallery, London.

18 Portrait from John Ray, *Stirpium Europaearum* (London, 1694). Library of the Arnold Arboretum.

19 Library of the Arnold Arboretum.

20 From J. C. Lettsom, *The Works of John Fothergill, M.D.* (London, 1784). Library of the Arnold Arboretum.

21 Library of the Arnold Arboretum.

22 Engraving from J. C. Lettsom, *Memoirs of John Fothergill, M.D.* (London, 1786). Library of the Arnold Arboretum.

23 Illustration from R. H. Semple, *Memoirs of the Botanic Garden at Chelsea belonging to the Society of Apothecaries of London* (London, 1878). Library of the Arnold Arboretum.

24 Drawings by Robin S. Lefberg.

25 Courtesy of the Waltham, Massachusetts, Public Library.

26 Private collection, by permission.

27 Courtesy of the Library of the Gray Herbarium.

28 Cartoon from *Chronica Botanica.* Library of the Arnold Arboretum.

29 From Sir William Chambers, *Plans, Elevations, Sections, and Perspective Views of the Gardens and Buildings at Kew, in Surrey* (London, 1763). By permission of the Houghton Library, Harvard University.

30 Photograph by Stephen A. Spongberg.

31 Courtesy of the Harvard University Art Museums (Fogg Art Museum, bequest of Grenville L. Winthrop).

32 Courtesy of the British Museum (Natural History).

CHAPTER TWO

33 Courtesy of the Archives of the Gray Herbarium.

34 From *The Horticulturist and Journal of Rural Art and Rural Taste* (1856). Library of the Arnold Arboretum.

35–37 Courtesy of the Archives of the Gray Herbarium.

38 From *Description des principaux parcs et jardins de l'Europe* (Germany, 1812). Library of the Arnold Arboretum.

39, 40 From F. A. L. von Burgsdorf, *Versuch einer vollstandigen Geschichte Vorzuglicher Holzarten*, 2 volumes (Berlin, 1783). Courtesy of the Library of the Gray Herbarium.

41 Archives of the Arnold Arboretum.

42 Drawing by Anthony Tyznik, courtesy of the Morton Arboretum.

43 Library of the Arnold Arboretum.

44 Drawings by Dorothy H. Marsh, courtesy of C. E. Wood, Jr., and the Generic Flora of the Southeastern United States project, Arnold Arboretum.

45 Courtesy of the American Philosophical Society.

46 From *Companion to the Botanical Magazine* (1836). Library of the Arnold Arboretum.

47 Drawings by Robin S. Lefberg.

48 Drawing by Nancy S. Hart. Courtesy of the Morton Arboretum.

49 Courtesy of the Library of the Gray Herbarium.

50 From *Gleason's Pictorial Drawingroom Companion,* volume 6 (1854). Courtesy of the Harvard College Library.

51 Painting courtesy of The Historical Society of Pennsylvania.

52 From *Woods and Forest,* volume 1 (1884). Library of the Arnold Arboretum.

53 Drawings by Anthony Tyznik. Courtesy of the Morton Arboretum.

54 Drawings by Robin S. Lefberg.

55 Portrait courtesy of the Massachusetts Historical Society.

56 Illustration from *Garden and Forest,* volume 2 (1889). Library of the Arnold Arboretum.

57 Painting courtesy of the Society for the Preservation of New England Antiquities.

58 Courtesy of the Archives of the Gray Herbarium.

59 From *Hortus Elginensis* (New York, 1811). Courtesy of the Library of the Gray Herbarium.

60, 61 Courtesy of the Archives of the Gray Herbarium.

62 Drawing by Isaac Sprague for Asa Gray's *Genera of North American Plants* (Boston, 1848, 1849). Library of the Arnold Arboretum.

63 From Charles Wilkes, *Narrative of the United States Exploring Expedition during the Years 1838,1839, 1840, 1841, 1842* (Philadelphia, 1845). Library of the Arnold Arboretum.

64, 65 Drawings by Robin S. Lefberg.

66 Photograph by Stephen A. Spongberg.

CHAPTER THREE

67, 68 From N. H. N. Mody, *A Collection of Nagasaki Prints and Paintings* (Tokyo: Charles E. Tuttle Co., Inc., 1969). Used by permission.

69 From *Ballou's Pictorial Drawing Room Companion,* volume 9 (1855). Courtesy of the Harvard College Library.

70 Library of the Arnold Arboretum.

71 Photograph by Stephen A. Spongberg.

72 From Siebold and Zuccarini, *Flora Japonica.* Library of the Arnold Arboretum.

73 From *Mémoires de Mathématique et de Physique* (Paris, 1786 [1788]). Library of the Arnold Arboretum.

74 Drawings by Karen Stoutsenberger.

75 From Johan Nieuhof, *An Embassy from the East-India Company of the United Provinces, to the Grand Tartar Cham Emperour of China* (London, 1669). By permission of the Houghton Library, Harvard University.

76 Frontispiece of A. Kircher, *China Illustrated* (Amsterdam, 1667). By permission of the Houghton Library, Harvard University.

77 From Louis Le Comte, *Beschryvinge van het machtige keyserryk China* (The Hague, 1698). Library of the Arnold Arboretum.

78 Drawings by Robin S. Lefberg.

79 Frontispiece to the reprint edition (arranged by Burman) of Plumier's *Plantarum Americanum* (Leiden, 1755). Courtesy of the Library of the Gray Herbarium.

80 Photograph by Albert Bussewitz.

81 Photograph by D. E. Boufford.

82 Photograph by Christopher Burnett.

83 Courtesy of The Trustees of The Royal Botanic Gardens, Kew. Reproduced with permission.

84 Photograph by Camillo Schneider. Photographic Archives of the Arnold Arboretum.

85 Photograph by Albert Bussewitz.

86, 87 Drawings by Robin S. Lefberg.

88 Courtesy of the Hunt Institute for Botanical Documentation, Carnegie Mellon University, Pittsburgh, Pennsylvania.

89 Courtesy of the Archives of the Gray Herbarium.

90 From George Vancouver, *A Voyage of Discovery to the North Pacific Ocean and Round the World* (London, 1798). Library of the Arnold Arboretum.

91 Courtesy of the Harvard University Art Museums (Fogg Art Museum, bequest of Grenville L. Winthrop).

92 Courtesy of the Royal Horticultural Society, London.

93 Etching by W. J. Hooker's sister-in-law, 1813. Courtesy of the Archives of the Gray Herbarium.

94 Photograph by Barth Hamberg, Arnold Arboretum.

95 Library of the Arnold Arboretum.

96 From Charles Wilkes, *Narrative of the United States Exploring Expedition during the Years 1838, 1839, 1840, 1841, 1842* (Philadelphia, 1845). Library of the Arnold Arboretum.

97 Cartography by Herb Heidt, Mapworks.

CHAPTER FOUR

98 From *Mr. Punch's Victorian Era* (London, 1887); the cartoon originally appeared in the issue of *Punch, or the London Charivari*, for September 4, 1858. Courtesy of the Harvard College Library.

99 From J. Zorn, *Icones Plantarum Medicinalium*, volume 3 (Nuremberg, 1779). Library of the Arnold Arboretum.

100 Library of the Arnold Arboretum.

101 Photograph by Christopher Burnett.

102 Courtesy of the Archives of the Gray Herbarium.

103 From *Journal of Horticulture, Cottage Gardner, and Home Farmer*, III, volume 38 (1899). Library of the Arnold Arboretum.

104 Courtesy of the Library of the Gray Herbarium.

105 From *Century Magazine* (1892). Archives of the Arnold Arboretum.

106 From *Gardeners' Chronicle* (1854). Library of the Arnold Arboretum.

107 Photograph by Stephen A. Spongberg.

108 Library of the Arnold Arboretum.

109 From *Hortus Veitchii* (London, 1906). Library of the Arnold Arboretum.

110 From J. Otis Williams, *Mammoth Trees of California* (Boston, 1871). Library of the Arnold Arboretum.

111 From *Century Magazine* (1892). Archives of the Arnold Arboretum.

112, 113 From N. H. N. Mody, *A Collection of Nagasaki Prints and Paintings* (Tokyo: Charles E. Tuttle Co., Inc., 1969). Reprinted with permission.

114 Library of the Arnold Arboretum.

115 From *Narrative of the Expedition of an American Squadron to the China Seas and Japan* (Washington, D.C., 1856). Library of the Arnold Arboretum.

116 Courtesy of the Archives of the Gray Herbarium.

117 Cartography by Herb Heidt, Mapworks.

118, 119 Photographs by Stephen A. Spongberg.

120 From *Century Magazine*, volume 45 (1893). Courtesy of the Harvard College Library.

121 Photographic Archives of the Arnold Arboretum.

122 From *Hortus Veitchii* (London, 1906). Library of the Arnold Arboretum.

123 Library of the Arnold Arboretum.

124 From *Botanical Magazine*, volume 61 (1834). Courtesy of the Library of the Gray Herbarium.

125 From *Narrative of the Expedition of an American Squadron to the China Seas and Japan* (Washington, D.C., 1856). Library of the Arnold Arboretum.

126 Courtesy of Mrs. George Gifford and the *Harvard Medical School Alumni Bulletin*.

127 Photographic Archives of the Arnold Arboretum.

CHAPTER FIVE

128 Library of the Arnold Arboretum.

129, 130 From *Hortus Veitchii* (London, 1906). Library of the Arnold Arboretum.

131 Photograph by G. R. King. Photographic Archives of the Arnold Arboretum.

132 Courtesy of the Archives of the Gray Herbarium.

133 Photograph by P. Bruns. Photographic Archives of the Arnold Arboretum.

134 Courtesy of the Archives of the Gray Herbarium.

135 Archives of the Arnold Arboretum.

136 From *Century Magazine*, volume 54 (1897). Courtesy of the Harvard College Library.

137, 138 Photographic Archives of the Arnold Arboretum.

139 Library of the Arnold Arboretum.

140 From *Hortus Veitchii* (London, 1906). Library of the Arnold Arboretum.

141 From *The Garden*, volume 24 (1883). Library of the Arnold Arboretum.

142 From *Garden and Forest*, volume 6 (1893). Library of the Arnold Arboretum.

143 Photograph by Albert Bussewitz.

144 Photographic Archives of the Arnold Arboretum.

145 Photograph by Barth Hamberg, Arnold Arboretum.

146 Archives of the Arnold Arboretum.

147 Photograph by Stephen A. Spongberg.

148 Photograph by E. H. Wilson, 1911. Photographic Archives of the Arnold Arboretum.

149 Photograph by David E. Boufford.

150 Portrait courtesy of the National Botanic Gardens, Glasnevin, Dublin, Ireland.

151 From *Journal of Horticulture, Cottage Gardener, and Home Farmer* III, volume 11 (1885). Library of the Arnold Arboretum.

152 Photographic Archives of the Arnold Arboretum.

153 Photograph of a portrait by H. G. Rivière. Photographic Archives of the Arnold Arboretum.

154 From L. Barron, "Chinese Wilson, Plant Hunter," *World's Work*, volume 27 (1913). Archives of the Arnold Arboretum.

155 Archives of the Arnold Arboretum.

156 Photographic Archives of the Arnold Arboretum.

157 Library of the Arnold Arboretum.

158–160 Photographs by Albert Bussewitz.

161, 162 Photographic Archives of the Arnold Arboretum.

CHAPTER SIX

163, 164 Photographic Archives of the Arnold Arboretum.

165 Photograph by Albert Bussewitz.

166, 167 Photographs by E. H. Wilson. Photographic Archives of the Arnold Arboretum.

168 Photograph by Christopher Burnett.

169 Photographic Archives of the Arnold Arboretum.

170 Archives of the Arnold Arboretum.

171 Photographic Archives of the Arnold Arboretum.

172, 173 Photographs by E. H. Wilson. Photographic Archives of the Arnold Arboretum.

174 Archives of the Arnold Arboretum.

175, 176 Photographs by E. H. Wilson. Photographic Archives of the Arnold Arboretum.

177 Photographic Archives of the Arnold Arboretum.

178 Drawings by Karen S. Stoutsenberger.

179 Cartography by Herb Heidt, Mapworks.

180, 181 Photographic Archives of the Arnold Arboretum.

182 Archives of the Arnold Arboretum.

183 Photographic Archives of the Arnold Arboretum.

184 Drawings by Robin S. Lefberg.

185 Courtesy of Dr. H. G. Hers.

186, 187 Photographic Archives of the Arnold Arboretum.

188 Photograph by Peter S. Ashton. Photographic Archives of the Arnold Arboretum.

189 Photographic Archives of the Arnold Arboretum.

190 From *Bulletin of the Fan Memorial Institute of Biology*, II, volume 1. Library of the Arnold Arboretum.

191 Archives of the Arnold Arboretum.

192 Photographic Archives of the Arnold Arboretum.

193 Drawings by Robin S. Lefberg.

194 Photographic Archives of the Arnold Arboretum.

195 From *Journal of the Arnold Arboretum*, volume 64 (1983).

196–198 Photographs by Stephen A. Spongberg.

199 From the collection of Stephen A. Spongberg.

200 Art by Amy Eisenberg.

COLOR PLATES

1 From Franz Schmidt, *Osterreichs Allgemeine Baumzucht* (Vienna, 1792). Library of the Arnold Arboretum.

2 From Franz Schmidt, *Osterreichs Allgemeine Baumzucht* (Vienna, 1792). Library of the Arnold Arboretum.

3 From Franz Schmidt, *Osterreichs Allgemeine Baumzucht* (Vienna, 1792). Library of the Arnold Arboretum.

4 Photograph by Stephen A. Spongberg.

5 Photograph by Stephen A. Spongberg.

6 From Joseph Jacobi Plenck, *Icones Plantarum Medicinalium, Centuria III* (Vienna, 1790). Library of the Arnold Arboretum.

7 From Jakob Trew, *Plantae Selectae* (Nuremberg, 1750–1773). Library of the Arnold Arboretum.

8 From Friedrich Guimpel, *Abbildung der fremden, in Deutschland ausdauernden Holzarten* (Berlin, 1819–1825). Library of the Arnold Arboretum.

9 From Franz Schmidt, *Osterreichs Allgemeine Baumzucht* (Vienna, 1792). Library of the Arnold Arboretum.

10 From Siebold and Zuccarini, *Flora Japonica* (Leiden, 1835–1870). Library of the Arnold Arboretum.

11 From Siebold and Zuccarini, *Flora Japonica* (Leiden, 1835–1870). Library of the Arnold Arboretum.

12 Library of the Arnold Arboretum.

13 From E. J. Ravenscroft, *The Pinetum Britannicum* (Edinburgh and London, 1863–1884). Library of the Arnold Arboretum.

14 From *Horticulturist and Journal of Rural Art and Rural Taste*, volume 4 (1854). Library of the Arnold Arboretum.

15 *Paeonia suffruticosa* from *Botanists Repository*, volume 6 (1840). *Rhododendron fortunei* from *Botanical Magazine*, volume 92 (1866). Library of the Arnold Arboretum.

16 Photograph by Stephen A. Spongberg.

17 From E. J. Ravenscroft, *The Pinetum Britannicum* (Edinburgh and London, 1863–1884). Library of the Arnold Arboretum.

18 From a painting by Alice Tangerini, Department of Botany, Smithsonian Institution. Courtesy of Alice Tangerini and Richard Eyde.

19 Photographs of *Magnolia stellata* and *Exochorda racemosa* by Stephen A. Spongberg. Photograph of *Taxus cuspidata* by Albert Bussewitz.

20 Photograph of *Parthenocissus tricuspidata* by Albert Bussewitz. Photograph of *Cercidiphyllum japonicum* by David E. Boufford.

21 From *Botanical Magazine,* volume 155 (1929). Library of the Arnold Arboretum.

22 From M. A. Franchet, *Plantae Davidianae,* volume 2 (Paris, 1888). Library of the Arnold Arboretum.

23 Photograph by Stephen A. Spongberg.

24 Photograph of *Pyrus pyrifolia,* by Albert Bussewitz. Photograph of *Albizia julibrissin,* by Stephen A. Spongberg.

INDEX

Page numbers are listed in roman type. Those that refer to pages on which illustrations appear are set in italics; plate numbers are also set in italic type. Numbers in bold face refer to locations on the end-paper map of the Arnold Arboretum.